Beginning SQL Server 2005 Express for Developers

From Novice to Professional

Robin Dewson

Apress®

Beginning SQL Server 2005 Express for Developers: From Novice to Professional

Copyright © 2007 by Robin Dewson

ISBN-13 (pbk): 978-1-59059-720-0

ISBN-10 (pbk): 1-59059-720-6

Printed and bound in the United States of America 9 8 7 6 5 4 3 2 1

Lead Editor: James Huddleston
Technical Reviewer: Vidya Vrat Agarwal
Editorial Board: Steve Anglin, Ewan Buckingham, Gary Cornell, Jason Gilmore, Jonathan Gennick, Jonathan Hassell, James Huddleston, Chris Mills, Matthew Moodie, Dominic Shakeshaft, Jim Sumser, Matt Wade
Project Manager: Kylie Johnston
Copy Edit Manager: Nicole Flores
Copy Editor: Ami Knox
Assistant Production Director: Kari Brooks-Copony
Production Editor: Kelly Winquist
Compositor: Lynn L'Heureux
Proofreader: April Eddy
Indexer: Broccoli Information Management
Cover Designer: Kurt Krames
Manufacturing Director: Tom Debolski

Distributed to the book trade worldwide by Springer-Verlag New York, Inc., 233 Spring Street, 6th Floor, New York, NY 10013. Phone 1-800-SPRINGER, fax 201-348-4505, e-mail orders-ny@springer-sbm.com, or visit http://www.springeronline.com.

For information on translations, please contact Apress directly at 2560 Ninth Street, Suite 219, Berkeley, CA 94710. Phone 510-549-5930, fax 510-549-5939, e-mail info@apress.com, or visit http://www.apress.com.

The source code for this book is available to readers at http://www.apress.com in the Source Code/Download section.

Contents at a Glance

Contents

CHAPTER 8 **Working with the Data** . 209

About the Author

ROBIN DEWSON has been hooked on programming ever since he bought his first computer in 1980, a Sinclair ZX80. His first major application of his own was a Visual FoxPro program that could be used to run a Fantasy League system. It was at this point he met up with a great help in his PC development life, Jon Silver at Step One Technologies, where in return for training he helped Jon with some other Visual FoxPro applications. From there, realizing that the marketplace for Visual FoxPro was limited, he decided to learn Visual Basic and SQL Server.

Starting out with SQL Server 6.5, Robin soon moved to SQL Server 7 and Visual Basic 5, and became involved in developing several applications for clients both in the UK and in the United States. From there, he moved to SQL Server 2000 and now SQL Server 2005, and Visual Basic 6 and now VS .NET, specializing in C# and VB .NET. Robin is also the coauthor of *Pro SQL Server 2005 Assemblies* along with other Apress books, and also contributes to ASP Today (www.asptoday.com). Robin can be contacted at robin@fat-belly.com or at www.fat-belly.com.

About the Technical Reviewer

VIDYA VRAT AGARWAL is a Microsoft .NET purist and an MCTS, MCPD, MCT, MCSD.NET, MCAD.NET, and MCSD. He works with Lionbridge (Nasdaq: LIOX), and his business card reads Subject Matter Expert (SME). He is also a life member of Computer Society of India (CSI). He started working on Microsoft .NET with its beta release. He has been involved in software development, evangelism, consultation, corporate trainings, and T3 programs on Microsoft .NET for various employers and corporate clients.

Vidya lives with his beloved wife and lovely daughter Vamika (nicknamed Pearly). He believes that nothing will turn into a reality without them. He is the follower of the concept "No Pain, No Gain" and believes that his wife is his greatest strength. He is a bibliophile, and when he is not working on technical stuff, he likes to play with his one-year-old daughter and also likes reading spiritual and occult science books. He blogs at http://dotnetpassion.blogspot.com.

Acknowledgments

Once again so many people to thank, so much like an Oscar-winning speech (without the tears): From the great Damian Fisher for teaching me how to play the drums, to Andrew and all at Host-It.co.uk, my ISP, for putting up with my incessant hassling over SQL Server and DotNetNuke; my bosses Bill Cotton and Richard Lowe at Lehman Brothers; and Robert McMillan (Toad), a great mate from college with whom I got back in touch after many years—each in their own way have helped me get through the last 12 months. Cheers also to Simon Collier for whipping me at table tennis week in and week out, and Paul Goodwin and Andrew Lockwood for letting me snore on the train and waking me up at the right stop. I also would like to thank Simon and all at Waitrose Steak and Oyster Bar for cooking my tea on a Wednesday night. They make great steak dinners! Finally, to Alan Grosvenor, Richard Porter, and Rob Robson for giving me the odd Sunday off from rugby training to get this and other books out on time. Talking of rugby, thanks to all at Bedford Rugby club who make my Saturdays, well, exciting.

Then, of course, there are my mum and dad who put up with not watching television while I commandeered it for my personal computer in 1980 and onward. I owe them so much and love them to bits. Thanks also to my mother-in-law, Jean, for being so brilliant when things needed doing and keeping my wife sane.

Finally, of course, thanks to my family. To my three kids, Cameron, Ellen, and Scott, who have been brilliant and wonderful and whom I am very, VERY proud of in many ways. And they are such great kids because they have what can only be termed the best mother kids can have, right there, helping, loving, and, yes, screaming at them when needed. Julie, I love you more than I love my Irn-Bru, rugby, and lemon meringue pie . . . and you know how much that means to me! It is to them this book is wholly dedicated.

Introduction

Beginning SQL Server 2005 Express for Developers is for those people who see themselves as becoming either developers, database administrators, or a mixture of both but have yet to tread that path with SQL Server 2005. This edition of my book *Beginning SQL Server 2005 for Developers* is for readers who wish to learn and develop on the free version of SQL Server 2005, SQL Server 2005 Express.

Whether you have no knowledge of databases, or have knowledge of desktop databases such as MS Access, or even come from a server-based background such as Oracle, this book will provide you with the insight to get up and running with SQL Server 2005 Express.

Right from the start, your basic knowledge will be expanded, and you will soon be moving from a perceived beginner through to a competent and professional developer. It is the aim of this book to cater to a wide range of developers, from those who prefer to use the graphical interface for as much work as possible to those who want to become more adept at using SQL Server 2005 Express's programming language, T-SQL. Where practical, each method of using SQL Server 2005 Express is demonstrated, explained, and expanded so that you can evaluate what works best in your situation.

There are plenty of examples within the book of every action along with details about the security of your data. You will also learn the best way to complete a task and even learn how to make the correct decision when there are two or more choices that could be made.

Once you reach the end of this book, you will be able to design and create solid and reliable database solutions competently and proficiently.

Who This Book Is For

This book is ideal for anyone starting out with SQL Server who prefers to use the free but very powerful Express Edition of SQL Server 2005.

How This Book Is Structured

Beginning SQL Server 2005 Express starts with downloading and installing the product. It then moves on to an overview of the product and the GUI tool, SQL Server Management Studio Express, that can be used to develop database solutions. From there it covers all the fundamental concepts and techniques you need to develop databases and complete database solutions.

Prerequisites

None. Everything you need to become proficient with SQL Server 2005 Express is fully covered in this book.

Downloading the Code

The source code for this book is available to readers at www.apress.com in the Source Code/Download section. Please feel free to visit the Apress web site and download all the code there. You can also check for errata and find related titles from Apress.

Contacting the Author

You can contact Robin at robin@fat-belly.com or visit his web site, www.fat-belly.com.

SQL Server 2005 Express Overview and Installation

Welcome to *Beginning SQL Server 2005 Express*. This book is centered around the Express Edition of SQL Server 2005, which is a free download from Microsoft. As you are reading this book, I assume that you are interested in learning how to create solutions with Microsoft SQL Server 2005 Express (SSE), but have no prior knowledge of SSE and probably no experience with any database. The aim of this book is to bring you quickly to a level at which you are developing competently with SSE. This book is specifically designed for beginners or those who at this stage wish to use only SSE. But you'll find this book useful for understanding the basics of any relational database, so moving from SQL Server to Oracle, DB2, MySQL, PostgreSQL, etc., will be relatively easy.

This chapter covers the following topics:

- What is SSE?

- How do I know if my hardware meets the requirements?

- Can I just confirm that I have the right operating system?

- What can I do with SSE?

We will also then look at installing SSE—this section of the chapter covers the following:

- Installing SSE on a Windows XP platform

- Options not installed by default

- Where to install SSE physically

- Multiple installations on one computer

- How SSE runs on a machine

- How security is implemented

- Logon IDs, especially the sa (system administrator) logon

What Is SSE?

SSE is a functionally-reduced version of SQL Server 2005 aimed at several market areas. These include students who cannot afford to purchase the full SQL Server 2005 and any users who wish to run SQL Server on lower-spec hardware. For example, if you have a web site based on a SQL Server host, you can develop your system on SSE and then migrate it to your web host without buying the full version.

SSE limits database size to 4GB and memory to 1GB.

Note If you have more than 1GB of memory, SSE will not use the excess.

SSE can support a wide variety of applications, both for small users and the largest corporations. However, SSE falls squarely in the smaller application category, appropriate for a small number of users and probably one or two developers. It also targets applications that require a database without a GUI built in like Microsoft Access, and it has the ability to scale up in size if required. If you start out on SQL Server Express, you can scale your application up to other SQL Server versions with potentially no, or very little, work. As you will see, it is easy to install, as it comes as one complete package for most of its functionality, with a simple install to be performed for the remaining areas if required.

Evolution of SQL Server

SQL Server has evolved over the years into the product it is today. Table 1-1 gives a summary of this process.

Table 1-1. *The Stages in the Evolution of SQL Server*

Year	Version	Description
1988	SQL Server	Joint application built with Sybase for use on OS/2.
1993	SQL Server 4.2 A desktop database	A low-functionality, **desktop** database, capable of meeting the data storage and handling needs of a small department. The concept of a database that was integrated with Windows and had an easy-to-use interface proved popular.
1995	SQL Server 6.05 A small business database	Major rewrite of the core database engine. First "significant" release, occurring a year after Microsoft splits from Sybase. Improved performance and significant feature enhancements. Still a long way behind in terms of the performance and feature set of later versions, but with this version SQL Server became capable of handling **small e-commerce and intranet** applications, and was a fraction of the cost of its competitors.
1996	SQL Server 6.5	SQL Server was gaining prominence such that Oracle brought out version 7.1 on the NT platform as direct competition.
1998	SQL Server 7.0 A web database	Another significant rewrite to the core database engine. A defining release, providing a reasonably powerful and feature-rich database that was a truly viable (and still cheap) alternative for **small-to-medium businesses**, between a true desktop database such as MS Access and the high-end enterprise capabilities (and price) of Oracle and DB2. Gained a good reputation for ease of use and for providing crucial business tools (e.g., analysis services, data transformation services) out of the box, which were expensive add-ons with competing databases.

Year	Version	Description
2000	SQL Server 2000 An enterprise database	Vastly improved performance scalability and reliability sees SQL Server become a major player in the **enterprise database** market (now supporting the online operations of businesses such as NASDAQ, Dell, and Barnes & Noble). A big increase in price (although still reckoned to be about half the cost of Oracle) slowed initial uptake, but the excellent range of management, development, and analysis tools won new customers. In 2001, Oracle (with 34% of the market) finally ceded its No. 1 position in the Windows database market (worth $2.55 billion in 2001) to SQL Server (with 40% of the market). In 2002 the gap had grown, with SQL Server at 45% and Oracle slipping to 27%. *Source: Gartner Report 5/21/2003*
2005	SQL Server 2005	Many areas of SQL Server have been rewritten, such as the ability to load data via a utility called Integration Services, but the greatest leap forward is the introduction of the .NET Framework. This will allow .NET SQL Server–specific objects to be built, giving SQL Server the flexible functionality that Oracle has with its inclusion of Java. This functionality is also available in SSE.

Hardware Requirements

Now that you know a bit about SQL Server, the next big question on your list may well be "Do I have a powerful enough computer to run SQL Server Express on? Will this help me refine my decision?"

Judging by today's standards of minimum specification hardware, even the low-cost solutions, the answer will in most cases be "Yes." However, you may have older hardware (technology moves so fast that even hardware bought a couple of months ago can be deemed below minimum specification), so let's take a look at what the minimum recommendations are, and how you can check your own computer to ensure that you have sufficient resources.

CPU

The minimum recommended CPU that SQL Server will run on is a Pentium III 500 MHz processor, a compatible processor, or similar processing power. However, as with most minimums listed here, Microsoft wholly recommends a faster processor, 1 GHz in fact. The faster the processor, the better your SQL Server will perform, and from this the fewer bottlenecks that could surface. Many of today's computers start at 2 GHz or above, and 500 MHz has not been the standard installation for a couple of years now. If you have a lower-speed processor, try to invest in upgrading it. You will find your development time reduced for it.

However, it is not processor alone that speeds up SQL Server. A large part is also down to the amount of memory that your computer has.

Memory

Now that you know you have a fast enough processor, it is time to check whether you have enough memory in the system. Microsoft recommends 512MB or above, although the minimum RAM is 192MB.

The more memory, the better: I really would recommend a minimum of 512MB. If a process can be held in memory, rather than swapped out to hard drive or another area while you are running another process, you are not waiting on SQL Server being loaded back into memory to start where it left off. This is called **swapping**, and the more memory, the less swapping could, and should, take place.

Together as a whole, CPU speed and memory are the two items crucial to the speed that the computer will run, and having sufficient speed will let you develop as fast as possible.

Hard Disk Space

You will need quite a lot just to install SQL Server and the help file called Books Online. Then you will have your data files on top of that. For SQL Server alone, ignoring any data files that you are then going to add on top, you will need over 350MB of space, then a further 425MB of space for Books Online and the sample databases, if you plan on downloading and installing them. You can reduce this by opting not to install certain options; however, even most notebooks these days come with a minimum 40GB, and 80GB is not uncommon. Hard disk space is cheap as well, and it is better to buy one disk too large for your needs than have one hard drive that suits now, and then have to buy another later, with all the attendant problems of moving information to clear up space on the original drive.

Again, you will need spare space on the drive for the expansion of SQL Server and the databases, as well as room for temporary files that you will also need in your development process. So think big—big is beautiful!

Operating System Requirements

You will find that SSE will run on Windows 2000 Professional Edition and above with Service Pack 4, or all editions of Windows XP Service Pack 2 or above. It will also work on the 64-bit operating systems for Windows XP, as well as the 64-bit editions of Windows Server 2003. So there is plenty of scope for running SQL Server on many operating systems.

The Example

In order to demonstrate SQL Server 2005 Express, together we will develop a system for a financial company that will have features such as banking, share purchase, and regular buying, such as a unit trust savings plan, etc. This is an application that could fit into a large organization, or with very minor modifications could be used by a single person to record banking transactions.

The book builds on this idea and develops the example, demonstrating how to take an idea and formulate it into a design with the correct architecture. It should be said, though, that the example will be the bare minimum to make it run, as I don't want to detract from SSE. The book will give you the power and the knowledge to take this example, expand it to suit your financial application needs, and give it the specifics and intricacies that are required to make it fully useful for yourself.

But before we can get to this point, we need to install SSE.

Installation

This chapter will guide you through the installation process of the Express Edition with Advanced Services. Although double the size of the vanilla Express Edition and SQL Server Management Studio Express, it does have that extra functionality that we look at in Chapter 14, which discusses creating reports.

Installation covers a great many different areas:

- Security issues
- Different types of installation—whether this is the first installation and instance of SQL Server Express or a subsequent instance, for development, test, or production
- Custom installations
- Installing only some of the products available

Most of these areas will be covered so that by the end of the chapter you can feel confident and knowledgeable to complete any subsequent installations that suit your needs.

A Standard Installation

Let's now take the time to install SSE on our machines. Microsoft offers SSE download options from http://msdn.microsoft.com/vstudio/express/sql/register/default.aspx.

Preparing to Install

First of all, ensure that you have logged on to your machine with administrative rights so that you are allowed to create files and folders on your machine, which is obviously required for installation to be successful. You will also have to have Microsoft .NET Framework 2.0 installed on your machine, which can be downloaded from the Microsoft web site.

Once you start the setup and after you accept the SQL Server End User Agreement, SQL Server then installs some support files. These files are part of SQL Server that will be included in service packs and form part of the installation process. The main files are setup files and the required .NET Framework version if it is not already installed. .NET is a framework that Microsoft created that allows programs written in VB, C#, and other programming languages to interoperate. SQL Server 2005 uses .NET for some of its own internal work, but also, as a developer, you can write .NET code in any of Microsoft's .NET languages, and include this within SQL Server.

■**Note** Including .NET code is an advanced topic and outside the scope of this book. For more information, try *Pro SQL Server 2005 Assemblies*, by Robin Dewson and Julian Skinner (Apress, 2006).

You are then presented with the welcome screen shown in Figure 1-1. Click Next.

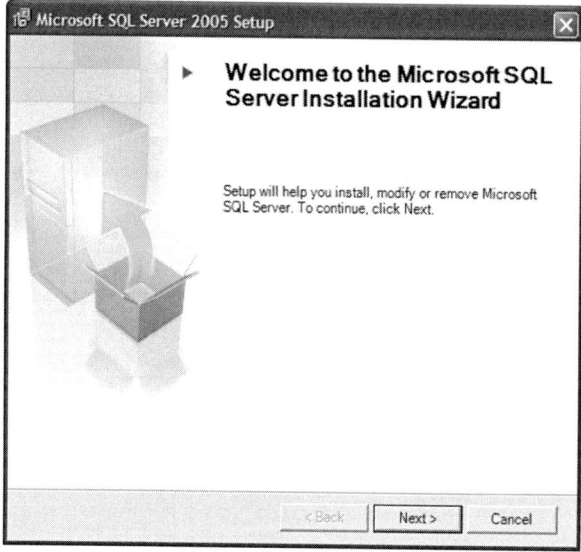

Figure 1-1. *Beginning the install*

We then come to the System Configuration Check, as you see in Figure 1-2. Its main function is to check that the PC meets the hardware and software requirements. There are certain requirements for certain parts of the installation; for example, SQL Server Reporting Services, a tool for producing reports from SQL Server that was an add-on with SQL Server 2000, requires Internet Information Services (IIS). IIS is a process that runs on Windows XP Professional that provides the ability to run a web server. If you are on Windows XP Home, there will be a warning message indicating IIS has not been installed and later in the process reporting services will not be able to be installed. SQL Server Reporting Services is web based.

Figure 1-2. *System configuration tool with IIS warning*

■**Tip** You may well see the Minimum Hardware Requirement warning come up whether you meet the require-
ments or not. Providing you meet the requirements detailed in this book, you should be fine.

You are then required to enter your registration information, as shown in Figure 1-3. Uncheck
the tick box detailed as Hide Advanced Configuration Options. This allows you to deal with looking
at settings such as how SQL Server runs among other items.

Figure 1-3. *Registration page*

Choosing the Components to Install

We now come to the Components to Install screen, where we have to make some decisions. As you
see in Figure 1-4, this installation will have everything installed, because this will be my development
instance where developers will be testing every aspect of SQL Server away from any development of
projects taking place. This is therefore going to be more of a training environment. Select every item
listed.

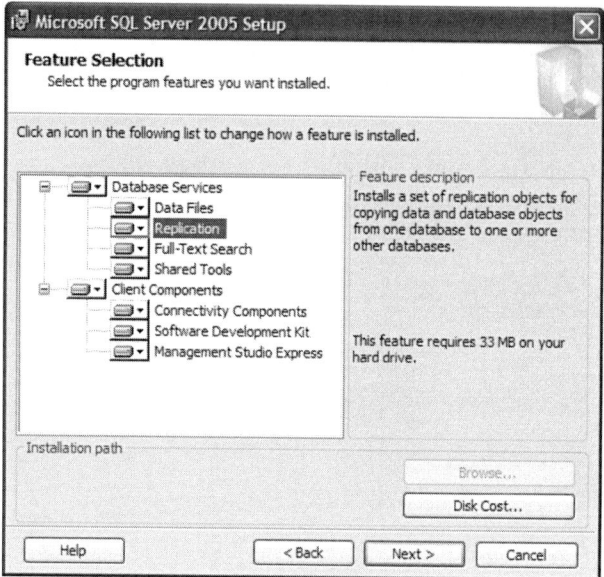

Figure 1-4. *Selecting every component to install*

Let's briefly take a look at what each of these components are:

- *Data Files*: This is the core of SSE, and it installs the main engine, data files, etc., to make SQL Server run.

- *Replication*: This component, which will not be discussed at length in this book, allows you to replicate changes in one database to another automatically.

- *Full Text Search*: This allows you to search text-based data types quickly. We don't look at this in this book.

- *Shared Tools*: These are tools used by every instance of SQL Server Express.

- *Connectivity Components*: This component allows importing and exporting data using as the data source not only SQL Server, but also Oracle, Excel, and so on.

- *Software Development Kit*: These are tools for working on the workstation. This installs the GUI we use with SQL Server, and it can also install the help feature, Books Online.

- *Management Studio Express*: This is the graphical tool used to work with SSE.

Naming the Instance

As you know, SSE is installed on a computer. It is possible to install SQL Server more than once on one computer. This could happen when you have a powerful server and it has enough resources such as memory, processor, etc., to cope with two or three different applications running. These different applications want to have their own SQL Server. Each install is called an **instance**. We are now at the stage that we can name the instance of the install. Each instance must have a unique name attached to it, although "no name," known as a **Default Instance**, is also classified as a unique name.

Naming instances is important as the first step to organizing your environments. For example, you may have an instance for development, another instance for system testing, and finally one for user testing.

Default Instance is available, which is what is selected when you are not giving the install a specific name; once you come to install SSE outside of a learning environment, you should avoid this, as it gives you an installation with no name and therefore no hint as to its use. As you are still learning, the easiest option to understand is to use the Default Instance, so select Default Instance as shown in Figure 1-5 and then click Next.

Figure 1-5. *Naming the install instance*

Service Accounts

SSE requires you to log in to Windows before starting, just as you need to log in to Windows before using the system. SQL Server, Reporting Services, etc., can run without you or anyone being logged in to the computer the install took place on. They can run just so long as the computer has fired up successfully. This is normal when SQL Server is installed on a server that is held in a remote location like a server room.

However, nothing can run on Windows without having some sort of login. If you do log in to the computer, as you likely will while working through this book because SSE will be running on a home or local system, you can use this Windows user ID for SQL Server to also use to log in and start its service. This is known as a **local system account**.

On the other hand, you can create a Windows login that exists purely for SSE. This could exist for several reasons. For example, your Windows account should be set up so that the password expires so many days after being set, or locks out after a number of incorrect password attempts. This is to protect your computer and the network, among many other areas. However, SSE should use a separate account that also has an expiring password and the ability to lock the account after a number of successful attempts. This kind of non-user-specific, "generic" account removes the link between SSE and a person within an organization.

Where you are installing your SSE will have an effect on the decision you make. As you are just learning SSE, let's not make it complicated at this point; I assume that this install will be for you to use to learn SSE. Therefore, just select Use the Built-in System Account/Local System as you see in Figure 1-6. You can also define what services start when the computer is started up. Keep the defaults at the moment as you can always change these later via the Services icon within the Control Panel. Click Next.

Figure 1-6. *Service account selection*

Authentication Mode

We now come to how we want to enforce security on our SSE installation. As Figure 1-7 shows, there are two choices: Windows authentication mode and mixed mode. You will learn more about modes later in the chapter but very, very simply, Windows authentication mode denotes that you are going to use Windows security to maintain your SQL Server logins, whereas mixed mode uses either Windows security or an SSE-defined login ID and password. We also need to define a password for a special login ID, called sa. Again, you will learn more about this in a moment, but for now you must enter a valid password. Use a meaningful and impossible-to-guess password.

Figure 1-7. *Authentication choices*

Collation Settings

Collation settings specify how sorting and comparison of rows of data are dealt with within SSE. For example, a collation setting will inform SSE of details such as whether the system is case sensitive. It is possible to have different collation settings for each type of SQL service, shown in the earlier setup process, so you could sort one way for Analysis Services different from that defined for your main SSE installation. It would only be in exceptional circumstances that you would do this, as it will cause extra processing complications when using the same processing in more than one service. Figure 1-8 shows that Windows collation has been chosen.

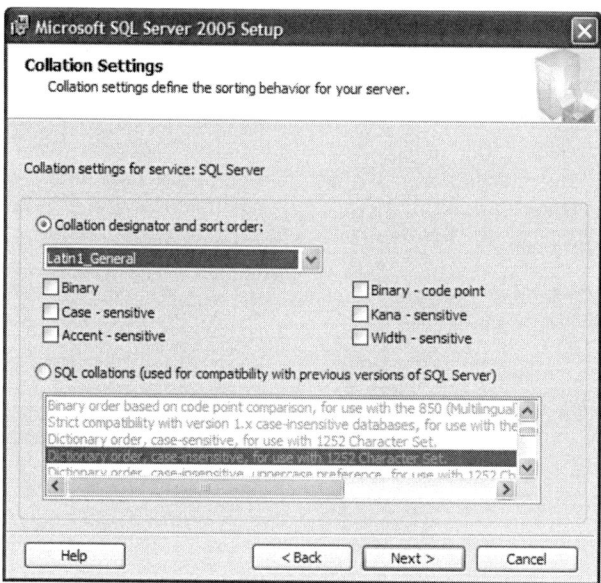

Figure 1-8. *Choosing the collations*

The Reporting Services Database

If you were on an operating system that had IIS, you may have selected Reporting Services to be installed. It is necessary to create a database for the reporting server to use. Depending on your requirements and how heavily used your SQL Server installation is, you may wish your reports to be run out of a separate and purpose-built SQL Server installation. For the moment, we will install Reporting Services on the same SQL Server (see Figure 1-9).

Figure 1-9. *Installing as default Reporting Services*

User Instances

When installing SQL Server Express, we can give the install, also known as an instance, a name. In this book, we are using SQL Server Express with no name, and therefore we are using the Default Instance. If we want to have one instance for development, one for testing, and one for user acceptance testing, we would need to run the SSE setup process again. Then at the instance screen shown previously in Figure 1-5, we would name an instance at that point. However, SSE also allows for databases that have been created to be attached to the server when an SQL client connection is made, if the Enable User Instances option is enabled, as shown in Figure 1-10. What this allows is for developers to attach a database through their code in, for example, C# at run time.

This option allows databases to exist in SQL Server only at run time, which would allow SSE to run a large number of databases. However, if you do go down this route, you'll encounter problems involving authority and the connecting user, whether they have enough permissions to attach the database and what they can execute. Unless you have a real need to attach and detach at run time, leave this option disabled.

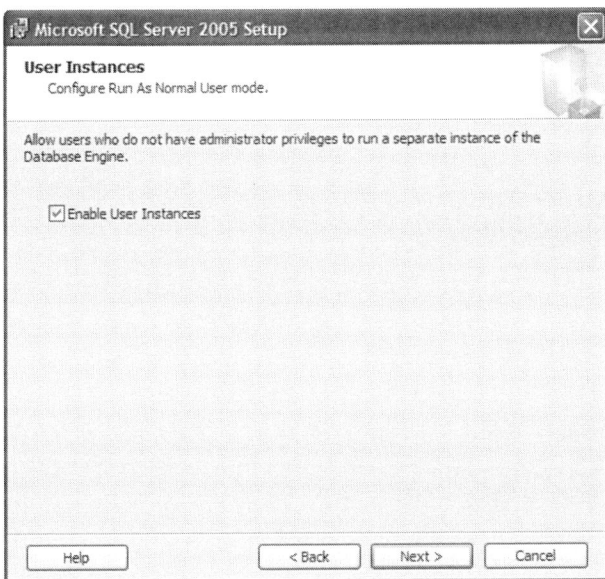

Figure 1-10. *Enabling user instances*

Error and Usage Report Settings

Within SSE, it is possible for any errors to be automatically reported and sent to Microsoft. These include fatal errors where SSE shuts down unexpectedly. It is recommended that you keep the error settings shown in Figure 1-11 enabled. No organizational information will be sent, therefore your data will still be secure. This is similar to sending reports when Excel crashes, for example. It is better to have these options enabled.

The final screen (see Figure 1-12) is displayed when the setup is complete. You can click the Summary Log link to check the install log. There is also a list of recommendations and information in a scrollable text box at the bottom of the screen. In between these two areas is a link to a new tool for SSE, the Surface Area Configuration tool. This tool, which we look at in Chapter 2, deals with enabling or disabling features, services, etc. You don't have to click the link now, as we will access this tool from the Start menu later.

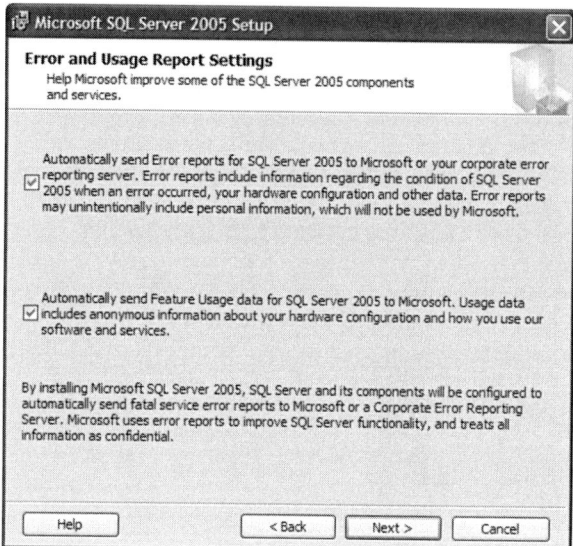

Figure 1-11. *Error and usage report settings*

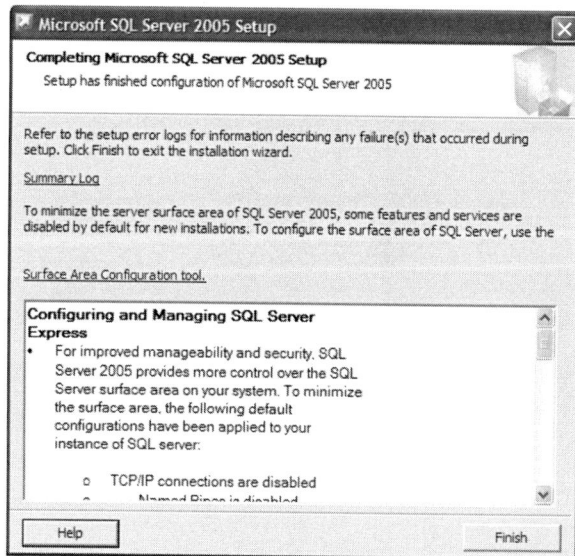

Figure 1-12. *Setup complete details*

And that is it, you are now ready to install.

Security

To discuss the Service Account dialog box that we came across in the installation properly (refer back to Figure 1-6), we need to delve into the area of Windows security.

In this section, we will first examine the concept of Windows services as opposed to programs, and then move on to discussing different types of authentication we can choose when installing SQL Server.

Services Accounts

SSE runs as a Windows service. So what is a service? A good example of a service is any antivirus software that runs continuously from when the user restarts a computer to the point that the computer shuts down. A program, on the other hand, is either loaded in memory and running, or not started. So what is the advantage of running a service? When you have a unit of work that can run as a service, Windows can control a great deal more concerning that process. A service can be set to start automatically before any user has even logged on; all other programs require a user to be logged in to Windows in order for him or her to start.

A service also has absolutely no user interface. There will be no form to display and no user input to deal with at run time. The only interaction with the process runs either through a separate user interface, which then links to the service but is a totally separate unit of work (for example, SQL Server Management Studio Express), or from Windows management of that service itself. Any output that comes from the service must go to the Event Log, which is a Windows area that stores any notification from the services that Windows runs.

Having no interface means that the whole process can be controlled without human intervention. Providing the service is designed well, Windows can take care of every eventuality itself, and can also run the service before anyone has even logged in to the computer.

In most production environments, SSE will be running on a remote server, one probably locked away in a secure and controlled area, possibly where the only people allowed in are hardware engineers. There probably isn't even a remote access program installed, as this could give unauthorized access to these computers. SSE will run quite happily and, with any luck, never give an error. But what if one day there is an error? If SSE was running as a program, some sort of decision has to be taken. Even if SSE crashed, there at least has to be some sort of mechanism to restart it. This means another process needs to be run, a monitoring process, which in itself could result in a whole ream of problems. However, as a service, SSE is under Windows control. If a problem occurs, whether with SSE, Windows, or any outside influence, Windows is smart enough to deal with it through the services process.

It's time to move on to the options we are given during installation regarding authentication mode.

Looking at the Authentication Mode

Probably the most crucial information in the whole setup process, and also the biggest decision that has to be made, concerns the authentication mode you wish to apply to your server. As we saw earlier in the setup process, there are two choices: **Windows authentication mode** and **mixed mode**.

Windows Authentication Mode

To log on to a Windows 2000/2003/XP machine, a username must be supplied. There is no way around this (unlike in Windows 9x/ME where a username is optional). So, to log on to Windows, the username and password have to be validated within Windows before the user can successfully log in. When this is done, Windows is actually verifying the user against username credentials held within the domain controller, or, if you are running SQL Server on a standalone machine at home, the credentials held locally. These credentials check the access group the user belongs to (the user rights). The user could be an administrator, who has the ability to alter anything within the computer, all the way down to a basic user, who has very restricted rights. This then gives us a trusted connection; in other words, applications that are started up after logging in to Windows can trust that Windows has verified that the account has passed the necessary security checks.

Once we have logged in to Windows, SQL Server uses a trusted connection. This means SQL Server is trusting that the username and password have been validated as we just mentioned. If, however, the username does not exist, you won't be able to log on to that machine. Someone else can log on to your machine with his or her user ID and password, and although he or she might be able to get to SQL Server by finding the executable on the C drive, SQL Server will first of all check to see whether that user has a valid login within SQL Server. If the login isn't valid, it will check the Windows group that the user belongs to and check its security to see whether that group is set up to access SQL Server. If that user has administration rights to your computer, the user may well be able to at least connect to SQL Server.

SSE on Windows XP Home works slightly differently with its security compared to Windows XP Pro, Windows Server 2003, and so on. We look at this in detail in Chapter 4, but for now, in the upcoming text, it is important to compare the two different scenarios in general terms.

We are also in a bit of a catch-22 situation here. You need to know about security for your install process, but to demonstrate it fully means working with SQL Server Management Studio Express, which the next chapter covers. We will keep the discussion simple here, so let's look at an example involving security now.

Try It Out: Windows Authentication Mode

1. Ensure that you are logged on to your machine as an administrator. If you are on a local computer, chances are that your login is in fact an administrator ID. If this computer is on a network and you are unsure about your access rights, ask your PC support desk to help you out with the ID and password.

2. Select Start ➤ Control Panel ➤ User Accounts.

3. When the Users and Passwords dialog box comes up, click Create a New Account.

4. Once the Name the New Account dialog box comes up, enter the username *AJMason*, as shown in Figure 1-13. When done, click Next.

5. Ensure that the account type specified is Limited (see Figure 1-14). This means that it will not have administrator privileges.

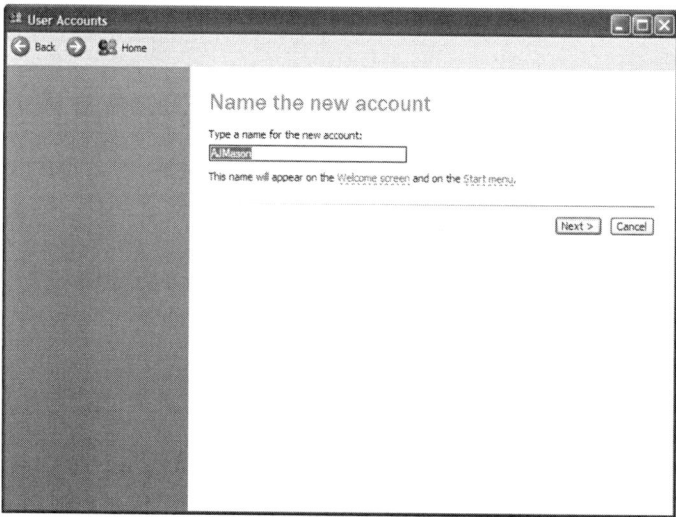

Figure 1-13. *Creating a new user account*

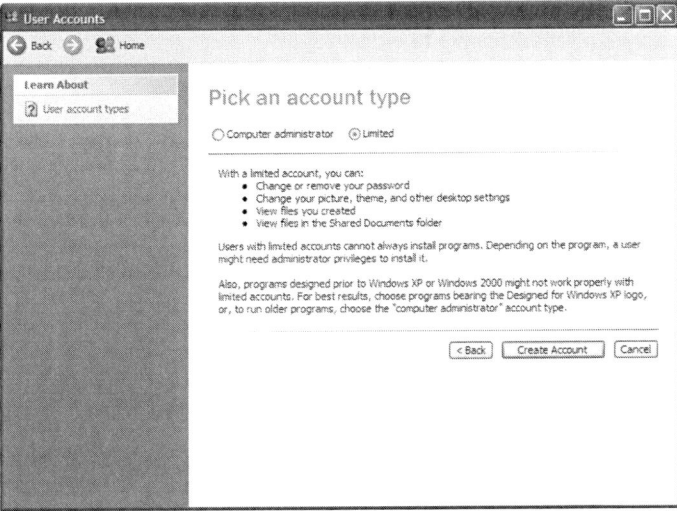

Figure 1-14. *Selecting the new user's account type*

6. Stay in the User Accounts dialog box, as you want to add a second username. Repeat the preceding process using the following details:

 Username: VMcGlynn

 Account type: Computer Administrator

7. Log off from Windows and then log on using the first ID that you created: AJMason.

8. Once logged on, start up SQL Server Management Studio Express by selecting Start ➤ All Programs ➤ Microsoft SQL Server 2005 ➤ SQL Server Management Studio Express. This is where Windows XP Home differs from other operating systems. The remaining steps will only be seen on Windows XP Professional.

9. Examine the error message that appears, which should resemble what you see in Figure 1-15. AJMason as a login has not been defined within SQL Server specifically and does not belong to a group that allows access. The only group at the minute is a user who is in the `Administrators` Windows group. Recall that AJMason is a Limited user.

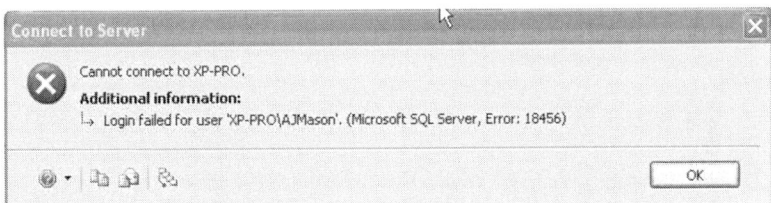

Figure 1-15. *Failed login to server*

10. We will now try out the other user we created. Close down SQL Server, log off Windows, and log on using the second ID—VMcGlynn. Once logged in, start up SQL Server Management Studio Express and connect to your server. This time the login will work.

We have created two usernames: one that has restricted access (AJMason), the other with administration rights (VMcGlynn). However, neither of these specific usernames exist within SQL Server itself: after all, we haven't entered them and they haven't appeared as if by magic. So why did one succeed and one fail?

The Windows security model has ensured that both IDs are valid. If the ID or password were incorrect, there is no way that you could be logged in to Windows. Therefore, when you try to connect to SQL Server, the only check that is performed is whether the user has access to SQL Server either via membership of an operating system group or through the specific logged-in user account. As you can see in Figure 1-16, neither AJMason nor VMcGlynn exist.

Figure 1-16. *Object Explorer for SQL Server*

However, you can see that there is a Windows group called BUILTIN\Administrators. This means that any username that is part of the Administrators group will have the capacity to log on to this SQL Server. Hence avoid if possible setting up users as administrators of their own PCs.

In a production environment, it may be advisable to remove this group from the system if you do allow users to be administrators. As VMcGlynn is a member of the Administrators group, then this username will also be a member of the BUILTIN\Administrators group.

Windows XP Home

Windows XP Home will place both AJMason and VMcGlynn in the same group, BuiltIn\Users. This group exists within SQL Server, and therefore all users you ever define for your computer will be able to connect. They will not be able to do anything on any database created that they have not been given access to, but you should instantly realize that the security is slightly more lax. The main reason behind this is the lesser complexity of Windows Security in Windows XP Home compared to versions of the software that are more likely to be running on networks and so on. I discuss this more in Chapter 4.

Mixed Mode

If we installed SQL Server with mixed mode, this means we can use either Windows authentication, as has just been covered, or SQL Server authentication.

How does mixed mode differ from Windows authentication mode? To start with, you need to supply a user ID and password to connect rather than Windows taking the ID of the logged-in account. There is no assumption that the username supplied is a valid ID. Using mixed mode is also appropriate in many cases when working with Internet Service Providers (ISPs). To clarify this, if you are working on your remote data at a local client machine, the remote machine would need to know the credentials of your login, and the easiest method is to use SQL Server authentication.

There is also another argument for mixed mode. We may wish for some reason (for example, for auditing purposes) that the user log on to SQL Server using a different username from that of his or her Windows account. We could be working on a large SQL Server development project that will have developers joining and leaving the team as the need arises. In this case, it might be necessary to create temporary usernames, as opposed to permanent IDs linked to the developers' Windows usernames. In SQL Server, we could create usernames of Developer1, Developer2, etc. These usernames can have different access rights within SQL Server. Another situation would be the case of an Internet-based application where there is just no way we could create a username for every visitor to our site. Therefore, we would create a generic login ID using a specific ID created for the web site. Whatever the reason, there is a need to have usernames not linked with the Windows usernames.

You will learn how to add usernames to SQL Server (as opposed to adding Windows users) when I talk about security in Chapter 4.

This leaves one area of security that needs to be discussed here: the sa login.

The sa Login

The sa login is a default login that has full administration rights for SQL Server. You saw during the installation process that you would be forced to include a password for this account if you were installing with SQL Server Authentication enabled because it is such a powerful login that exists in every SQL Server installation. If you logged in to SQL Server as sa, you will have full control over any aspect of SQL Server. SQL Server inserts this ID no matter which authentication mode you install. If you have a Windows account defined as sa, for example, for Steve Austin, then this user will be able to log in to the server if you have set the server up to implement Windows authentication mode without any further intervention on his part. Try to avoid login IDs of sa.

In a mixed mode installation, sa will be a valid username and validated as such. As you can guess, if any user gets ahold of this username and the password, that user will have full access to view and amend or delete any item of data. At worst, the user could corrupt any database, as well as corrupt SQL Server itself. He or she could even set up tasks that e-mail data to a remote location as it is being processed.

It is essential to set up a strong password on the sa account in the Authentication Mode screen if you choose mixed mode. It is a major improvement in SQL Server 2005 that you are now forced to enter a password, although it is possible to set up a very easily guessed password. Do not use passwords such as password or adminpwd, for example. Always keep the password safe, but also make a note of it in a safe place. If you forget the sa password and this is the only administration ID that exists, you will need to reinstall SQL Server to get out of this problem. A good password is one that mixes numbers and letters, but doesn't always include letters than can be made into numbers and numbers into letters in all cases. For example, pa55word is just as easy to guess as password. Or 4pr355 for Apress.

There is also another reason not to log on to SQL Server with the sa username. At times it will be essential to know who is running a particular query on a SQL Server database. In a production database, someone may be running an update of the data, which is filling up the disk space or filling up the transaction log. We need to contact that person to check whether he or she can stop the process. If that person logs in as sa, we will have no idea who he or she is. However, if that person logs on with an identifiable name, he or she will have an ID in SQL Server, which we can track. By

restricting the sa login so that people have to use their own accounts, we can ensure a much higher degree of system monitoring and integrity.

There will be times that we want mixed mode authentication; it is perfectly acceptable to wish this. Internet providers use mixed mode, as many applications may be on one web server. If this ISP is a reseller (in other words, many people around the globe use the one computer), you will not want these people to have the ability to see your data. We have also decided not to have sa as an administration logon at this point. So what do we do? Well, we create a logon ID that will have the access privileges we wish; in other words, the ability to just see the data and work with the data that we need, and no more. The ISP may require you to supply a user ID and password that it uses to create an account on its SQL Server instance. You will encounter more about this in Chapter 4.

■**Note** Regardless of the authentication mode, it is important you always supply a strong password.

Summary

By this point, you should understand the small differences between each version of SQL Server. You should also know how to check your computer to see whether it is suitable for a SQL Server installation.

By following the steps given earlier, you should have a successful installation of SQL Server on your computer. You may even have completed the installation twice so that you have a development server installation as well as a test server installation. This is a good idea, and something to consider if you have only one installation so far. Whether you are working in a large corporation or are a "one-man band," keeping your production and development code separate leads to greatly reduced complications if, when developing, you need to make a production fix.

This chapter introduced you to security in SQL Server so that you can feel comfortable knowing which way you want to implement this and how to deal with different usernames. You may not have any data yet, but you want to ensure that when you do, only the right people get to look at it!

You are now ready to explore SQL Server 2005 Express Edition. One of the best ways of managing SQL Server is by using SQL Server Management Studio Express, which will be discussed in the next chapter.

SQL Server Management Studio Express

Now that SQL Server Express is successfully installed on your machine, it is time to start exploring the various areas that make this an easy and effective product to use. With SSE, all the administration interfaces now reside in **studios**, such as SQL Server Management Studio Express, which features tools for working with developing database solutions, and if you move to the full version of SQL Server 2005, you'll see SQL Server Business Intelligence Development Studio, designed for analyzing data with Analysis Services. This chapter concentrates on the SQL Server Management Studio Express tool, also known as SSMSE. It also looks at the Surface Area Configuration (SAC) tool.

SSMSE is the graphical user interface (GUI) you will use to build your database solutions. This is an easy-to-use and intuitive tool, and before long, you will feel confident in using it to work with SQL Server quickly and efficiently.

SQL Server Management Studio Express is crucial to your success as a developer. By the end of this chapter, you should be proficient in the following areas:

- How to use SQL Server Management Studio Express

- How to configure SQL Server Management Studio Express

- How to secure the surface area of SQL Server from vulnerabilities

A Quick Overview of SSMSE

SSE runs as a separate Windows process as touched on in Chapter 1, whether on a standalone desktop machine, on a server, or on a network (with the more advanced versions of SQL Server). If you open Task Manager and move to the Processes tab, you will see **sqlservr.exe**. This process (or service) runs in its own process space, and is isolated from other processes on the machine. SQL Server should not be affected by any other piece of software that does not talk to any SQL Server component. If you have to kill any other component's process, the SSE engine should continue to run.

SSE runs as a service that is controlled and monitored by Windows itself. Windows ensures that it is given the right amount of memory, processing power, and time, and that everything is working well. Because SSE runs as a service, it has no interface attached to it for a user to interact with. As a result, there needs to be at least one separate utility that can pass commands and functions from a user to the SSE service, which then passes them through to the underlying database. The GUI tool that accomplishes this is SQL Server Management Studio Express.

SSMSE can be used to develop and work with several installations of SQL Server in one application. These installations can be on one computer, or on many computers connected through a local area network (LAN), a wide area network (WAN), or even the Internet. Therefore, it is possible to deal with your development, system testing, user testing, and production instances of SQL Server

from one instance of SSMSE. SSMSE helps you in the development of database solutions, including creating and modifying components of a database, amending the database itself, and dealing with security issues. Getting to know this tool well is crucial to becoming a successful professional SQL Server developer, as well as a database administrator.

One of the tools within SSMSE that we will use for completing tasks is Query Editor. This tool allows program code to be written and executed, from objects to commands that manipulate data, and even complete tasks such as backing up the data. This program code is called **Transact-SQL** (T-SQL). T-SQL is a Microsoft proprietary language, although it is strongly linked to the ANSI-92 SQL standard of the American National Standards Institute (ANSI).

Query Editor allows us to programmatically build the same actions as dragging and dropping or using wizards. However, using T-SQL within Query Editor can give developers more control over certain aspects of certain commands. Note that the name "Query Editor" comes from the fact that it sends **queries** to the database using T-SQL. Don't worry if you don't quite grasp this—all will become clear very soon.

Let's spend some time taking a look at SSMSE in more detail.

Try It Out: Touring SQL Server Management Studio Express

1. To start up SQL Server Management Studio Express, select Start ➤ Programs ➤ Microsoft SQL Server 2005 ➤ SQL Server Management Studio Express.

2. Click the Options button to bring up a Connect to Server dialog box similar to the one in Figure 2-1. Note the following items in this dialog box:

 - SQL Server Express only allows Database Engine to be selected in the Connect to Server dialog box unless you install the toolkit for Business Intelligence.

 - The second combo box contains a list of SQL Server installations that the Connect to Server dialog box can find or knows about. In the dialog box in Figure 2-1, you will see the name of the computer that the local install is on. If you open the Server Name combo box, you can search for more servers locally or over a network connection.

 - The final combo box specifies which type of connection you wish to use. We installed SQL Server with Windows authentication in Chapter 1, therefore this is the only available option for connecting with our local server.

Figure 2-1. *SQL Server Management Studio Express Connect to Server dialog box without expanded options*

3. Click Options, which will switch you to the Connection Properties tab. Here you will see specific properties for this connection, as shown in Figure 2-2:

- The first area to take a look at is the Connect to Database combo box, which provides a list of databases based on the server and login details in the Login tab. Clicking the down button for this combo box allows you to browse for and select a database on the server to which you wish to connect. Only the databases that the Windows Account or SQL Server login can connect to will populate this list. Also, any error in the login details will cause an error message to be displayed here instead of listing databases.

- The Network area details how we want this connection to be made with SQL Server. At the moment, there is no need to change the current settings.

- The third area, Connection, deals with connection timeouts. The first item, Connection Time-Out, defines how long the connection should wait before returning an error. For local installs and even most network installs, a setting of 15 seconds should be more than enough. The only situation that may require you to increase this setting is if you were connecting over a WAN or to a SQL Server installation at an ISP. The second option, Execution Time-Out, details the timeout value for any T-SQL code that you execute. A setting of 0 means that there is no timeout; there should be few, if any, occasions when you would want to change this setting.

Figure 2-2. *SQL Server Management Studio Express connection properties*

4. Once you are happy with all the items in the Connection Properties tab, click Connect. This brings you to SSMSE itself. If you have ever used Visual Studio .NET, you will notice that SSMSE has a reasonably similar layout. This is deliberate on Microsoft's part, as the company is bringing SQL Server a tool more integrated with .NET. Your layout should look like the one in Figure 2-3, with only minor name changes based on the server you have connected to and the connection you have used. This figure shows I have connected to XP-HOME SQL Server using the Windows account XP-HOME\rdewson.

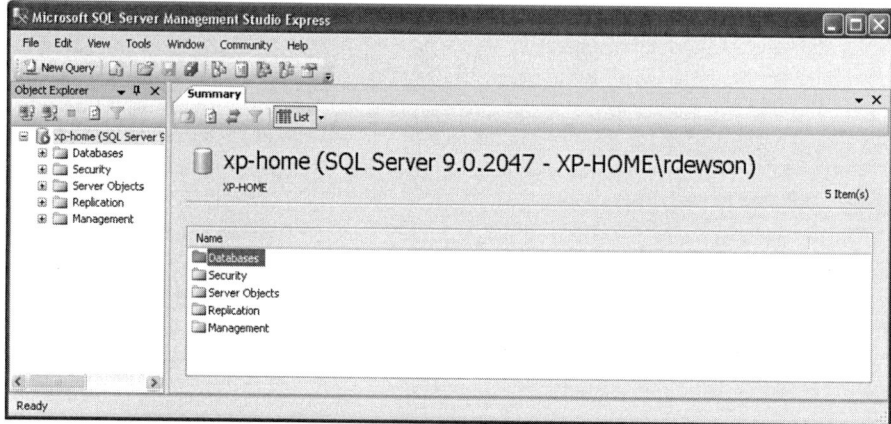

Figure 2-3. *SQL Server Management Studio Express*

5. The first area of SSMSE we will look at is the Registered Servers Explorer. Access this explorer, shown in Figure 2-4, by selecting View ➤ Registered Servers or by pressing Ctrl+Alt+G. This area details all SQL Server servers that have been registered by you on this SSMSE installation. At present, there will only be the server just registered. This explorer will also show registered services for other services such as Reporting Services (which is covered in detail in Chapter 14).

Figure 2-4. *A list of registered servers*

6. If you need to register another server, right-click the Database Engine node and select New ➤ Server Registration to bring up a dialog box very similar to the Connect to Server dialog box shown earlier. Go ahead and do this now to familiarize yourself with the New Server Registration dialog box, shown in Figure 2-5.

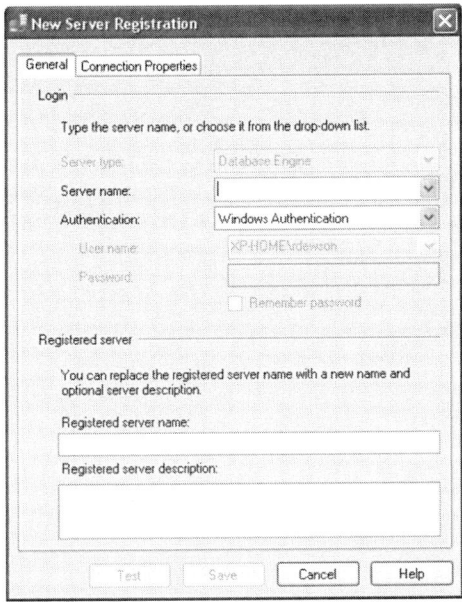

Figure 2-5. *The New Server Registration dialog box*

7. As you can see, the only real difference from the Connect to Server dialog box is that the Server Name combo box is empty and there is a new section called Registered Server. In this area, you can give a registration a different name, such as Development Server or User Testing Region, and on top of this give the registration a description. You don't have a server to register, so just click Cancel now.

8. Moving back to SSMSE's explorer window below the registered servers, take a look at the Object Explorer, which should have been present when you first brought up SSMSE. If it isn't there or if it disappears, you can redisplay it by selecting View ➤ Object Explorer or by pressing F8. You will likely use this explorer the most, as it details every object, every security item, and many other areas concerning SQL Server Express. You can see that SSMSE uses nodes (which you expand by clicking the + signs) to keep much of the layout of the Object Explorer (the hierarchy) compact and hidden until needed. Let's go through each of the nodes you see in Figure 2-6 now.

Figure 2-6. *Object Explorer nodes*

- *Databases*: Holds the system and user databases within the SQL Server you are connected to.

- *Security*: Details the list of SQL Server logins that can connect to SQL Server. You will see more on this in Chapter 4.

- *Server Objects*: Details objects such as backup devices and provides a list of linked servers, where one server is connected to another remote server.

- *Replication*: Shows the details involving data replication from a database on this server to another database (on this or another server) or vice versa.

- *Management*: Details maintenance plans, which you will learn more about in Chapter 7, and provides a log of informational and error messages that can be very useful when troubleshooting SQL Server.

9. Select the topmost node in the Object Explorer to see a summary page similar to the one in Figure 2-7. This area is known as the documents area. You don't have to be on the top node for this page to be of use, as it will provide a summary of any node within this explorer. This works a bit like folders within Windows Explorer, where you can navigate through each item to get a summary of details of objects within the node.

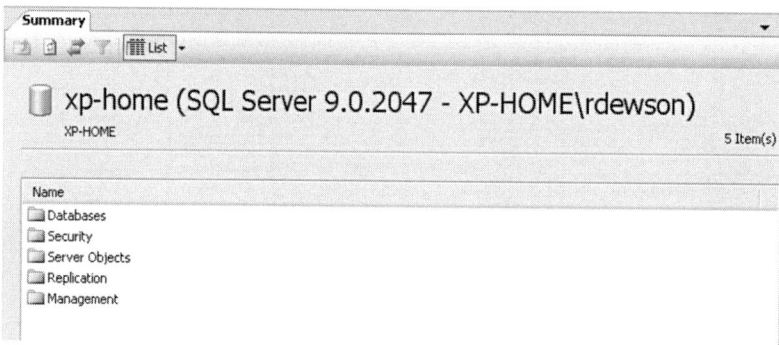

Figure 2-7. *Summary page*

10. Moving to the menu bar of SSMSE, the first item of interest is the View menu option. The first three options on the View menu, shown in Figure 2-8, bring up the two explorer windows, Object Explorer and Registered Servers Explorer, and the summary page we encountered previously. Therefore, if you ever need to close these items to give yourself more screen space, you can reopen them from the menu or with the shortcut keys you see defined. The other options on the View menu are as follows:

- *Template Explorer*: Provides access to code templates. In the examples in this book, we will be building objects using T-SQL. Rather than starting from scratch, we can use code templates that contain the basic code to create these objects.

- *Properties Window*: Displays the set of properties for each object.

- *Bookmark Window*: Allows you to create **bookmarks**, which you place into various locations in your code to allow you to jump quickly to those locations.

- *Web Browser*: Brings up a web browser within SQL Server, ideal for searching the web for answers to SQL Server problems for which you may require information.

- *Other Windows*: Allows you to access other windows generated when running T-SQL from Query Editor, which may hold error messages or results from queries.

- *Toolbars*: Brings up toolbars for Query Editor, diagramming the database, and integration with Visual SourceSafe for source control, if they are not opened by default.

- *Full Screen*: Removes title bars and explorer windows, and then maximizes SSMSE to show as much of the main pages as possible.

Figure 2-8. *The View menu options*

11. SQL Server Express has no built-in tools unlike other versions of SQL Server, although there is the capability to include other tools when they are launched. These can be accessed through the Tools menu, shown in Figure 2-9, along with the means to customize keyboard commands, show or hide toolbar buttons, etc., as is the case with other Microsoft products such as Word. In paticular, note the following Options item, which lets you access different options you can use to set up your SSMSE as you like. We will take a look at this area of the Tools menu in the next section.

Figure 2-9. *The Tools menu options*

12. The final part of SSMSE that we will take a look at is the main SSMSE toolbar, as you can see in Figure 2-10. Some of the icons, such as the Save icon, will be instantly recognizable, but let's go through each button, from left to right, so that it is clear what they all do.

Figure 2-10. *Main toolbar*

- In the next section, you will see how you can write code to add objects, work with data, and so on. Clicking the New Query button will open up a new query window, which allows you to do this using the connection already made with SQL Server.

- Similar to the New Query button, the New Database Engine Query button will also create a new query window. However, this will give you the option of having a different connection to SQL Server through which to run your code. This is a good way of testing code with a different connection to ensure that you cannot see data that should be secure, such as wages, via that connection.

- As with every other Windows-based product, it is possible to open and save files. The Open button will allow you to search for a T-SQL file. The next two buttons change their function depending on what you are doing, but in the main, the Single Save button will give you the option to save the details of the window that is active in the main documents area of SSMSE. The Multiple Save button will give you the option to save all the open tabs in the documents window.

- The last set of buttons open up explorers and document tabs that we have covered already. From left to right, these buttons access the Registered Servers Explorer, Summary page, Object Explorer, Template Explorer, and Properties window.

Now that we've covered the main areas of SSMSE, we'll next take a closer look at the Options area off the Tools menu, as it warrants a more detailed discussion.

Examining SSMSE's Options

As you saw earlier, the Tools menu has an Options menu choice. This allows you to choose what options you would like to set as part of the setup for SSMSE. We will go through each node and option in this area one at a time, except for the options dealing with Analysis Services, which are not of interest to us just now.

Environment Node

The first node we'll look at is the Environment node, which covers the SSMSE environment and how you would like it to look and feel. This contains the General, Fonts and Colors, Keyboard, and Help nodes, which you'll see next.

General Node

The General node, shown in Figure 2-11, contains the following options:

At Startup: This controls how SSMSE behaves when it is started. You have a choice of four options here. It is possible to open Object Explorer after prompting for a connection, open a new query window after prompting for a connection, open both of these after prompting for a connection, or open with an empty SSMSE and no connection.

Hide System Objects: In SQL Server, system objects are hidden. This is a good option to have enabled unless you are a database administrator.

Environment Layout: The layout can either be defined as tabbed documents (a bit like Excel) or as MDI (a bit like Word).

Docked Tool Window—Close Button: If checked, when you click the close button, only the active document is closed. Unchecked means that all windows will be closed.

Docked Tool Window—Auto Hide Button: You can pin toolboxes or unpin explorers to hide them. Unpinning windows affects only the currently active document.

Display NN Files in Recently Used List: This indicates the number of recently opened files to place under the File menu option.

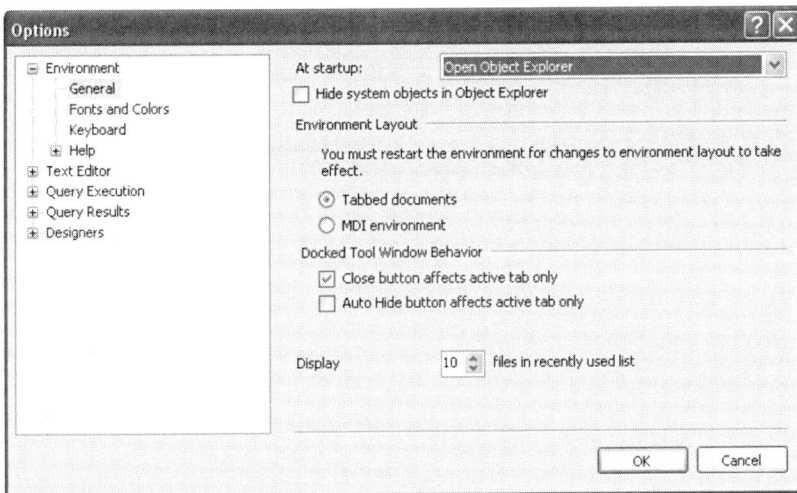

Figure 2-11. *Environment area, General options*

Fonts and Colors Node

As you might guess, the Fonts and Colors node options, shown in Figure 2-12, affect the fonts and colors for different areas of SSMSE. The Display Items list box contains a list of all the different areas for which fonts and colors can be set. By selecting one of these items, you can define the color of the foreground and background as well the font type and size.

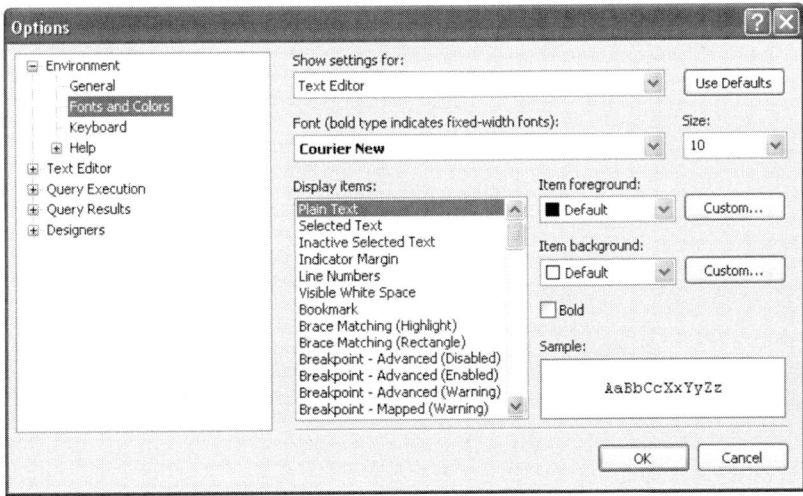

Figure 2-12. *Environment area, Fonts and Colors options*

Keyboard Node

The Keyboard node section, shown in Figure 2-13, allows you to define fast keys for commands you run often. Any T-SQL stored procedure can be defined. The examples in this book assume you are using the standard keyboard scheme.

■**Note** A **stored procedure** is a set of code that is stored within SQL Server, a bit like a program.

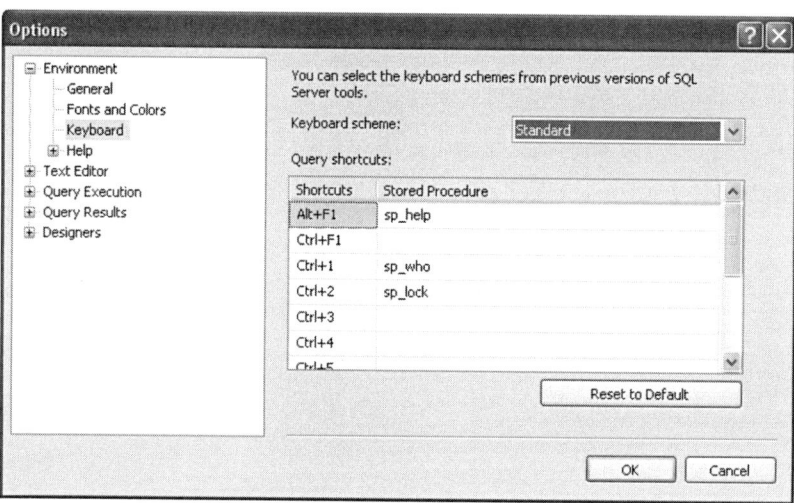

Figure 2-13. *Environment area, Keyboard options*

Help Node

The help system for SQL Server Express as a whole has been altered: you now have the ability to use not only the help installed on the computer, but also the online help; thus you have access to the most up-to-date information. Configure the help system through the Help node options shown in Figure 2-14.

Figure 2-14. *Environment area, Help options*

Text Editor Node

The Text Editor node contains options that affect how you work with text.

File Extensions

Files specific to a particular Microsoft product have their own unique file extension so that they are instantly recognizable to users and can then be linked to the relevant product. These products have different filtering when accessing them through Open from the File menu so that you will only see files with the relevant extension. This also holds true for SQL Server Express, but it is possible to alter these extensions within the File Extension option, as you see in Figure 2-15, although I strongly recommend that you don't. You'll come across a few of these extensions throughout the book, although the majority are for more advanced work.

Figure 2-15. *Text Editor, File Extension defaults*

All Languages ➤ General

Taking a look at the second option within the Text Editor node, you can see how different text editors' options can be set. The All Languages node sets the options from both the Plain Text and XML nodes below it, as shown in Figure 2-16. Of the general options discussed here, the first three are for the XML editor:

Auto List Members: Lists the members, properties, and values available to you when typing.

Hide Advanced Members: Shows more commonly used items.

Parameter Information: Displays the parameters for the current procedure.

Enable Virtual Space: Adds spaces so that comments are placed at a consistent location when using a text editor.

Word Wrap: Specifies text be wrapped to the next line once you type past the end of the viewing area.

Show Visual Glyphs: Enabled when word wrap is checked. Shows text glyphs when a line has been word wrapped. This is a logical character that doesn't physically exist, and therefore will not appear on any printouts.

Apply Cut or Copy Commands to Blank Lines: If this is checked, and you "copy" a blank line, a blank line will be pasted. If this is unchecked, nothing is inserted.

Line Numbers: Displays line numbers only against the code. This will help when an error occurs and is reported back, because the error message will mention the line number.

Enable Single-Click URL Navigation: When working with data and a URL is displayed, as the cursor moves over it, the cursor will change to a hand to indicate a URL, and clicking it will open up a browser.

Navigation Bar: Displays a navigation bar at the top of the code editor.

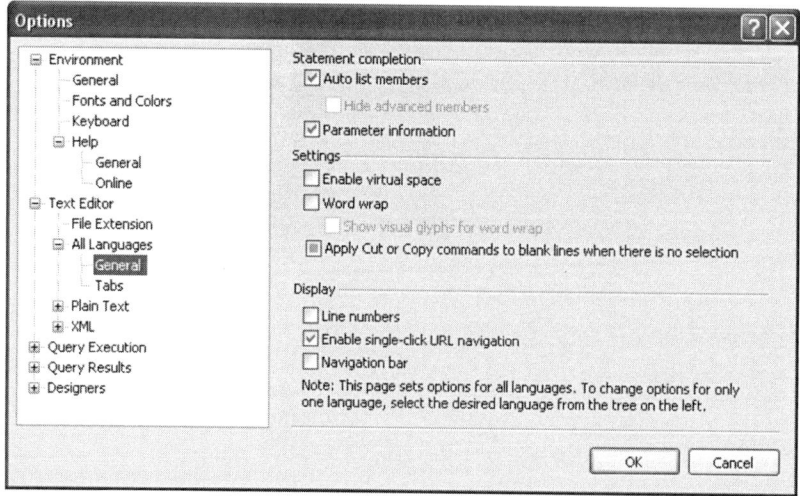

Figure 2-16. *Text Editor, All Languages, General options*

All Languages ➤ Tabs

The Tabs node deals with tabs within editors. As you can see in Figure 2-17, there are only two sections:

Indenting: The first two options are for plain text and XML. When you press Enter, these set whether the new line starts at the leftmost point (None) or at the same point as the preceding line (Block). The Smart option is for XML only and determines whether the new line is tabbed, depending on the context of the XML element.

Tab: This sets the number of characters for a physical tab (via Tab Size) and the number of characters in an intelligent tab, or an indent (via Indent Size). If you want spaces in the tabbing or indentation, click the Insert Spaces option; otherwise tabbing will use tabulation characters.

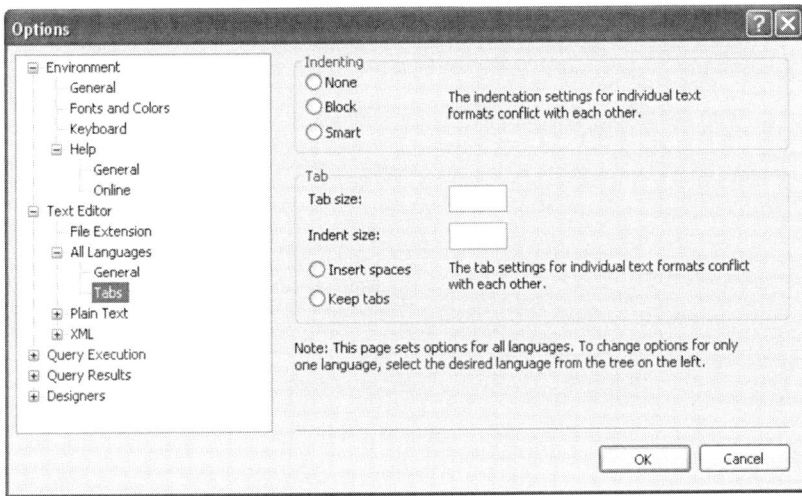

Figure 2-17. *Text Editor, All Languages, Tabs options*

Query Execution Node

The Query Execution node contains options that affect your T-SQL code. You can change the environment in which you write T-SQL and how SSMSE interacts with SQL Server when running T-SQL.

SQL Server ➤ General

When we come to running T-SQL code within Query Editor, a number of options affect how it will run, and these are shown in Figure 2-18. The settings are only for SSMSE and don't apply to other connections such as a .NET program connecting to the data.

SET ROWCOUNT: Defines the maximum number of rows to return before stopping. A setting of 0 defines that every row should be returned. This option is more often defined at the top of your T-SQL code to reduce the number of rows for that query—for example, if you have a large table and you want to see only a few rows as examples.

SET TEXTSIZE: Sets the maximum size of text data that will be seen in the results editor.

Execution Time-Out: Specifies how long you are prepared for the query to run before forcing it to stop. This can be useful especially in a production environment, when you don't wish a query to take up a large amount of processing time.

Batch Separator: Separates batches of code by a word or character. At present this is set to GO. Although you could change this, it is best not to, because GO is known universally as a batch separator.

By Default, Open New Queries in SQLCMD Mode: Selecting this option will open a command prompt utility for creating queries based on SQLCMD code rather than T-SQL code. This will allow you to create code that has extensions within it that can then be used to run as a batch file via the SQLCMD utility. We don't look at SQLCMD within this book, but if you want to run code as a batch file, then Books Online can show you how.

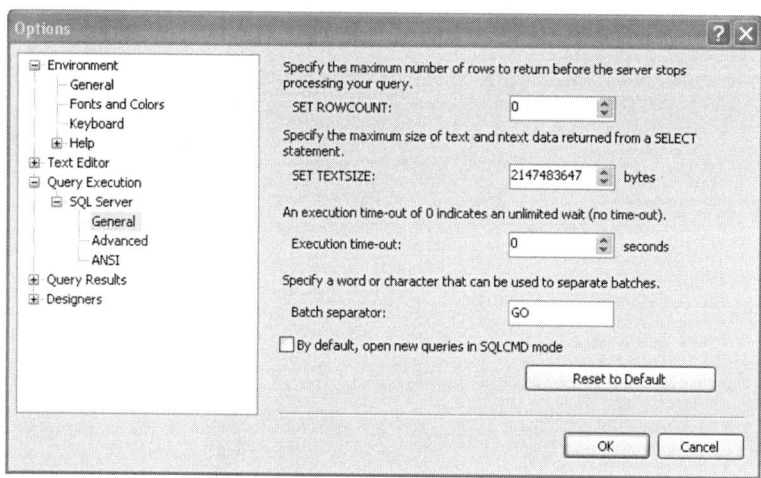

Figure 2-18. *Query Execution, SQL Server, General options*

SQL Server ➤ Advanced

This area deals with how SQL Server Express executes T-SQL code within SSMSE, and the options available here are shown in Figure 2-19. We go through those options relevant to someone learning SQL Server when creating a database in Chapter 3. The only two options not covered in that chapter that you should know are the following:

Suppress Provider Message Headers: Status messages about the query that is running will not show the data provider. Therefore, by selecting this option, you will be suppressing the data provider for SQL Server being displayed (.NET SqlClient Data Provider).

Disconnect After the Query Executes: After your query has completed, disconnect the connection. This is ideal for situations where you have a limited number of connections or you want to keep the connection count down.

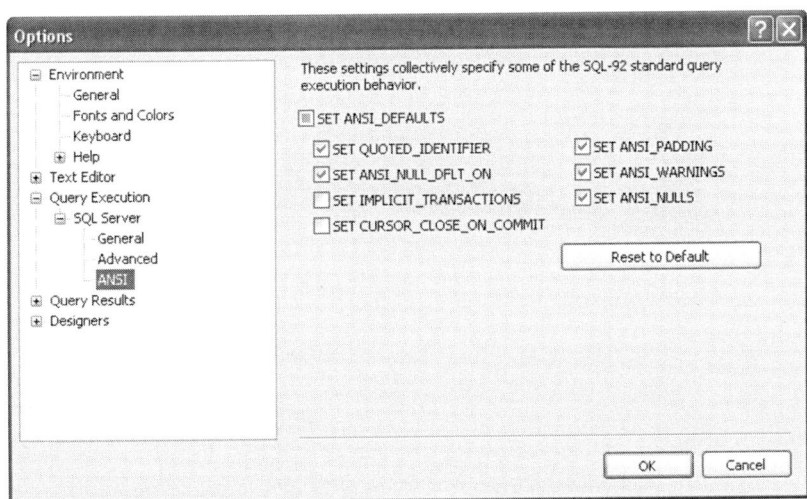

Figure 2-19. *Query Execution, SQL Server, Advanced options*

SQL Server ➤ ANSI

Like the options for the previous area, the options for the ANSI area are discussed in Chapter 3. For now, note the default settings shown in Figure 2-20.

Figure 2-20. *Query Execution, SQL Server, ANSI standard options*

Query Results Node

When you run T-SQL code, the database will return the results to SSMSE. The Query Results node is where you can modify how these results will look.

SQL Server ➤ General

The options in this area, shown in Figure 2-21, detail how results will be displayed and where they will be saved.

Default Destination for Results: This option defines how you would like to see the results of a query that returns some data.

Default Location: This option specifies the default directory for saved results.

Play the Windows Default Beep: If you wish SQL Server Express to beep you at the end of a query, check this option, and if you run a lot of queries, be prepared to lose a lot of friends. I would leave this unchecked *unless* you are going to be running a long-term query, which will allow for notification when the query finishes rather than you having to sit and watch it.

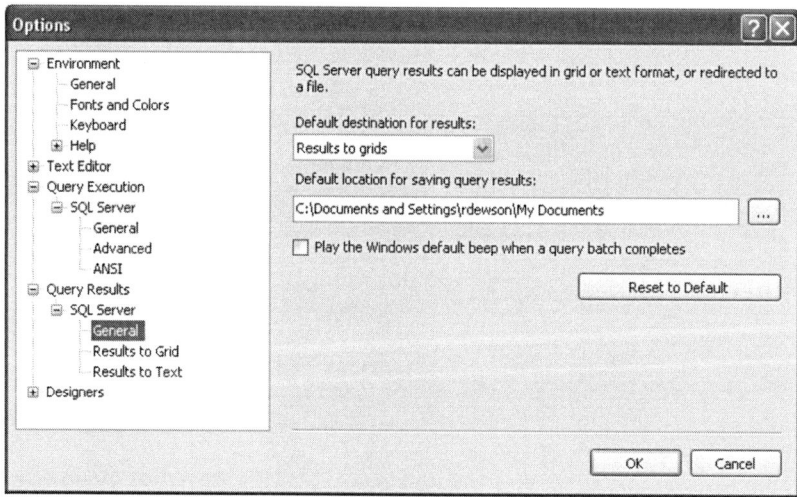

Figure 2-21. *Query Results, SQL Server, General options*

SQL Server ➤ Results to Grid

When we run T-SQL to retrieve data, SSMSE can place it within a grid, a bit like in Excel (although it will be read-only), or represent it as text, like in Notepad (also read-only). You can also save the data to a file, which is based on Results to Text options. The Results to Grid options, shown in Figure 2-22, cover how the results will look if we are outputting to a grid.

Include the Query in the Result Set: The T-SQL used to run the query is placed prior to the results.

Include Column Headers: If you want to copy information from the results, for example, to place it within an e-mail, selecting this option will include the column headers as well as the results.

Discard Results After Execution: Once the query executes, any results displayed will be immediately discarded at the end, therefore leaving nothing to display.

Display Results in a Separate Tab: Instead of the results appearing below the query, they can be in their own tab, giving more space for a larger set of results to be displayed.

Maximum Characters Retrieved: This defines the maximum amount of data to be displayed in a single cell for results.

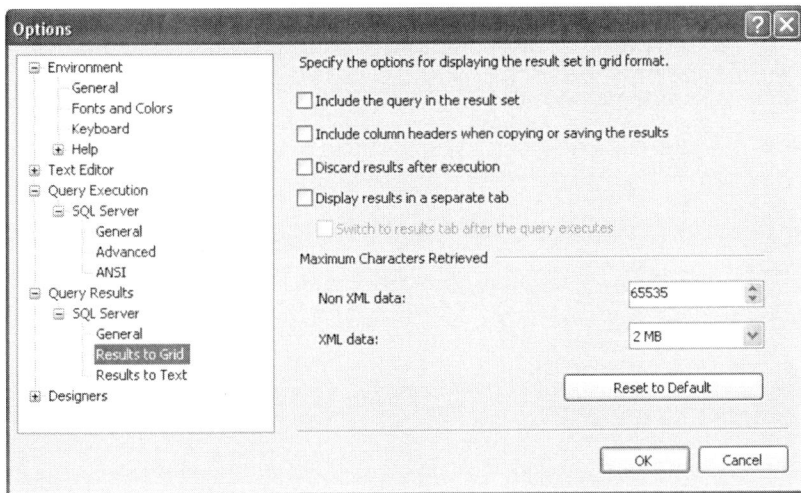

Figure 2-22. *Query Results options when the results are to grids*

SQL Server ➤ Results to Text

The other results options, shown in Figure 2-23, affect how results are displayed when they are in text format.

Output Format: This combo box presents you with five different formatting options: Column Aligned, Comma Delimited, Tab Delimited, Space Delimited, and Custom Delimiter. These different options allow you to set your output delimiter so that you can import your data into other systems.

Include Column Headers in the Result Set: Uncheck this if you just wish the results. Again, this is ideal for when you are passing data on to other systems.

Include the Query in the Result Set: The T-SQL used to run the query is placed prior to the results.

Scroll As Results Are Received: As rows are returned, if they extend past the end of the page, the results are scrolled so that the last row of data is displayed.

Right Align Numeric Values: Any numeric values are aligned to the right instead of the left.

Discard Results After Query Executes: Once the query executes, any results displayed will be immediately discarded at the end, therefore leaving nothing to display.

Display Results in a Separate Tab: Instead of the results appearing below the query, they can instead be in their own tab, giving more space for a larger set of results to be displayed.

Maximum Number of Characters Displayed in Each Column: This defines the maximum amount of data to be displayed in a single cell for results.

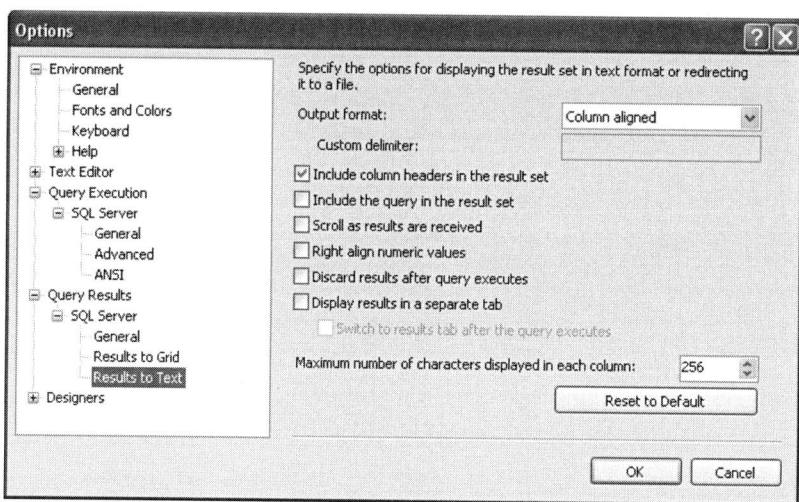

Figure 2-23. *Query Results options when the results are to text*

That completes our look at the options that are of relevance to us. The next section discusses the Query Editor within the documents area of SSMSE.

Query Editor

As you progress through the book, the creation of objects, the manipulation of data, and the execution of code will be shown either by using the graphical interface and options that Object Explorer provides or by writing code using T-SQL. To write code, we need a free-form text editor so that we can type anything we need. Luckily, SSMSE provides just such an editor as a tabbed screen within the document view on the right-hand side. This is known as Query Editor, and it can be found when you click New Query of the main toolbar or by selecting File ➤ New ➤ Database Engine Query.

We discussed some of the options that affect Query Editor, such as how text is entered and how results from running the T-SQL code are displayed, in the preceding section. There is not a great deal to say about the editor itself, as it really is a free-form method of entering commands and statements for SQL Server to execute. However, Query Editor has a toolbar that is worth covering at this point in time. Figure 2-24 shows this toolbar.

Figure 2-24. *Query Editor toolbar*

The first three buttons, as shown in Figure 2-25, work with connections to the server. The first button requests a connection to the server if one doesn't currently exist, the second disconnects the current server connection, and the third allows you to change the connection you are using.

Figure 2-25. *SQL Server connection buttons*

The next item, shown in Figure 2-26, is a combo box that lists all the databases in the server you are currently connected to. If you wish a query to run against a different database, you can select it here. The database the code will execute against, providing you have permissions, is the database that is displayed.

Figure 2-26. *Available database list*

The next buttons, shown in Figure 2-27, are concerned with executing the code entered in Query Editor. The red exclamation mark and the Execute button execute the code. The green tick will parse the code, but will not actually run it. Parsing the code will not find every error that could occur but will ensure that the syntax is correct. The last option is a grayed-out button that turns red when code is executing. If you would like to send a cancel command to SQL Server, press this button. This may not always instantly cancel the query, depending on what is executing and whether your server is local or remote. There will be a delay in sending the command and SQL Server "pausing" to receive the command.

Figure 2-27. *Query Editor execution buttons*

The next button, shown in Figure 2-28, helps you analyze the T-SQL query for optimization. We won't look at this subject within this book.

Figure 2-28. *Query optimization analysis button*

Rather than typing T-SQL code by hand, we can use a type of wizard that allows a query to be built up by selecting tables and columns via check boxes, and so on. Pressing the button shown in Figure 2-29 brings up this wizard, known as the Query Design Editor, which you will see in action in Chapter 9.

Figure 2-29. *Query Design Editor button*

The button in Figure 2-30 allows you to work with code templates. Templates feature the basics of commands or actions. They have options that act as default values. Pressing this button brings up a dialog box to change the values in each of a template's parameters.

Figure 2-30. *Code templates variable values button*

The next set of buttons, shown in Figure 2-31, deals with the query. The first two place details on how your code was executed and statistics about the code within the output. The third button runs your code as a command prompt as if it were run by SQLCMD, which is a command-line utility for executing SQL batches. These options are not covered within this book.

Figure 2-31. *Code execution buttons*

The first two buttons shown in Figure 2-32 affect how the results from the query are displayed, either as text or as a grid, respectively. The third button sends results to a file.

Figure 2-32. *How output will be displayed*

Finally, we can comment out lines of code by clicking the first button shown in Figure 2-33, or we can uncomment code by clicking the second button. The third and fourth buttons will place or remove indentations of code. All of these buttons work only on currently selected lines of code.

Figure 2-33. *Code formatting and commenting option buttons*

Now that you have seen the Query Editor toolbar, I'll turn your attention to a security tool you need to know about.

Surface Area Configuration Tool

The last section in this chapter is our first look at locking down security within SQL Server Express. The Surface Area Configuration tool, or SAC, which is new to SQL Server Express, is not, as the name may imply, a tool for configuring what parts of SSMSE are displayed on your screen or how. This tool is a method for reducing the number of services and components of SQL Server Express that run to help protect SQL Server Express from a security breach. You can also change how SQL Server runs to the same end.

This tool does have a number of areas that can be configured, but we will take a look at only four of these services, which will give you good grounding in how to work with the other services and components when the need arises.

Every item within this tool can be dealt with from another means within your computer. For example, with services, you can also do the same actions we will be performing from Control Panel ➤ Administrative Tools ➤ Services. The benefit of the Surface Area Configuration tool, though, is that it offers a unified interface and brings together all the items required to keep security of SQL Server at its optimum, in one place.

Try It Out: Using the Surface Area Configuration Tool

1. To start up the Surface Area Configuration tool, select Start ➤ Programs ➤ Microsoft SQL Server 2005 ➤ Configuration Tools ➤ SQL Server Surface Area Configuration. This will bring up the tool, as shown in Figure 2-34. You then have two choices on which area you wish to work with. Click the first choice, Surface Area Configuration for Services and Connections.

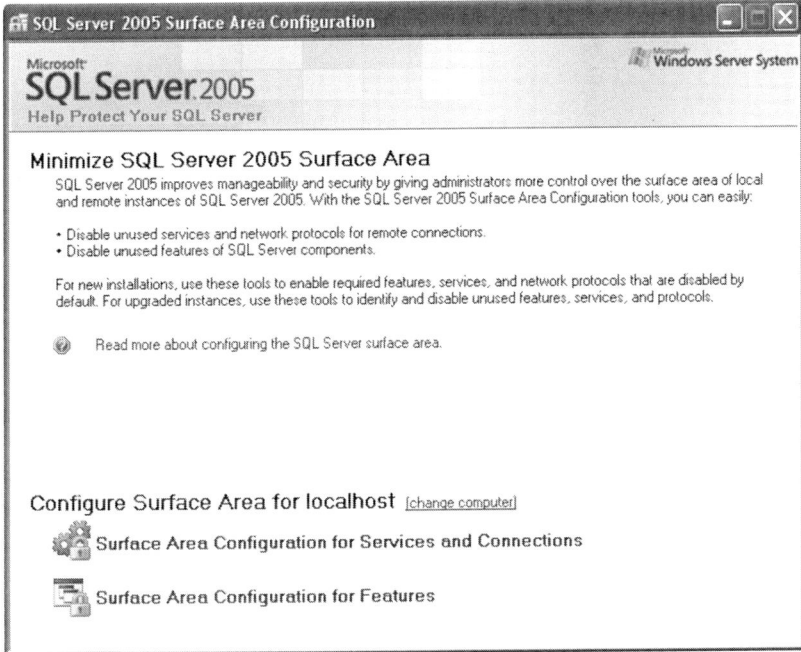

Figure 2-34. *SAC tool selection*

2. This brings you to the dialog box shown in Figure 2-35. Notice in the title bar that there is a suffix of localhost. This shows that the tool is working with a local install of SQL Server. If you wish to work with a different install, you need to return to the screen shown in Figure 2-34 and select the Change Computer option. The first item we'll discuss, as shown in Figure 2-35, is under Database Engine ➤ Service. Here you will see how the SQL Server database engine service is defined on this computer, how it starts, and its status:

 - Our first area of interest is Startup Type, which could have a setting of Automatic, as you see in Figure 2-35, which means that SQL Server will start when your machine reboots; Manual, which means that SQL Server will only start when you start the service; and Disabled, which means that the service cannot be started.

 - The Service Status box shows us what the state of the service is (whether it is running or not). You can change the state as required with the four buttons underneath the status box.

3. The Remote Connections node governs client connections. Figure 2-36 shows the default settings. Your installation will probably only allow local connections, which means that if your standalone computer is switched to being on a network, no other computer on the network will be able to connect to it until you select the Local and Remote Connections radio button option. By keeping Local Connections Only selected, you know that your SQL Server will be safe from other users on your network who may be defined as administrators on your computer, for example, as this would allow them to connect to SQL Server. Once you have specified all your options, click Apply or OK, which will then take you back to the screen in Figure 2-34.

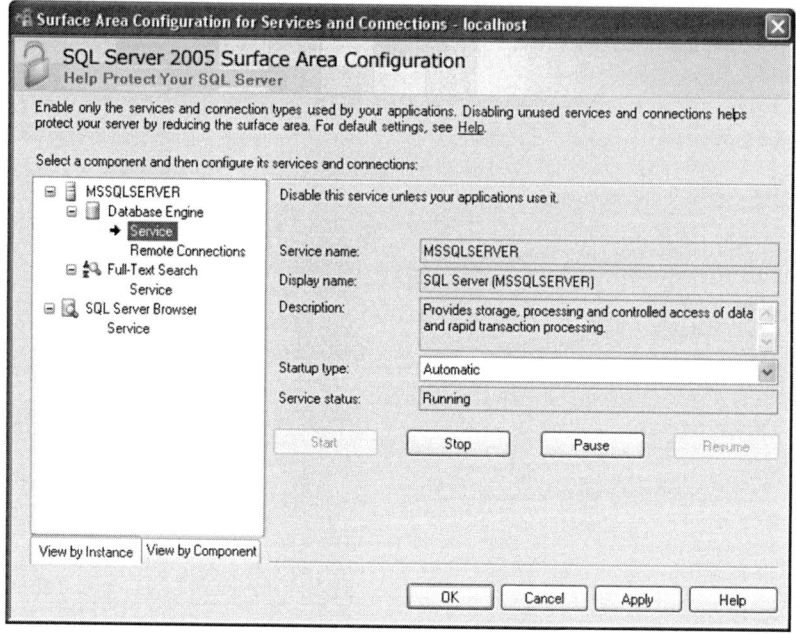

Figure 2-35. *SAC tool for configuring a component's services*

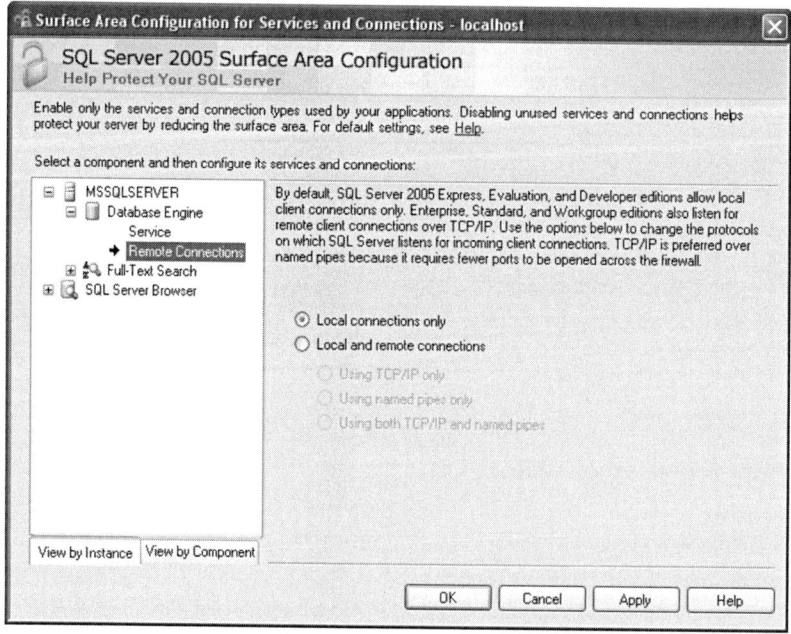

Figure 2-36. *SAC tool for configuring a component's connections*

4. Now that you are back at the main screen, click Surface Area Configuration for Features. The only option we will take a look at here is the option that affects the dedicated administrator connection (DAC), shown in Figure 2-37. There may be times that SQL Server is under stress or has a problem and you cannot connect to it via SSMSE. This option allows for a separate unique and specialized connection to SQL Server that will allow you to connect and perform administrative functions to help get SQL Server working again. If you wish to allow this connection to be enabled from a remote computer, you can switch it on here. However, it should only be enabled if your SQL Server is behind a secure firewall; otherwise, it could lead to a connection to your server that is undesired.

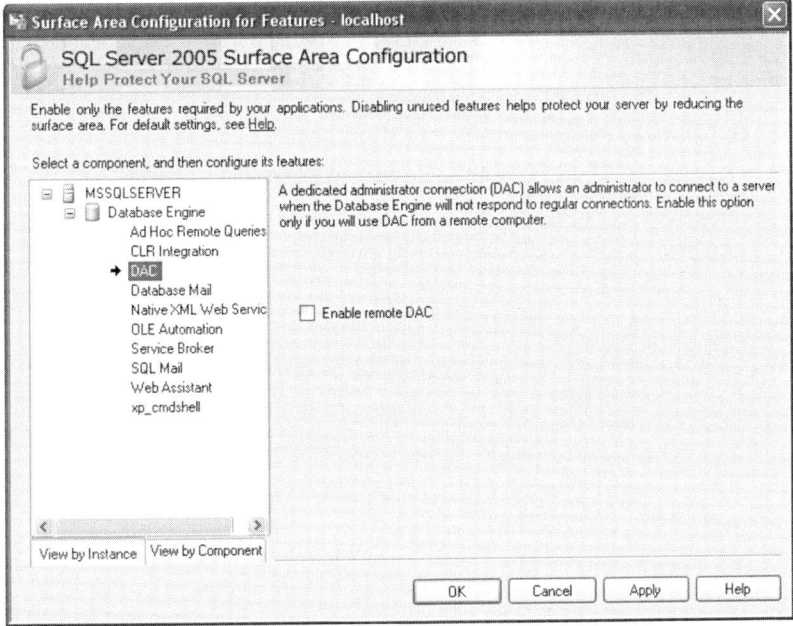

Figure 2-37. *SAC tool for DAC*

■**Caution** If you have a network install, you may wish to consider disabling Analysis Services, Reporting Services, etc., if they are not being used.

This concludes our look at the Surface Area Configuration tool.

Summary

SQL Server Management Studio Express (SSMSE) is a tool for working with SQL Server that you will see in action throughout this book, whether we work with the graphical interface or use Query Editor for writing T-SQL code. As you learned in this chapter, the main areas of the tool are the Registered Servers Explorer, shown in Figure 2-4, the Object Explorer, shown in Figure 2-6, and the main documents window that will contain graphical representations of objects in the database.

Now that you have installed SSE and seen the GUI, time to look at the example and the design process of getting the database right for the right type of system, which you will learn about in Chapter 3.

CHAPTER 3

■ ■ ■

Database Design and Creation

Now that you've installed SQL Server 2005 Express Edition (SSE) and examined the main tools you'll use as an SSE developer or administrator, it's almost time to start building the `ApressFinancial` database solution. However, you can't do this yet because you still don't know what the database will hold. At this point in time, all the information you have so far is that you'll be building a database to hold some financial transactions for a personal or corporate financial tool. You'll gear this database toward a financial tool that a corporation might use, because SSE can be scaled from one user to thousands of users very easily. The next step is to gather more information about the requirements of the solution and about what information is required to be stored in the database. Once the information gathering stage is complete, you'll then be able to create the database in SSE.

The design of a database solution is not a simple task; it requires a great deal of work. This chapter will provide you with insight into the vast area that is database design. Armed with this information, you'll proceed through arranging the data so that retrieval is as efficient as possible and ensuring that data duplication is minimal, or ideally that no data duplication exists. Part of this process will include **normalizing** your data, where different normalization forms will aid in having an efficient database. You'll also need to know how the system and the data within it will be used on a day-to-day basis. Finally, you need to know what type of system is being built—for instance, whether it will receive instant data updates or just be used to analyze already-defined data. Once the design is complete, building a database solution is a much smoother process. A good design ensures you've gathered all the information you need to build the correct tables with the correct information without duplication.

Although the methods and processes involved with the design may not meet the needs of every organization, this chapter presents an overview of the processes involved, and also shows how to build up information and ensure that the design is well thought out. This chapter covers the following topics:

- What a database is, what it consists of, and where it is stored

- How to define the type of system: transactional or analytical

- How to collect data about the current system and seek out information about the new system

- How to create a database through SQL Server Management Studio Express, a wizard, or a Query Editor window, and how to set database options in a Query Editor window

- How to review the database details

- How to remove a database using SQL Server Management Studio Express and a Query Editor window

■**Note** No specific formal design techniques will be used in this chapter's exercises, as this is not a book specifically on database design. However, the processes—both physical and logical—to get to the final design of the database will be the same.

Defining a Database

A **database** is a container for objects that not only store data, but also enable data storage and retrieval to operate in a secure and safe manner. An SSE database can hold the following objects (among others):

- **Table** definitions
- **Columns** within those tables, which make up **rows** of data
- **Programs** (either **stored procedures**, **triggers**, or **assemblies**) used to access or manipulate the data
- **Indexes**, which are used to speed up the retrieval of data
- **Views**, which are specialized ways of looking at the actual data
- **Functions**, which are repetitive tasks that can be applied to rows of data

The preceding list contains a fair number of technical terms, so let's take a few moments to look at their definitions:

Tables: These are where data is kept within the database. A database must contain at least one table to be of use, although you can have a database with no user tables and only system tables. **System tables** are special tables that SSE uses to help it work with the database. These tables contain information within rows and columns, much like in Excel, but they have a great deal more power than cells within Excel. **Temporary tables**, another type of database table, can take several different forms. These will be discussed at several points within the book.

Columns: These provide a definition of each single item of information that builds up to a table definition. A column is made up of cells that all hold data, much like a column in Excel. Unlike in Excel, though, where each cell can hold a different type of data, a column within an SSE table is restricted as to what the data within it relates to, the type of data it will hold, and how much information can be stored in it. Each table must have at least one column, although the column doesn't need to contain any information.

Rows: A row is made up of one cell from every column defined for the table. There can be any number of rows in a table; you are limited only by your disk space or the amount of disk space that you defined as the maximum in your database creation definition, or the amount of disk space on your server. A row will define a single unit of information, such as a user's bank account details or a product on an e-commerce site. Rows are also called **records**.

Stored procedures: When it comes to requiring a program to manipulate or work with data, or perform the same data-intensive task repeatedly, it's often better to store this code in a stored procedure. Stored procedures contain one or more T-SQL statements, which are compiled and ready to be executed when required. Stored procedures are permanently stored in the database, ready for use at any time.

T-SQL statement: This is a program statement that SSE can use to work with your data.

Triggers: This is a special program that executes when either a modification to data takes place or when an action happens against a data definition. Triggers allow further processing to happen when a given action that the trigger has been defined for happens.

Assemblies: These are new to SSE. Assemblies are similar to stored procedures in that they can be used to manipulate or work with data, but they are used more for procedural logic, as you might find in a .NET program. An assembly can be more than a replacement for a stored procedure and can take on many different guises—for example, you can also build data types using an assembly.

Indexes: These can be regarded as predefined lists of information that can inform the database how the data is physically sorted and stored, or they can be used by SSE to find rows of data quickly using information supplied by a T-SQL query and matching this information to data within columns. An index consists of one or more columns from the table it is defined for, but it is not possible for an index to cover more than one table. An index in SSE is very much like the index of a book, which is used to locate a piece of information faster than looking through the book page by page.

Views: These can be thought of as virtual tables. Views can contain information combined from several tables and can present a more user-friendly interface to the data. Views can also add a great deal of security to an application, but they do give reduced functionality over the use of stored procedures or direct access to the tables. Views can also be indexed to speed processing of data within.

Functions: A function is similar to a stored procedure, but it takes information one row at a time or produces information one row at a time as you work through the rows of data you are processing. For example, you would use a stored procedure to produce output to create a statement, but you would use a function to go through each transaction one at a time to calculate interest on a daily basis.

Also within every database is a set of system tables that SSE uses to maintain that database. These tables hold information about every column, information about every user, and many other pieces of information (i.e., metadata). Unlike in previous versions of SQL Server, system-table security in SSE has been increased and improved, and you cannot access these tables directly—only through views. There is no need to investigate system tables at this point, as their data can't be modified, and the information they produce is useful only for working with advanced functionality.

Databases Within SSE

Several databases are installed and displayed when SSE is first installed. This section explores each of these databases, so that you'll know what each does and feel comfortable when you come across them outside of this book.

Let's first look at the most important database in SSE: the `master` database. I'll then cover the `tempdb`, `model`, `msdb`, and `AdventureWorks/AdventureWorksDW` sample databases.

■**Note** The `AdventureWorksDW` database is used with Analysis Services, which we haven't installed.

master

`master` is the most important database in SSE, so I must start off with a warning:

Caution Directly alter the `master` database at your own peril!

There should be no reason to go into any of the system views within this database and alter the records or column information directly. There are system functions that allow a constructive alteration of any of the data in an orderly fashion, and these are the only approaches you should use to alter the `master` database.

The `master` database is at the heart of SSE, and if it should become corrupted, there is a very good chance that SSE will not work correctly. The `master` database contains the following crucial information:

- All logins, or roles, that the user IDs belong to
- Every system configuration setting (e.g., data sorting information, security implementation, default language)
- The names of and information about the databases within the server
- The location of databases
- How SSE is initialized
- Specific system tables holding the following information (this list is not exhaustive):
 - How the cache is used
 - Which character sets are available
 - A list of the available languages
 - System error and warning messages
 - Special SSE objects called assemblies (tables within every database that deal with SSE objects and therefore are not specific to the `master` database)

The `master` database is the security guard of SSE, and it uses the preceding information to ensure that everything is kept in check.

Note It is crucial that you regularly back up the `master` database. Ensure that doing so is part of your backup strategy. Backups are covered in more detail in Chapter 7.

tempdb

The `tempdb` database is—as its name suggests—a temporary database whose lifetime is the duration of an SSE session; once SSE stops, the `tempdb` database is lost. When SSE starts up again, the `tempdb` database is re-created, fresh and new, and ready for use. There is more to this process, but before we delve into that, you first need to know what the `tempdb` database is used for.

As you know, a database can hold data, and that data can be held in many tables. You use commands and functions to retrieve and manipulate that data. However, there may be times when you wish to temporarily store a certain set of data for processing at a later time—for example, when you pass data from one stored procedure to another that is going to run right after the first one. One option is to store that data within the `tempdb` database. Any temporary table created within a stored procedure or query will be placed within the `tempdb` database. This is fine, as long as the `tempdb` database is not refreshed. If it is, your data will be gone, and you will need to rebuild it.

You may be thinking that this is not an ideal solution. After all, wouldn't it be wonderful if temporary information could be stored somewhere outside of the database? Well, that's not really where `tempdb` would be used. It really should be thought of only as transitional storage space.

Another reason `tempdb` is refreshed is that not only is it available for a developer to use, but also SSE itself uses `tempdb`. Actually, SSE uses `tempdb` all the time, and when you reinitialize SSE, it will want to know that any temporary work it was dealing with is cleaned out. After all, there could have been a problem with this temporary work that caused you to restart the service in the first place.

Being just like any other database, `tempdb` has size restrictions, and you must ensure that it is big enough to cope with your applications and any temporary information stored within it generated both by yourself and by SSE itself. An example of SSE using `tempdb` is when it has to hold data while it completes a sort of data for output.

As you read through the next sections, you will see that a database has a minimum and a maximum size. `tempdb` is no exception to this, and you should ensure that its settings provide for expansion so it can grow as required.

Caution Because `tempdb` has a limited size, you must take care when you use it that it doesn't get filled with records in tables from rogue procedures that indefinitely create tables with too many records. If this were to happen, not only would your process stop working, but also the whole server could stop functioning and therefore impact everyone on that server!

As indicated in the first paragraph of this section, there's more to say about `tempdb`'s refresh process, which we'll examine in the next section.

model

Whenever you create a database, as you'll do shortly in this chapter, it has to be modeled on a predefined set of criteria. For example, if you want all your databases to have a specific initial size or to have a specific set of information, you would place this information into the `model` database, which acts as a template database for further databases. If you wanted all databases to have a specific table within them, for example, you would put this table in the `model` database.

The `model` database is used as the basis for creating the `tempdb` database as well as any other database created by you. Thus, you need to think ahead and take some care if you decide to alter the `model` database, as any changes will be mirrored within the `tempdb` database.

msdb

`msdb` is another crucial database within SSE, as it provides the necessary information to run jobs to **SSE Agent**.

SSE Agent is a Windows service in SSE that runs any scheduled jobs that you set up (e.g., jobs that contain backup processing). A **job** is a process defined in SSE that runs automatically without any manual intervention to start it.

As with `tempdb` and `model`, you should not directly amend this database, and there is no real need to do so. Many other processes use `msdb`. For example, when you create a backup or perform a restore, `msdb` is used to store information about these tasks.

AdventureWorks/AdventureWorksDW

`AdventureWorks` and `AdventureWorksDW` are the example databases that can be downloaded and installed from Microsoft at `http://msdn.microsoft.com/vstudio/express/sql/register/`

`default.aspx`. This link may change if any updates are done, but you should be able to find it under SSE downloads. Once these databases have been downloaded and installed on the computer, you have to attach these databases to SSE. This will be demonstrated in Chapter 7 when we look at how to attach databases. You don't need it attached for this part of the chapter.

These databases are based on a manufacturing company that produces bicycles. They exemplify the new features in SSE, such as reporting services, Common Language Runtime (CLR) functionality, and many others, in a simple, easy-to-follow way.

The following excerpt from the Microsoft documentation provides a concise overview of what the `AdventureWorks` databases are about:

> *Adventure Works Cycles, the fictitious company on which the AdventureWorks sample databases are based, is a large, multinational manufacturing company. The company manufactures and sells metal and composite bicycles to North American, European, and Asian commercial markets. While its base operation is located in Bothell, Washington, with 290 employees, several regional sales teams are located throughout their market base.*

The example databases are not meant for novice SSE developers, although you'll have no problems with them after you learn the basics of SSE. They are also very good for practicing what you learn about SSE from this book with different data to check out your learning.

Now that you know what databases are in SSE, let's start building one! We'll start by deciding what type of database to create, depending on what we'll use it for.

Choosing the Database System Type

Before we can design a database, we have to decide whether the system will be an **Online Transaction Processing** (OLTP) system or an **Online Analytical Processing** (OLAP) system. We could find this out prior to our first meeting with the users, or even during the first meeting, but the choice of OLTP or OLAP will probably be indicated in the initial proposal.

Before we make the decision, we need to understand these two key types of systems.

OLTP

An OLTP system provides instant updates of data. There is a good chance that an OLTP database system has a separate user front end written in a .NET language such as Visual Basic .NET (VB .NET), C#, or ASP.NET. This user front end calls through to the database and instantly updates any changes a user has made to the underlying data.

OLTP systems require many considerations to ensure they're fast and reliable, and can keep the data integrity intact. When you design an OLTP system, it's crucial that you get not only the database structure right, but also where the data physically resides. It's common to find that OLTP systems are normalized to third normal form (more on what this term means later in the chapter), although this may not happen in every case. By normalizing your data, you will aid the achievement of one of the main goals of an OLTP system: keeping data updates as short as possible. When you normalize your data by removing redundant or duplicate columns, you should ensure that the data to be written is as compact as possible. In many OLTP systems, normalization is critical.

Backups

Many OLTP systems are in use 24 hours a day, 7 days a week. The high frequency of changes in such a systems' data means that backing up the database is a necessary and mandatory task.

It is possible to back up a database while SSE is in use, although it is best to perform a backup when SSE is either not in use or when there will be a small amount of activity updating the data taking place. The ideal timeframe might be in the middle of the night or even during a break period.

Whenever you decide to perform a backup, it's crucial that you constantly monitor and check it within an OLTP system to see that the system is still performing as desired. You would not be the first person to find that what you thought was a valid backup that could be restored in a disaster situation was in fact corrupt, incomplete, or just not happening. Therefore, periodically take a backup from production and reload it into a secure development area just to confirm that it works.

Indexes

Speed is essential to a successful OLTP system. You should see a higher number of indexes within an OLTP system as compared to an OLAP system, with these indexes used not only to help relate data from one table to another, but also to allow fast access to rows within tables themselves.

Note Chapter 6 covers how to build indexes, how indexes work, and how to manage indexes within your solutions.

OLAP

When considering an OLTP system, you must keep in mind that an update to the database could happen at any moment in time, and that update must be reflected within the database instantly. It is also crucial that the system performs many updates simultaneously, and that it does not corrupt any data when it does so.

An OLAP system is designed with the premise that the data remains fairly static with infrequent updates. These updates could be every night, weekly, monthly, or any other time variant as long as updates aren't happening on a frequent basis, like in an OLTP system. As the name "Online Analytical Processing" suggests, in this system a large amount of the processing involves analysis of existing data. There should be little or no updating of that data—ostensibly only when the data within the analysis is found to be incorrect or, as mentioned previously, when more data is being applied for analysis. Backing up the data will probably take place only as a "final action," after the database has had changes applied to it. There is no need to make it a regular occurrence.

Systems designed for OLAP sometimes do not follow any design standards or normalization techniques, and most certainly have fewer indexes than an OLTP system. You tend to see no normalization in an OLAP system, as it is easier to take data and to slice and dice it without having to bring in data from a normalized table. There will be few or no updates taking place in an OLAP system, so performing transactions and keeping them compact aren't concerns. Most OLAP systems will contain no normalization. Quite often, you'll find one or two large flat tables—rather than several tables related together—and therefore as there are fewer relationships, there will be fewer indexes.

Note OLAP systems are often called **data warehouses**, although data warehousing is only one aspect of the overall OLAP system design. Put simply, a data warehouse is a database to hold organized data that is not constantly changing like an OLTP system. This data is then used within the OLAP system.

Example System Choice

So, when you take into consideration all of the information presented in the preceding sections, it is fairly obvious that although the data updates will be relatively infrequent in our example system (in other words, only when a financial transaction occurs or a statement is generated), there will be updates occurring online with instant results expected. Therefore, our system will be an OLTP system.

Gathering the Data

Among the first things you need to do before building a database is find out what information the database system has to hold and then decide how that information should be stored (e.g., numerical or text, length, etc.). To achieve this, you'll perform a data-gathering exercise, which could involve talking with those people who are the owners of the system and those who will be using the system.

For larger systems, you would hold several meetings, each of which would pinpoint one area of the system to discuss and research. Even then, several meetings may be spent going back and discussing each area. You could also conduct interviews, distribute questionnaires, or even just observe any existing processes in action, all in an effort to gather as much information as possible about the database and how it will be used.

The key indicator of whether or not a database solution is successful is not found so much in the building of the system, but rather in the information gathering process before the first line of code is actually written. If you're working off of an incorrect piece of information, or you're missing an element that might be crucial to the final solution, already the system is flawed. Involving the users as much as possible at the earliest stage and then including them as the design progresses should result in a significant reduction of errors and missing parts in the final product.

For our example financial program, the first people we want to talk to are the owners of the data that will be displayed. These are the people in the banking department who handle checks, cash withdrawals, credit card transactions, and so forth, and also those people who handle the purchase and sale of stock shares, unit trusts, life insurance policies, and so on. These people will know what documentation or statements are sent to which customers, as well as the information those statements contain. In addition, these people will likely have a good idea about how customers may want to reconcile these statements, and also what data will be allowed to be downloaded and inserted into your SSE database.

At the first meeting, we examine all the documentation to see what information is stored in the current system. We find out at the meeting that the current system sends out different statements, one statement for the customer's current account and a separate statement for each financial product the customer owns. When we examine the statements, we focus on the information each contains that we need to capture to produce a similar statement. This could be not only customer-related information, but also regulatory statements.

With the information from all the documentation in hand, we can start discussions about what is required from the system we are to build. Obviously, a great deal of information will be discussed in these meetings, some of it useful and some not. Make sure that the discussions are recorded in the order in which the people present make points, and not in "logical" order. This simulates meetings where people "remember" items that have to be catered to, where one point raised may make someone remember a point elsewhere.

Out of our initial discussions, we note the following points:

1. The software must be able to handle working with more than one product. The main product is a current checking account that a bank or a single user might use to hold banking details. The system also must be able to manage by-products such as loans, life insurance policies, and car insurance policies, and it should be able to record any trading of shares on the stock market.

2. Statements are produced on a monthly basis or at any time the customer requests them from the system. If a customer requests a statement within the month, there will still be a statement produced for that month.

3. Interest accrues daily on accounts that are in credit and is subtracted daily from overdrawn accounts.

4. Annual, monthly, or single-premium products can be held for a customer or held by a customer in a standalone version of the system. We need to know when to start and stop collecting payments, and we also need to determine which products we send out a reminder for (e.g., a notice to let a customer know her car insurance policy is up for renewal).

5. If a collection for a product failed, the system needs to recognize this so the amount can be collected the next time a collection is due.

6. Each product will have its own statement, but only banking statements will be produced on request.

7. Trading of stock shares is allowed, so the system needs to record and display the current value for a customer's specific share at any time.

Notice how the information in this list is in no set order, as this is how information tends to come out. Also notice that there is also a little bit of duplication of information in points 2 and 6; if this is not realized and understood, it could cause problems.

■**Note** This is the only data-gathering exercise performed for our example database. The information gathered here should be cross-checked later in the design phase with the users, although this is beyond the scope of this book.

Determining the Information to Store in the Database

Using the notes we took in the previous section, we'll now try to find every area each point has an interest in. Looking at the list of areas that require information to be recorded and stored within our database, it's clear we need to arrange them in some sort of order. We're still looking at our solution from a logical viewpoint, and we're not ready to start building anything in SSE yet.

First off, let's scan through the points listed and try to group the information into specific related areas. The list items are numbered, so we'll be able to easily demonstrate the groupings. The following list shows some initial groupings and reasons for them:

- *Financial products*
 - *Point 1*: We are dealing with more than one product. We need to record each product.
 - *Point 2*: Statements will be produced for each product, and specific product information for those statements would be recorded here, such as the name of the product.
 - *Point 4*: We need to record what type of premium is associated with this product.
 - *Point 5*: This point deals with a financial products premium collection.
 - *Point 6*: This point deals again with statement production.
- *Customers*
 - *Point 2*: Customers can request statements.
 - *Point 3*: We need to record the amount of interest for a customer.
 - *Point 4*: A list of the different products associated with each customer is required.
 - *Point 7*: For each share, we need to keep a current value.
- *Customer addresses*
 - *Point 2*: We need each customer's address in order to send a statement.
 - *Point 6*: As with point 2, we need the customer's address to send a statement.
- *Shares*
 - *Point 1*: We trade shares on the stock market, therefore we need to record share information.
 - *Point 7*: We need to keep a given share's value.
- *Transactions*
 - *Point 2*: A list of transactions is required for statement production.
 - *Point 4*: Regular and ad hoc premiums have to be recorded.
 - *Point 5*: We need to record failed transaction collection.
 - *Point 6*: Statements will be produced for each product.

These five distinct groups could translate into five distinct tables within our proposed database. At this point in the logical design process, we would base our design on these five tables. From here, it is then possible to start examining the information that should go into these logical tables. There might be duplication of data with columns in the "wrong" table, and the potential for multiple columns to hold "similar" information or the same column in more than one table.

Let's look at the list points in turn in the following sections and examine what information should be stored in each "table." The information listed in the sections that follow is taken from the discussion with the users, and a list of the columns is supplied that may initially form the basis of the tables. Each column has a description, so when we go back to the users they'll understand the purpose of the columns. Also at this stage, we'll add columns to hold any identifiers for finding the records; in the following sections, these are denoted with (**K**). These are potentially our keys, which we'll cover in more detail later in the "Building Relationships" section of this chapter.

Financial Products

The aim of this table is to hold the different products the company sells. From bank accounts to life insurance, all products will be held here. This table will be used when producing statements and creating transactions when the user's balance changes, for example, when buying further shares.

- *Financial Product ID (K)*: This is a unique identifier.
- *Financial Product Name*: This is the name of the product, such as checking account, share, loan, and so forth.
- *Frequency of Payment*: For each product, this indicates how often payments are collected for those products that require it, such as a loan, a regular savings account, and so on.

Customers

This table will hold customer details, such as the customer's products and balances. To clarify and reiterate, there will be items currently within this table that will no longer reside within it once we normalize the data. For example, you will see an attribute for "Account Numbers for Each Product." When we proceed through normalization, you will see how attributes such as this are "moved."

- *Customer ID (K)*: This is a unique ID for each customer.
- *Financial Product Balance*: This is the current balance of each product.
- *Title*: This is the customer's title (Mr., Ms., etc.).
- *First Name*: This is the customer's first name.
- *Last Name*: This is the customer's last name.
- *Address*: This is the customer's address.
- *Account Numbers for Each Product*: This is the account number of each product the customer owns.
- *Financial Products Details*: These are the details of each financial product the customer owns.

Customer Addresses

As part of the analysis, we will not hold our own address information and we will instead get this from a third-party system. This table will therefore not exist within our database.

Shares

This table holds the details of each stock share, such as its current price and its price history.

- *Share Price ID (K)*: This is a unique ID for each share.
- *Share Name*: This is the name of the share.
- *Current Price*: This is the current price of the share.
- *Previous Price*: This contains previous prices of the share.
- *Price Date*: This is the date the price was set at this level.
- *Stock Market Ticker ID*: This is the ID for this share on the stock market.

Transactions

This table holds the details of each financial transaction that takes place for each product.

- *Financial Transaction ID (K)*: This is a unique ID for each financial transaction.
- *Customer ID*: This is the customer's unique identifier, as defined in the "Customers" section earlier.
- *Date of the Transaction*: This is the date the transaction took place.
- *Financial Product*: This is a link to the financial products table.
- *Amount*: This is the amount the transaction is for.
- *Debit/Credit*: This flag denotes whether the transaction is a debit or a credit.

External and Ignored Information

At this point, we have a first draft of the logical tables and attributes, but there are still no relationships between these tables. There is one more piece of information that we need to know, which concerns information not recorded, as it won't be included within this database.

The example database will not hold every item of information that is required to make the system complete. This is to keep the example simple and to avoid having extra tables that will not be used within the book's exercises. However, there may be other times when you may wish to implement only some of the tables—for example, when performing a viability study (in other words, when you're building part of a system to prove the viability of an idea). Or perhaps there are third-party tools available that can fill in the gaps.

For example, a system might use an external addressing system, and instead of holding all customer addresses within the system, it may use a cross-reference ID. A table could also exist to hold all of the financial transactions for products not covered where specialized tables are required, such as for company pension plans.

Next, let's move on to consider relationships between the tables in the database.

Building Relationships

Much like people, databases can be temperamental creatures and need a bit of TLC. Good relationships can provide this kind of care and attention.

At the moment, the tables in our example database are essentially single, unrelated items. Of course, they have columns with the same name in different tables, but there is nothing tying them together. This is where defining relationships between the tables comes in. Binding the tables together in this way ensures that changes in one table do not cause data in another table to become invalid.

Using Keys

A **key** is a way of identifying a record in a database table. We can use keys to build relationships between tables because a key refers to a whole record—a property we can exploit when working with columns that, for example, have the same name in different tables. Using a key as a shortcut, we can make the link between the two very easily. Keys can also uniquely identify a record in a table when that is an important part of the database's design.

A key can be defined on a single column if that's enough to identify the record, or it can be defined on more than one column if not. The sections that follow introduce the three kinds of keys you can use in a database: primary, foreign/referencing, and candidate/alternate. We'll also look at using an SSE method called a constraint instead of a primary key.

Primary Key

The **primary key** is probably the most important key type. First and foremost, the column (or columns) on which the primary key is defined must only contain unique values. A primary key cannot be defined on a column, or a sequence of columns, that does not return a single row. To this end, it is not possible to define a primary key for any columns that allow NULL values. A further restraint is that a table may have only one primary key.

A primary key can be used to link data from one table to data from another. For instance, in our example database, we have two tables: one holding customers and another holding customer banking transactions. We define a primary key on the Customers table on the customer ID that is generated uniquely each time a new customer record is inserted. This is then used to link to the many records within the banking transactions table, to return all the transactions for that customer ID. The link between these two tables is the customer ID, which as previously mentioned is defined as a primary key in the Customers table.

Later on, you'll see how to join tables together and define a relationship between them. A join and a relationship essentially mean the same thing: a logical link between two or more tables that can be defined through a specific column or set of columns between the tables.

Foreign/Referencing Key

There will be times when you have two or more tables linked together in a relationship, as demonstrated in the previous section's example, where the link between the Customers and Transactions tables is the customer ID column. This column returns a unique row in the Customers table; hence it is defined as the primary key of the Customers table. However, there has to be a corresponding **foreign** (or **referencing**) **key** in the Transactions table to link back to the Customers table, which is the customer ID column of the Customers table.

When it comes to creating relationships within our example database, you will later see how a foreign key is created that will create a link, or a relationship, between two columns. This link is created through a **constraint**, which is a method SSE uses to check the details built into the relationship. From the viewpoint of a foreign key, this constraint, or check, will ensure that the relationship follows the conditions set with it. We'll examine foreign keys in more depth in the "More on Foreign Keys" section.

Candidate/Alternate Key

As mentioned previously, a table can have only one primary key. However, there may be another key that could just as easily be defined as a primary key. This is known as a **candidate key**, as it is a candidate for being the primary key.

There is no logical difference at all between the definition of a candidate key and a primary key. For example, if we have a table that holds spare parts for a General Motors (GM) vehicle, we could have an internal GM part number to use when ordering parts at the head office for various GM branches. This part number would be unique and would likely be used as the primary key. However, a part number is also created by each of the manufacturers, which is unique to them. This, too, could be a primary key if we include the supplier identifier in the database. We can't have two primary keys, and we've chosen the GM part number as the primary key, but we could create a candidate key using the manufacturer identifier and part number.

A Unique Constraint Instead of a Primary Key

This is where having a constraint defined will ensure that unique values can only be entered into columns defined within the constraint. This sounds very much like the previous primary key definition, but there are differences.

A **unique constraint** is not a primary key, but the column or columns defined within the constraint could be a primary key. Also, a unique constraint can contain NULL values, but recall that a primary key cannot. However, NULL is treated as any other value within a column; therefore, the columns used to make a unique constraint must remain unique, including the NULL value, when you're looking to insert or update data. Finally, it is possible to have multiple unique constraints, but you can have only one primary key.

Creating Relationships

A **relationship** in an SSE database is a logical link between two tables. It is impossible to have a physical link; although, as you will see later, a physical line is drawn between two tables when designing the database. To have a physical link would mean the actual data linking the two tables would be stored only once in a central location, and that information within the keys linking the tables would be stored more than once, which is just not the case.

When defining a logical relationship, we're informing SSE that we'll be linking a primary key from the master table to a foreign key in another table. So already there is a need for two keys: one on each table.

The following sections present specific details about relationships, starting with a look at how relationships work with the concept of **referential integrity**.

Relationships and Referential Integrity

A relationship can be used to enforce data integrity. In other words, if you are expecting data in one table because there is data in another, you can place a relationship between these two tables to ensure that no SQL command breaks this rule. However, don't confuse referential integrity with other processes that are associated with maintaining data integrity, such as placing checks or default values on columns to ensure that values for a specific column are valid.

Referential integrity revolves around the idea that there are two tables in the database that contain the same information, and it requires that the duplicated data elements are kept consistent. For example, if you have a primary key in one table and a foreign key in another table that have data that matches exactly, it is important that both pieces of data either change together or don't change at all. Relationships are not the only way referential integrity can be enforced; you can also use triggers to ensure that data remains valid (we'll examine this further in Chapter 13).

For instance, our example banking system includes the Customers and Transactions tables. It is not possible to record customer transactions without a customer record. As a result, we have to use referential integrity to enforce data integrity between these two tables, so that a customer record can't be removed from the database while there are customer transaction records for that customer. Similarly, this rule should allow the removal of a customer record when there are no customer transaction records.

Another result of enforcing referential integrity is that it isn't possible for a customer transaction to be entered using a customer reference number that doesn't exist within the Customers table. Instead, to enter a customer transaction in this situation, we first have to create the customer record, and then we can carry out the transaction.

Finally, if we had a customer record and related customer transaction records, we couldn't alter the customer reference number in the customer record without first altering the customer transaction records and checking that the reference we're altering the customer transaction records to already exists.

So, there are a number of rules to follow if we want to maintain the integrity of our data. If we so desired, we could use referential integrity to enforce data integrity. However, a flip side to all of this to be aware of is that we can keep data integrity within a system and not use referential integrity. Instead, we can create stored procedures or triggers, which are types of programs within SSE, to do this task. We'll look at these topics in Chapters 10 and 13.

Using stored procedures and triggers is a possible but undesirable solution, because it leaves our system open to instances where data integrity is not kept, because of holes within the design of the system or perhaps because a developer doesn't have the correct processing sequence to ensure that all data is always valid. Not only that, but if someone adds data directly to a table, the referential integrity will be lost. That said, having the data integrity checks in an application does lead to less traffic flow over the network, as all the validation is done on the front end.

There is one more important point about referential integrity before we move on to discuss database relationship types: if you want to maintain referential integrity by creating a relationship between two tables, these two tables must be in the same database. It is not possible to have referential integrity between two databases.

Types of Relationships

Three main relationship types can exist in a database:

- One-to-one

- One-to-many

- Many-to-many

The sections that follow cover each type, so when it comes to creating a relationship, you'll know which one to create, when to create it, and why. We'll start off by looking at the one-to-one relationship, which is perhaps the easiest type of relationship to understand, although it is one of the least used.

One-to-One

This relationship type isn't very common within a working database. Typically, there is no real reason for one record in one table to match just one record in another. This scenario would really only exist, for example, if you were splitting a very large table into two separate tables.

To illustrate the one-to-one relationship, imagine that in our example bank database there is a table that holds PINs for ATM cards, keeping them completely separate from the remainder of the customer records (see Figure 3-1). In most cases, there would be one PIN record for each customer record, but there may be exceptions—for instance, a high-interest deposit account may not have a card, and therefore there would be no associated PIN record.

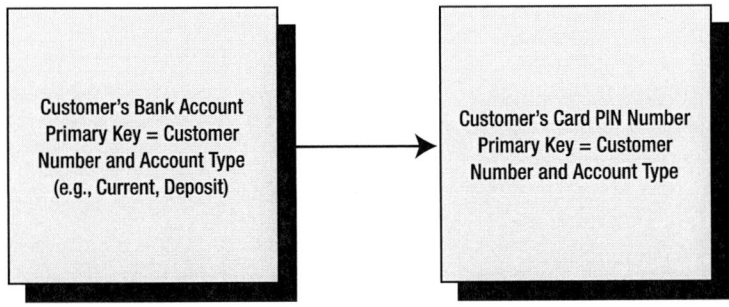

Figure 3-1. *One-to-one relationship*

One-to-Many

Perhaps the most common relationship found in a database is the one-to-many relationship. This is where one master record is linked with zero, one, or more records in a child table.

Using our banking example, say we have a customer master record along with any number of associated transaction records. The number of these transaction records could range from none, which corresponds to when a customer is new to the bank and hasn't made a deposit or performed a transaction, to one or more, which corresponds to when there has been an initial deposit in an account, and then further deposits or withdrawal transactions after that (see Figure 3-2).

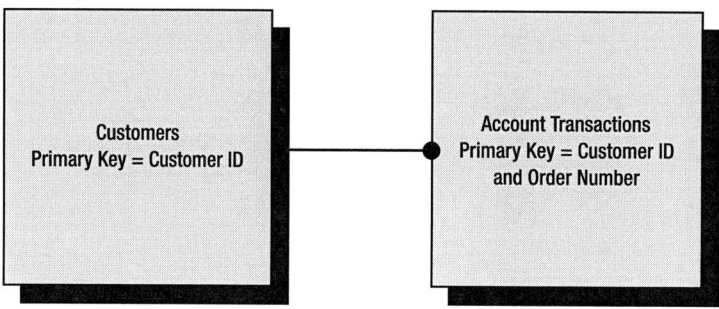

Figure 3-2. *One-to-many relationship*

You'll see this concept in action again in the customer-to-transactions relationship we'll build for our solution.

Many-to-Many

Many-to-many is the final relationship type that can exist in a database. This relationship can happen relatively frequently where there are zero, one, or indeed many records in the master table related to zero, one, or many records in a child table.

An example of a many-to-many relationship might be where a company has several depots for dispatching goods, presented in the master table, which then dispatch goods to many stores, presented in the child table (see Figure 3-3). The depots could be located and organized so that different depots could all supply the same store, and they could be arranged in groups of produce, frozen, perishables, and bonded. In order for a store to be supplied with a full complement of goods, it would need to be supplied by a number of different depots, which would typically be in different locations.

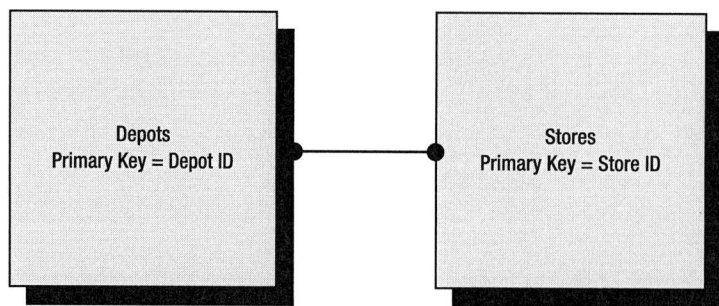

Figure 3-3. *Many-to-many relationship*

When building relationships within a database, it is necessary to have a foreign key. I covered foreign keys briefly earlier in the chapter; let's take a closer look at them in the next section.

More on Foreign Keys

A foreign key is any key on a child table where a column, or a set of columns, can be directly matched with exactly the same number and information from the master table. By using this foreign key, you can build up the data to return via a relationship.

However, a foreign key does not have to map to a primary key on a master table. Although it is common to see a foreign key mapped to a primary key, as long as the key in the master table that is being mapped to is a unique key, you can build a relationship between a master table and a child table.

The whole essence of a foreign key lies in its mapping process and the fact that it is on the child table. A foreign key will exist only when a relationship has been created from the child table to the parent table. But what exactly are the master table and the child tables? To demonstrate, let's refer back to our relationship examples. Take the one-to-many relationship. The master table would be on the left-hand side, or the "one" side of the relationship, and the child table would be on the right-hand side, or the "many" side of the relationship (see Figure 3-4).

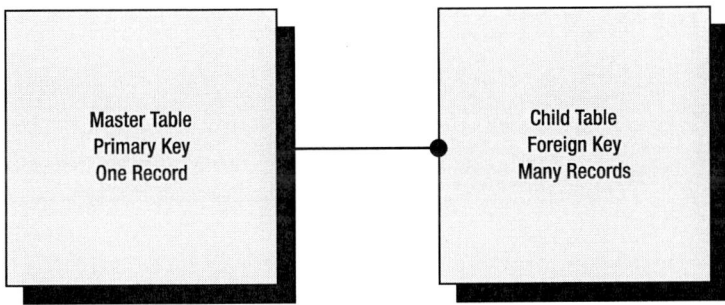

Figure 3-4. *Foreign key*

There is one final point to mention concerning foreign keys, relationships, and the master and child tables. It is totally possible for the master table and the child table to be the same table, and for the foreign key and the unique key to both be defined within the same table. This is called a **self-join** or a **reflexive relationship**. You don't tend to see this much within a database, as it is quite an unusual situation, although you could use it to ensure that the data in one column exactly matches the information in another column, just as in any other join.

For example, say you have a table built around customers, and you have two columns, one of which is a parent customer ID, which holds an ID for the head office and is used to link all the branches. If the head office is also seen as a valid branch of the conglomerate, the second column could be the specific branch ID, and you could put a link between these two columns so that there is still a valid link for the head office as a branch as well (see Figure 3-5). Another example is in an employees table where all employees reside, with a self-join from an employee back to his or her manager.

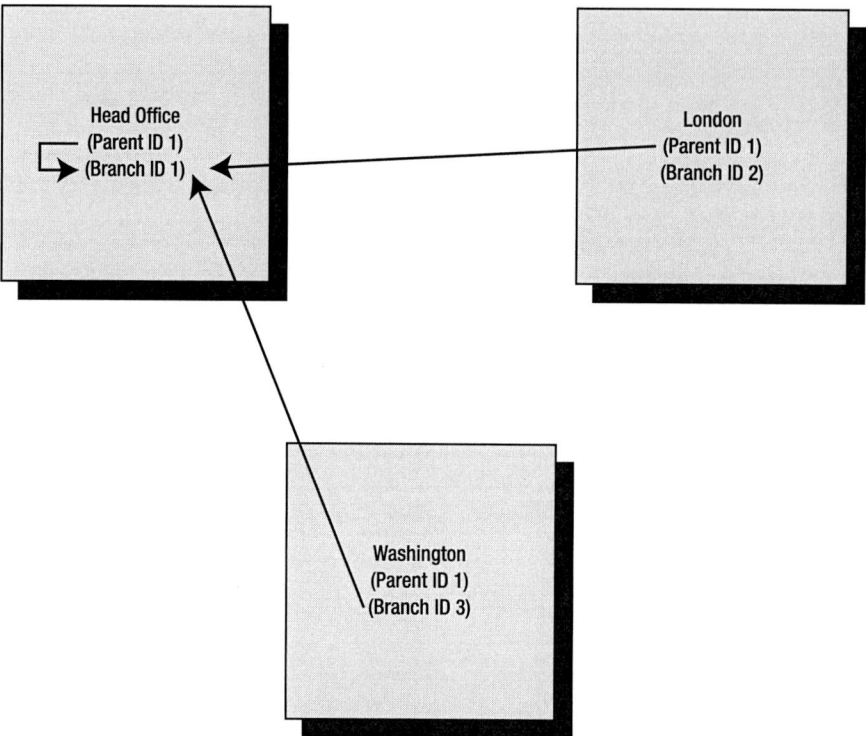

Figure 3-5. *Foreign keys in same table*

Now that we've looked at relationships, let's move on to see how to normalize the database.

Normalization

Normalizing a database means minimizing data redundancy to avoid update anomalies. The removal of duplicate data leads to smaller, more compact databases. This avoids confusion over which column holds data, and reduces the overhead of keeping data up to date.

Note Just a reminder that we're still in the logical phase of building our solution, and we're not ready to start building our database within SSE.

A database designer should not normalize with impunity, as this may negatively affect the data retrieval. Although normalization can make a database more efficient for updating, **overnormalizing** data may create too many relationships and too many slim, small tables, so that to retrieve one piece of information requires access to many tables and many joins between these tables. Database design is an art, and a knowledgeable designer knows when to stop normalizing. In fact, after normalizing a design, it's often wise to denormalize it by adding back some redundancy to make critical queries run faster. (See "Denormalization," later in this chapter.) This knowledge comes with experience and practice, but in our database example you'll learn where to "stop."

In this section of the chapter, we'll model our example in a method known as **logical modeling**. The purpose of the logical model is to show the data that the application must store to satisfy business requirements. It demonstrates how this data is related and explores any integration requirements with business areas outside the scope of the development project. It is created without any specific computer environment in mind, so no optimization for performance, data storage, and so forth is done.

In logical modeling, the term **entity** is used to mean a conceptual version of a table. As we're still in the logical modeling stage of designing our database, I'll use "entity" rather than "table" in this discussion, since it is less tied to implementation. Also within logical modeling, a column of data is referred to as an **attribute**. To build our logical model, we'll take the information gathered previously in the chapter and implement attributes in our entities. From that, we'll see how we need to alter our design.

The question remains, what should be contained in an entity? Three principles should govern the contents of an entity:

- Each entity should have a unique identifier.
- Only store information that directly relates to that entity.
- Avoid repeating values or columns.

The sections that follow provide more detail about each principle.

Each Entity Should Have a Unique Identifier

It must be possible to find a unique row in each entity. You can do this through the use of a unique identifying attribute or the combination of several attributes. However, no matter which method you use, it must be impossible for two rows to contain the same information within the unique identifying attribute(s).

Consider the possibility that there is no combination of attributes in an entity that can make a row unique, or perhaps you wish to build a single value from a single attribute. SSE has a special data type, called **unique identifier**, that can do this, but a more common solution is to build a column attribute with an integer data type, and then set this up as an **identity** column. You'll learn more about this technique when building the tables in Chapter 5.

Only Store Information That Directly Relates to That Entity

It can be very easy in certain situations to have too much information in one entity and therefore almost change the reason for the existence of the specific entity. Doing so could reduce efficiency in an OLTP system, where duplicate information has to be inserted. It could also lead to confusion when an entity that has been designed for one thing actually contains data for another.

Avoid Repeating Values or Columns

Having attributes of data where the information is an exact copy of another attribute within either the same entity or a related entity is a waste of space and resources. However, what tends to happen is that you have repeated values or attributes within two or more tables, and therefore the information is duplicated. It is in this scenario that you are expected to avoid the repeating values and move them elsewhere.

Normal Forms

Now that you know what should be contained within an entity, how do you go about normalizing the data? The normal forms addressed within this chapter are as follows:

- First normal form (1NF)
- Second normal form (2NF)
- Third normal form (3NF)

There are a other, "higher" normal forms, but they are mainly of theoretical significance, so they will not be covered here.

First Normal Form

To achieve 1NF within a database, it is required that you eliminate any repeating groups of information. Any groups of data found to be repeated will be moved to a new table. Looking at each table in turn, we find that we have two tables in our example database that potentially flout the first requirement of 1NF: customers and shares.

Customers

There are two columns with possible repeating values in this table:

Title: A customer's title will be Mr., Miss, Ms., or Mrs., all of which you could put in to a reference table. Some corporations do this; others don't. It all depends on whether you want to restrict what users can enter.

Address: The address should be split out into separate lines, one for each part of the address (e.g., street, district, etc.). It is probably well worth having a reference table for cities, states, and countries, for example.

Shares

There is one column that will possibly repeat: share name. This is really due to the Shares table actually doing two jobs: holding details about the share, such as its name and the market ticker, which really are unique; and holding a historical list of share prices. This table actually needs to be split into Shares details and SharePrices, which you'll see happening when we discuss 3NF.

Second Normal Form

To achieve 2NF, each column within the table must depend on the whole primary key. This means that if you look at any single column within a table, you need to ask whether it is possible to get to this information using the whole key or just part of the key. If only part of the key is required, you must look to splitting the tables so that every column does match the whole key. So, you would look at each column within the table and ask, "Can I reach the information contained within this column just using part of the key?" All of the tables use an ID as the primary key, and only one column will define that ID. Therefore, to break 2NF with this is almost impossible. Where you are more likely to break 2NF is a scenario in which the primary key uses several columns.

If we look at all the tables within our example, every column within each table does require the whole key to find it.

Third Normal Form

To achieve 3NF, you must now have no column that is not part of the primary key be dependent on any other column within the table.

The `Customers` table does have data derived from another table, with account numbers for each product the customer has bought and financial product details. This means that the account number plus details about the product such as the date opened, how much is paid with each payment, and the product type do not belong in the `Customers` table. If such information did remain in the table, `Customers` would have multiple rows for the same customer. Therefore, this table also now needs to be split into `Customers` details such as name and address, and `CustomerProducts`, such as a row for each product bought with the customer details about that product.

We have now reached full normalization to 3NF of the tables within our database. Let's take a moment to clarify where we are. Figure 3-6 shows that we're now moving from a logical model to a physical model, where we are physically defining what information is stored where.

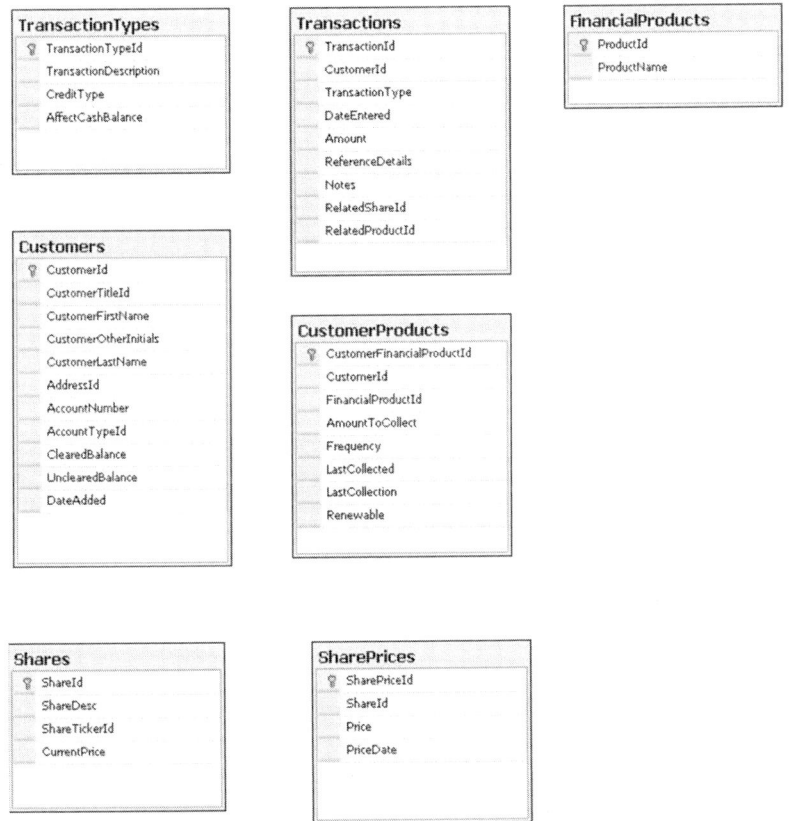

Figure 3-6. *Logical database model*

Denormalization

Despite having normalized our data to be more efficient, there will be times when denormalizing the data is a better option. **Denormalization** is the complete opposite of normalization: it is where you introduce data redundancy within a table to reduce the number of table joins and potentially speed up data access. Instances of denormalization can be found in production systems where the join to a table is slowing down queries, or perhaps where normalization is not required (e.g., when working with a system in which the data is not regularly updated).

Just because others say your data should be totally normalized, it is not necessarily true, so don't feel forced down that route. The drawback of denormalizing your data too far, though, is that you'll be holding duplicate and unnecessary information that could be normalized out to another table and then just joined during a query. This will therefore create performance issues as well as use a larger amount of data storage space. However, the costs of denormalization can be justified if queries run faster. That said, data integrity is paramount in a system. It's no use having denormalized data in which there are duplications of data where one area is updated when there's a change, and the other area isn't updated.

Denormalization is not the route we want to take in our database example, so now that we have all the data to produce the system, it's time to look at how these tables will link together.

Creating the Sample Database

Let's now begin to create our example database. In this section, we'll examine two different ways to create a database in SSE:

- Using the SSMSE graphical interface
- Using T-SQL code

Both methods have their own merits and pitfalls for creating databases, as you'll discover, but these two methods are used whenever possible throughout the book, and where you might find one method is good for one task, it may not be ideal for another. Neither method is right or wrong for every task, and your decision of which to use basically comes down to personal preference and what you're trying to achieve at the time. You may find that using T-SQL code for building objects provides the best results, as you will see instantly the different possible selections. However, if the syntax for the commands is not familiar to you, you may well choose to use a wizard or SSMSE. Once you become more comfortable with the syntax, then a Query Editor pane might become your favored method.

We'll also examine how to drop a database in SSMSE.

Creating a Database in SQL Server Management Studio Express

The first method of creating a database we'll look at is using SSMSE, which was introduced in Chapter 2.

Try It Out: Creating a Database in SSMSE

1. Before creating the database, you'll need to start up SSMSE. To do this, select Start ➤ All Programs ➤ Microsoft SQL Server 2005 ➤ SQL Server Management Studio Express.

■Tip Throughout the book examples, I'm working on a server called XP-HOME, using the default installed instance. You will have your own name for your host computer and possibly even your own name for the installed instance of SSE. Therefore, throughout the book, when you see XP-HOME, this will mean your host computer name.

2. Ensure that you have registered and connected to your server. If the SSE service was not previously started, it will automatically start as you connect, which may take a few moments.

3. In Object Explorer, expand the Databases node until you see the individual system databases you installed earlier in the book. If you are on a Windows XP Pro system setup that allowed you to install the reporting area of SSE, you will see the Reporting Services databases listed. You will find nodes for each of the system databases: master, msdb, and so forth. Ensure that the Databases folder is highlighted and ready for the next action, as shown in Figure 3-7.

Figure 3-7. *The Databases node in Object Explorer*

A minimum amount of information is required to create a database:

- The name the database will be given

- How the data will be sorted

- The size of the database

- Where the database will be located

- The name of the files used to store the information contained within the database

SSMSE gathers this information using the New Database menu option.

4. Right-click the Databases folder to bring up a pop-up menu with a number of different options. Select New Database, as shown in Figure 3-8.

Figure 3-8. *Creating a new database*

5. You are now presented with the New Database screen set to the General tab. First enter the name of the database you want to create—in this case, `ApressFinancial`. Notice as you type that the two file names in the Database Files list box also populate. This is simply an aid, and the names can be changed (see Figure 3-9). However, you should have a very good reason to not accept the names that the screen is creating, as this is enforcing a standard.

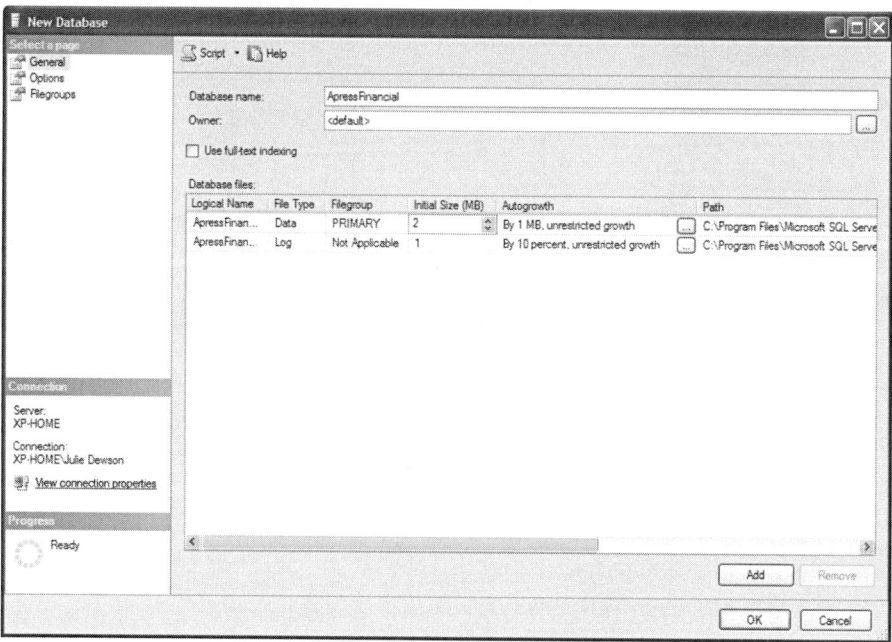

Figure 3-9. *General settings in the New Database screen*

The General screen within this page collects the first two pieces of information. The first piece of information required is the database name. No checks are done at this point as to whether the database exists (this comes when you click OK); however, there is some validation in the field so that certain illegal characters will not be allowed.

Note Illegal characters for a database name are as follows:

" ' */?:\<> -

Keep to alphabetic, numeric, underscore, or dash characters. Also, you may want to keep the database name short, as the database name has to be entered manually in many parts of SSE.

Below the database name is the owner of the database—in other words the login defined as having the ability to do anything within the database as well as determine what other logins can do within the database. This can be any login that has the authority to create databases. A server in many—but not all—installations can hold databases that belong to different development groups. Each group would have an account that was the database owner, and at this point you would assign the specific owner. For the moment, let it default to the <default> account, which will be the account currently logged in to SSE; you'll learn how to change this later. If you're using Windows authentication, your Windows account will be your user ID, and if you're using SQL Server authentication, it will be the ID you used at connection time.

The database owner initially has full administration rights on the database, from creating the database, to modifying it or its contents, to even deleting the database. It is normal practice for a database administrator–type account to create the database, such as a user that belongs to the Builtin\Administrators group, as this is a member of the sysadmin role, which has database creation rights.

Ignore the check box option for full-text indexing, which can be seen below the database owner in Figure 3-9. You would select this option if you wanted your database to have columns that you could search for a particular word or phrase. For example, search engines could have a column that holds a set of phrases from web pages, and full-text searching could be used to find which web pages contain the words being searched for.

The File Name entry (offscreen to the right in Figure 3-9) is the name of the physical file that will hold the data within the database you're working with. By default, SSE takes the name of the database and adds a suffix of _Data to create this name.

The database files are stored on your hard drive with an extension of .MDF—for example, ApressFinancial_Data.MDF. In this case, .MDF is not something used by DIY enthusiasts, but it actually stands for **Master Data File** and is the name of the **primary data file**. Every database *must* have at least one primary data file. This file holds not only the data for the database, but also the location of all the other files that make up the database.

It is also possible to have **secondary data files**. These would have the suffix .NDF. Again, you could use whatever name you wished, and in fact you could have an entirely different name from the primary data file. However, if you did so, the confusion that would abound is not worth thinking about. So do use the same name, and if you need third, fourth, and so on, then add on a numeric suffix.

You would place the file name for a secondary data file in the row below the ApressFinancial_Data entry in the Data Files list box, after clicking the Add button. I'll come back to why you can have a secondary data file when I talk about file properties later in this section.

The File Type column shows whether the file is a data file or a log file, as in a file for holding the data or a file for holding a record of the actions done to the data.

The next column in the grid is titled Filegroup. This allows you to specify the PRIMARY filegroup and any SECONDARY data filegroups for your database. Every database must have a PRIMARY filegroup. To designate secondary data files, click the Add button to add a new data or log file and then click the drop-down arrow on the third row. Select <new filegroup>.This will bring up a dialog box where you can enter **SECONDARY** into a Name field (see Figure 3-10). You may use any name for secondary data files, but it is advisable to choose a name that closely resembles the PRIMARY filegroup name. You can also make that new name the default filegroup, which will become the filegroup to which further additions of data files will be assigned. Also, if you make the new name the default filegroup, it will be in to this area that tables, views, and so on will be assigned unless otherwise specified.

Figure 3-10. *New filegroup*

The logic behind secondary data files is relatively straightforward. A PRIMARY filegroup will always—and must always—contain the system tables that hold the information about the database, the tables, the columns, and so on. If you have the Autogrowth file option (covered shortly) switched off, the PRIMARY filegroup is likely to run out of space at some point. If this happens, and no secondary data files are specified, the database will grind to a halt until some space is added. However, in most instances, especially when you're first starting out, you can leave the database with only a PRIMARY filegroup. Don't misunderstand filegroups and space, though: filegroups are there to help you organize your files within your database storage, and the files that make up the filegroup may span several disks for a performance issue. You will move files around filegroups for speed, efficiency, security, backups, and a number of other reasons. However, you can still hold all the files in one filegroup—the PRIMARY filegroup—which is what you'll do throughout this book.

6. Click Cancel in the New Filegroup for ApressFinancial dialog box and then click Remove in the New Database dialog box to remove the third line within the Database files grid.

■**Note** Remember that the PRIMARY filegroup may hold not only data, but also the system tables, so the PRIMARY filegroup could fill up purely with information about tables, columns, and so forth.

The next item in the New Database dialog box is the Initial Size (MB) column. The initial size of the database is its size when empty. Don't forget that the database won't be totally empty, and some of the space will be initially taken up with the system tables. It is impossible to say, "I'm creating a database, and the initial size must be *nn*MB"—the database size depends on many factors, such as the number of tables, how much static information is stored, to what size you expect the database to grow, and so on.

Let's move on to the next, and possibly most important, area: Autogrowth. This option indicates whether SSE will automatically handle the situation that arises if your database reaches the Initial Size limit. If you don't set this option, you will have to monitor your database and expand its size manually, if and when required. Think of the overhead in having to monitor the size, let alone having to increase the size! It is much easier and less hassle, and much less of a risk, to let SSE handle this when starting out.

■**Note** In a production environment, or even when you're developing in the future, it will be more common to switch Autogrowth off and fix the size. This prevents your hard drive from filling up and your server from being unable to continue. At least when you fix the size, you can keep some hard drive space in reserve to enable your SSE to continue running while the development team tries to clear out unwanted data.

While SSE handles increasing the size of the database for you, it has to know by how much. This is where the Autogrowth option comes in. You can let SSE increase the database either by a set amount each time in megabytes or by a percentage. The default is By Percent, and at this stage it doesn't really matter. In our example, the first increase will be 2MB; the second increase will be 2.2MB. For our example, this is sufficient, as there won't be a great deal of data being entered. However, the percentage option does give uneven increases, and if you like order, the option By MB is for you. If you want to change these options by selecting the autogrowth options button (the ellipsis) to the right of the current setting, you can disable autogrowth of your database in the dialog box that appears. You can also, as discussed, alter it to increase By MB rather than By Percent.

In the Autogrowth dialog box, the Maximum File Size option sets a limit on how large the database is allowed to grow. The default is "unrestricted file growth"—in other words, the only limit is the spare space on the hard drive. This is good, as you don't have to worry about maintaining the database too much. But what if you have a rogue piece of code entering an infinite loop of data? This scenario is rare, but not unheard of. It might take a long time to fill up the hard drive, but fill up the hard drive it will, and with a full hard drive, purging the data will prove troublesome. When it is time to start moving the database to a production environment, the Restricted File Growth (MB) option should be set to guard against such problems.

The final column that you will find in the New Database dialog box by scrolling to the right is Path (refer back to Figure 3-9). In this column, you define where the database files will reside on your hard drive. If SSE is installed on your C drive and none of the paths for the data were changed, you will find that the default is `C:\Program Files\Microsoft SQL Server\MSSQL.1\MSSQL\Data`. Figure 3-9 shows I'm working on a mapped drive that has been given the drive letter C. The command button with the ellipsis (. . .) to the right of the path brings up an explorer-style dialog box that allows you to change the location of the database files. For example, if you move to a larger SSE installation, moving the location of these files to a server relevant to the needs of your database will probably be a necessity.

The line that has a File Type setting of Log includes the same information as a Data File Type setting, with one or two minor exceptions. The File Name entry adds a suffix of `_Log` onto the database name, and there is no means to change the Filegroup column, since the Transaction Log doesn't actually hold system tables, and so would only fill up through the recording of actions. It is possible, however, to define multiple log file locations. Filling the transaction log and not being able to process any more information because the log is full will cause your SSE to stop processing. Specifying more than one log location means that you can avoid this problem. The use of a failover log file in larger production systems is advisable.

Let's now move on to discussing the Options area of the New Database dialog box (see Figure 3-11).

The first field in the Options area is labeled Collation. We discussed this option in Chapter 1 when installing SSE. If you need to alter a collation setting on a database, you can do so, but care is required. Note that altering the collation sequence on a server should only be undertaken by the SSE system administrator, who will be aware of the issues and have the authority to perform specific tasks.

Figure 3-11. *Options area of the New Database dialog box*

The next setting is Recovery Model. You'll learn about backing up and restoring your database in Chapter 7, and this option forms part of that decision-making process. In development, the best option is to choose the Simple backup mode, as you should have your most up-to-date source being developed and saved to your local hard drive. The three modes are as follows:

- *Full:* Allows the database to be restored to where the failure took place. Every transaction is logged; therefore, you can restore a database backup and then move forward to the individual point in time required using the transaction log.

- *Bulk-Logged:* Minimally logs bulk operations, so if you're performing a bulk operation such as bulk copying into SSE, or if you're inserting a bulk of rows of data, only the action is recorded and not every row is inserted. This will increase performance during these sorts of operations, but if a problem occurs, recovery can only occur to the end of the last log backup.

- *Simple:* After each database backup, the transaction log is truncated. This allows restores to be created to the last successful data backup only, as no transaction log backups are taken. You should not use this mode in a production environment.

The third item in the Options page is Compatibility Level. It is possible to build a database for previous versions of SSE, provided you are willing to sacrifice the new functionality. This will allow you to connect to a SQL Server 2000–defined database, for example, from SSE.

Among the next set of options, the ones of interest to us at the moment are the first four. We'll examine the remaining options when we build the database using T-SQL.

- *Auto Close*: If you want the database to shut down when the last user exits, set this option to True. The standard practice is a setting of False, and you should have a good reason to set this option to True, especially on a remote server.

- *Auto Create Statistics*: This option relates to the creation of statistics used when querying data. The standard practice is a setting of True; however, in a production environment, especially if you have a nightly or weekly process that generates statistics on your data, you would switch this to False. Creating and updating statistics while your system is being used does increase processing required on your server, and if your server is heavily used for inserting data, you will find a performance degradation with this option set to True. To clarify, though, it is necessary to balance your choice with how much your system will have to query data.

- *Auto Shrink*: Database and transaction logs grow in size not only with increased data input, but also through other actions, which we'll discuss in more detail in Chapter 7. You can shrink the size of the log file through certain actions, some of which can be instigated by T-SQL and some as a by-product of actions being performed.

- *Auto Update Statistics*: This is a more common option to have set to True, even on production servers, although there is still a performance degradation. This option will update statistics as data is inserted, modified, or deleted for tables for use in indexes, and it will also update statistics for columns within a table. We'll discuss indexes further in Chapter 6.

7. Click the OK button at the bottom of the screen to create the database.

SSE will now perform several actions. First, it checks whether the database already exists and, if so, you will have to choose another name. Once the database name is validated, SSE does a security check to make sure that the user has permission to create the database. This is not a concern here, since by following this book, you will always be logged on to SSE with the proper permissions. Now that you have security clearance, the data files are created and placed on the hard drive. Providing there is enough space, these files will be successfully created, and it is not until this point that SSE is updated with the new database details in the internal system tables.

Once this is done, the database is ready for use. As you can see, this whole process is relatively straightforward, and simple to complete. Congratulations!

■Tip You need not create the database at this point if you don't want to. There are several other options available to you to save the underlying T-SQL to a file, to the clipboard, or to Query Editor. The first two options are very useful as methods of storing actions you're creating to keep in your source code repository, such as Visual SourceSafe. The third option is ideal if you wish to add more options to your database than you have defined within the wizard setup. All of the options enable you to see the underlying code and understand what is required to create a database. We'll look at the code in a moment.

When you return to Object Explorer in SSMSE (you may need to refresh the contents), you will see the new database listed, as shown in Figure 3-12.

Figure 3-12. *The new database within Object Explorer*

SSMSE is simply a GUI front end to running T-SQL scripts in the background. As you progress through the book, you'll see the T-SQL generated for each object type we're using, as well as the objects being created graphically, as you've just seen. Once the database has been created, you can right-click and, as shown in Figure 3-13, have the details sent to one of three locations.

Figure 3-13. *Scripting the database from SSMSE*

Whichever method you choose to use, the script will be the same, with the exception of a comment line when you create the script in the second option. The script for generating the database from this option is listed here so we can go through what is happening.

First of all, SSE points itself to a known database, as shown in the following snippet. master has to exist; otherwise, SSE will not work. The USE statement, which instructs SSE to alter its connection to default to the database after the USE statement, points further statements to the master database.

```
USE [master]
GO
```

Next, the script builds up the CREATE DATABASE T-SQL statement built on the options selected. (We'll walk through the CREATE DATABASE syntax that could be used in the "Creating a Database in a Query Pane" section, as this statement doesn't cover all the possibilities.) Notice in the code that follows that the name of the database is surrounded by square brackets: []. SSE does this as a way of defining that the information between the square brackets is to be used similarly to a literal and not as a variable. Also it defines that the information is to be treated as one unit. To clarify, if we want to name the database Apress Financial (i.e., with a space between "Apress" and "Financial"), we need to have a method of knowing where the name of the database starts and ends. This is where the identifier brackets come into play.

■**Note** Recall the Set Quoted Identifier option that we encountered in Figure 2-20, which can also be defined with the T-SQL command SET QUOTED_IDENTIFIER ON/OFF. Instead of using the square brackets around master as you see in the preceding text, you can define identifiers by surrounding them with double quotation marks using this command. Therefore, anything that has double quotation marks around it is seen as an identifier rather than a literal, if this option is set to ON. To get around this requirement, you can use single quotation marks, as shown in the example, but then if you do have to enter a single quote mark—as in the word "don't"—you would have to use another single quotation mark. So as you can see, this situation can get a bit messy. I prefer to have QUOTED_IDENTIFIER set to OFF, to reduce confusion.

```
/****** Object:  Database [ApressFinancial]
Script Date: 05/29/2006 14:14:35 ******/
CREATE DATABASE [ApressFinancial] ON  PRIMARY
( NAME = N'ApressFinancial', FILENAME = N'C:\Program Files\Microsoft SQL
Server\MSSQL.1\MSSQL\DATA\ApressFinancial.mdf' , SIZE = 3072KB , MAXSIZE =
UNLIMITED, FILEGROWTH = 1024KB )
 LOG ON
( NAME = N'ApressFinancial_log', FILENAME = N'C:\Program Files\Microsoft SQL
Server\MSSQL.1\MSSQL\DATA\ApressFinancial_log.ldf' , SIZE = 1024KB , MAXSIZE =
2048GB , FILEGROWTH = 10%)
 COLLATE SQL_Latin1_General_CP1_CI_AS
GO
```

Have you noticed that every so often there is a GO command? This signals to SSE utilities that this is the end of a batch of T-SQL statements, and the utility should send the batch to SSE. Certain statements need to be in their own batch and cannot be combined with other statements in the same batch. To clarify, a GO statement determines that you have come to the end of a batch of statements and that SSE should process these statements before moving on to the next batch of statements.

■**Note** GO commands are used only in ad hoc T-SQL, which is what I'm demonstrating here. Later in the book, you'll build T-SQL into programs called stored procedures. GO commands are not used in stored procedures.

Next, we define the new database's compatibility level. This statement defines that the database's base level is SSE. It is possible to define SSE to an earlier level, as far back as SQL Server

version 6.0, by changing the version number in the parameter @new_cmptlevel. You'll learn more about this code in Chapter 10.

```
EXEC dbo.sp_dbcmptlevel @dbname=N'ApressFinancial', @new_cmptlevel=90
GO
```

■**Note** The values for this are 60 = SQL Server 6.0, 65 = SQL Server 6.5, 70 = SQL Server 7.0, 80 = SQL Server 2000, and 90 = SQL Server 2005.

We then can define the remaining database options. The statements have a GO statement separating them, but in this scenario, the GO statements are superfluous. So why are they included? When SSE is preparing the wizard, it is safer for it to place GO statements after each statement, as it then doesn't have to predict what the next statement is, and therefore whether the end of the batch of transactions has to be defined.

It is possible to initialize full-text indexing or remove all full-text catalogs from the current database. By default, all user-created databases in Microsoft SQL Server 2005 are enabled for full-text indexing.

```
EXEC [ApressFinancial].[dbo].[sp_fulltext_database] @action = 'disable'
GO
```

There will be times when columns have no data in them. This is known as a special value of NULL. Setting ANSI_NULL_DEFAULT to OFF means that a column's default value is NOT NULL. We look at NULL values in Chapter 5 during our table creation discussion.

```
ALTER DATABASE [ApressFinancial] SET ANSI_NULL_DEFAULT OFF
GO
```

Still with NULL values, the ANSI standard states that if you are comparing two columns that have this special NULL value in them, the comparison will fail. This is different from columns that do have data in them. Setting this value to OFF alters the standard, so when you do compare two NULL values, the comparison will pass.

```
ALTER DATABASE [ApressFinancial] SET ANSI_NULLS OFF
GO
```

There are columns of characters that can store variable-length data. We'll come across these when we build our table in Chapter 5. If set to ON, this option makes every column of data contain the maximum number of characters, whether you sent through just one character or many more. It is common to have this set to OFF.

```
ALTER DATABASE [ApressFinancial] SET ANSI_PADDING OFF
GO
```

If an ANSI standard warning or error occurs, such as divide by zero, switching the ANSI_WARNINGS setting to OFF will suppress these. A value of NULL will be returned in any columns that have the error.

```
ALTER DATABASE [ApressFinancial] SET ANSI_WARNINGS OFF
GO
```

If the ANSI_WARNINGS setting was ON, and you performed a divide by zero, the query would terminate. To change this, we tell SSE not to abort when there's an arithmetic error:

```
ALTER DATABASE [ApressFinancial] SET ARITHABORT OFF
GO
```

If you have a database that is only "active" when users are logged in, switching the AUTO_CLOSE setting to ON would close down the database. This is unusual, as databases tend to stay active 24/7, but closing unwanted databases frees up resources for other databases on the server to use if required. One example of when to switch this setting ON is for a database used for analyzing data by users through the day (e.g., one in an actuarial department, where death rates would be analyzed).

```
ALTER DATABASE [ApressFinancial] SET AUTO_CLOSE OFF
GO
```

SSE uses statistics when returning data. If it finds that statistics are missing when running a query, having the following option ON will create these statistics.

```
ALTER DATABASE [ApressFinancial] SET AUTO_CREATE_STATISTICS ON
GO
```

If the volume of data within your database grows smaller (e.g., if you have a daily or weekly archive process), you can reduce the size of the database automatically by setting the following option ON. It is standard to have the option OFF because the database size will simply increase as data is readded. It would be switched ON only if a reduction in the database is required—due to disk space requirements, for example—but it is never a good idea for this option to kick in when the database is in use, so really it is best to keep it OFF.

```
ALTER DATABASE [ApressFinancial] SET AUTO_SHRINK OFF
GO
```

When data is added or modified to SSE, statistics are created that are then used when querying the data. These statistics can be updated with every modification, or they can be completed via a T-SQL set of code at set times. There is a performance reduction as data is inserted, modified, or deleted, but this performance is gained back when you want to return data. Your application, being a pure insertion, a pure query, or a mix, determines whether you'll want this option ON. If you have a pure insertion application, you probably want this option switched OFF, for example, but this is an optimization decision.

```
ALTER DATABASE [ApressFinancial] SET AUTO_UPDATE_STATISTICS ON
GO
```

A *cursor* is an in-memory table built with T-SQL used for row-at-a-time processing. It's a temporary memory resident table, in essence. A cursor can exist for the lifetime of a program but, if you switch the following setting to ON, when a batch of data is committed or rolled back during a transaction, the cursor will be closed.

```
ALTER DATABASE [ApressFinancial] SET CURSOR_CLOSE_ON_COMMIT OFF
GO
```

A cursor can either exist locally or globally. This means that if GLOBAL is selected for this option, any cursor created in a program is available to any subprogram that is called. LOCAL, the other option, indicates that the cursor exists only within the program that created it.

```
ALTER DATABASE [ApressFinancial] SET CURSOR_DEFAULT  GLOBAL
GO
```

If you're concatenating character fields and if the following option is ON, then if any of the columns has a NULL value, the result is a NULL.

```
ALTER DATABASE [ApressFinancial] SET CONCAT_NULL_YIELDS_NULL OFF
GO
```

When you're working with some numeric data types, it is possible to lose precision of the numerics. This can occur when you move a floating-point value to a specific numeric decimal point location, and the value you're passing has too many significant digits. If the following option is set to ON, an error is generated. OFF means the value is truncated.

```
ALTER DATABASE [ApressFinancial] SET NUMERIC_ROUNDABORT OFF
GO
```

As mentioned earlier, when you're defining database names, if there is a space in the name or the name is a reserved word, it is possible to tell SSE to ignore that fact and treat the contents of the squared brackets as a literal. You are using **quoted identifiers** when you use the double quotation mark instead of square brackets. We'll delve into this further when inserting data in Chapter 8, as there are a number of details to discuss with this option.

```
ALTER DATABASE [ApressFinancial] SET QUOTED_IDENTIFIER OFF
GO
```

The following option relates to a special type of program called a **trigger**. A trigger can run when data is modified, and one trigger can call another trigger. A setting of OFF means that this cannot take place.

```
ALTER DATABASE [ApressFinancial] SET RECURSIVE_TRIGGERS OFF
GO
```

I mentioned statistics earlier with another option and how they can be updated as data is modified. The following option is similar to AUTO_UPDATE_STATISTICS. If this option is set to ON, the query that triggers an update of the statistics will not wait for the statistics to be created. The statistics update will start, but it will do so in the background asynchronously.

```
ALTER DATABASE [ApressFinancial] SET AUTO_UPDATE_STATISTICS_ASYNC OFF
GO
```

This option governs whether there is a relationship between datetime columns in related tables:

```
ALTER DATABASE [ApressFinancial] SET DATE_CORRELATION_OPTIMIZATION OFF
GO
```

The following option defines how the filegroups are set: READ_WRITE or READ_ONLY. The use of READ_ONLY is ideal where you have a backup database that users can use to inspect data. The database is an exact mirror of a production database, for example, so it has the security on it set to allow updates to it, but by setting this option to READ_ONLY you can be sure that no updates can occur.

```
ALTER DATABASE [ApressFinancial] SET READ_WRITE
GO
```

The next option determines how your data can be recovered when a failure such as a power outage happens. In other words, the following option defines the recovery model, as discussed earlier. We'll look at this in more detail when we discuss database maintenance in Chapter 7.

```
ALTER DATABASE [ApressFinancial] SET RECOVERY SIMPLE
GO
```

The following option defines the user access to the database. MULTI_USER is the norm and allows more than one user into the database. The other settings are SINGLE_USER and RESTRICTED_USER, where only people who have powerful privileges can connect. You would set your database to RESTRICTED_USER after a media or power failure, for example, when a database administrator needs to connect to the database to ensure everything is OK.

```
ALTER DATABASE [ApressFinancial] SET  MULTI_USER
GO
```

When you have an I/O error (e.g., a hard drive might be on its way to breaking down), this option will report an error if checksums don't match:

```
ALTER DATABASE [ApressFinancial] SET PAGE_VERIFY CHECKSUM
GO
```

Finally, the following line is used for controlling whether permissions checks are required when referring to objects in another database:

```
ALTER DATABASE [ApressFinancial] SET DB_CHAINING OFF
```

Dropping the Database in SQL Server Management Studio Express

To follow the next section properly and build the database using code, it is necessary to remove the database just created. It is also handy to know how to do this anyway, for those times when you have made an error or when you wish to remove a database that is no longer in use. Deleting a database is also known as **dropping** a database.

Try It Out: Dropping a Database in SSMSE

1. If SSMSE is not started, start it up now and expand the nodes until you see the database `ApressFinancial`.

2. Right-click `ApressFinancial` to bring up the context menu.

3. Click the Delete option, as shown in Figure 3-14.

Figure 3-14. *Deleting a database within SSMSE*

4. The dialog box shown in Figure 3-15 will display. Select Close Existing Connections and then click OK.

Figure 3-15. *Confirming database deletion in the Delete Object dialog box*

The first check box, Delete Backup and Restore History Information for Databases, gives you the option of keeping or removing the history information that was generated when completing backups or restores. If you want to keep this information for audit purposes, uncheck the box.

The second check box is very important. If there is a program running against a database, or if you have any design windows or query panes open and pointing to the database you want to delete, this option will close those connections. If you are deleting a database, there really should be no connections. This is a good check and will prevent accidents from happening, and it also allows any rogue databases to be removed without having to track down who is connected to them.

5. Click OK. The database is now permanently removed.

When you click the OK button, SSE actually performs several actions. First, a command is sent to SSE informing it of the name of the database to remove. SSE then checks that nobody is currently connected to that database. If someone is connected, through either SSE Query Editor or a data access method like ADO.NET, SSE will refuse the deletion. Only if you select Close Existing Connections will this process be overridden.

For SSE to refuse the deletion, it does not matter if anyone connected to the database is actually doing anything; all that is important is the existence of the connection. For example, if you selected ApressFinancial in Query Editor, and then returned to SSMSE and tried to drop the database, you would see the error shown in Figure 3-16.

Figure 3-16. *Failed database deletion*

■**Tip** Errors like the one shown in Figure 3-16 provide hyperlinks to documentation that can give you further help.

Once SSE has checked that nobody is connected to the database, it then checks that you have **permission** to remove the database. SSE will allow you to delete the database if it was your user ID that created it, in which case you own this database and SSE allows you do what you want with it. However, you are not alone in owning the database.

If you recall from Chapter 1, there was mention of the sa account when installing SSE. Since it is the most powerful ID and has control over everything within SSE, there were warnings about leaving the sa account without any password and also about using the sa account as any sort of login ID in general. This section also mentioned that the sa account was in fact a member of the sysadmin server role. A **role** is a way of grouping together similar users who need similar access to sets of data. Anyone in the sysadmin role has full administrative privileges—and this includes rights to remove any database on the server.

So whether you are logged in as yourself or as sysadmin, take care when using SSMSE to drop a database.

Creating a Database in a Query Pane

To use the second method of creating databases, you first need to drop the ApressFinancial database as described in the previous section. Then you can continue with the following steps.

Try It Out: Creating a Database in a Query Pane

1. From the standard toolbar of SSMSE, select New Query.

2. In the query pane, enter the following T-SQL script:

```
CREATE DATABASE ApressFinancial ON  PRIMARY
( NAME = N'ApressFinancial',
FILENAME = 'C:\Program Files\Microsoft SQL
Server\MSSQL.1\MSSQL\DATA\ApressFinancial.mdf' , SIZE = 3072KB ,
MAXSIZE = UNLIMITED, FILEGROWTH = 1024KB )
 LOG ON
( NAME = 'ApressFinancial_log',
FILENAME = 'C:\Program Files\Microsoft SQL
Server\MSSQL.1\MSSQL\DATA\ApressFinancial_log.ldf' ,
SIZE = 1024KB , MAXSIZE = 2048GB , FILEGROWTH = 10%)
 COLLATE SQL_Latin1_General_CI_AS
GO
```

3. Execute this code by pressing F5 or Ctrl+E, or by clicking the Execute Query toolbar button.

4. Once the code is executed, you should see the following result:

```
Command(s) completed successfully.
```

How It Works: Creating a Database in a Query Pane

The main focus of this section of the chapter is the code listed in the previous exercise: the CREATE DATABASE command.

When placing code in Query Editor, you're building up a set of instructions for SSE to act on. As you progress through the book, you will encounter many commands that you can place in Query Editor, all of which build up to provide powerful and useful utilities or methods for working with data. An in-depth discussion of Query Editor took place in Chapter 2, so if you need to refresh your memory, take a quick look back at the material covered in that chapter.

Before we actually look at the code itself, we need to inspect the syntax of the CREATE DATABASE command:

```
CREATE DATABASE <database name>
[ON
    ( [ NAME = logical_name, ]
        FILENAME = physical_file_name
      [, FILESIZE = size ]
      [, MAXSIZE = maxsize ]
      [, FILEGROWTH = growth_increment] ) ]
```

```
[LOG ON
    ( [ NAME = logical_name, ]
        FILENAME = physical_file_name
      [, FILESIZE = size ]
      [, MAXSIZE = maxsize ]
      [, FILEGROWTH = growth_increment] ) ]
[COLLATE collation_name ]
```

The parameters are as follows:

- `database name`: The name of the database that the `CREATE DATABASE` command will create within SSE.

- `ON`: The `ON` keyword informs SSE that the command will specifically mention where the data files are to be placed, as well as their name, size, and file growth. With the `ON` keyword comes a further list of comma-separated options:

 - `NAME`: The logical name of the data file that will be used as the reference within SSE.

 - `FILENAME`: The physical file name and full path where the data file will reside.

 - `FILESIZE`: The initial size, in megabytes by default, of the data file specified. This parameter is optional, and if omitted it will take the size defined in the model database. You can suffix the size with `KB`, `MB`, `GB`, or `TB` (terabytes).

 - `FILEGROWTH`: The amount that the data file will grow each time it fills up. You can specify either a value that indicates by how many megabytes the data file will grow or a percentage, as discussed earlier when we created a database with SSMSE.

- `LOG ON`: The use of the `LOG ON` keyword informs SSE that the command will specifically mention where the log files will be placed, and their name, size, and file growth:

 - `NAME`: The name of the log file that will be used as the reference within SSE.

 - `FILENAME`: The physical file name and full path to where the log file will reside. You must include the suffix `.LDF`. This could be a different name from the `FILENAME` specified earlier.

 - `FILESIZE`: The initial size, in megabytes by default, of the log file specified. This parameter is optional, and if omitted, it will take the size defined in the model database. You can suffix the size with `KB`, `MB`, `GB`, or `TB`.

 - `FILEGROWTH`: The amount by which the log file will grow each time the data file fills up, which has the same values as for the data file's `FILEGROWTH`.

- `COLLATE`: The collation used for the database. Collation was discussed earlier in the chapter when we created a database with SSMSE.

It's now time to inspect the code entered into Query Analyzer that will create the `ApressFinancial` database.

Commencing with `CREATE DATABASE`, you are informing SSE that the following statements are all parameters to be considered for building a new database within SSE. Some of the parameters are optional, and SSE will include default values when these parameters are not entered. But how does SSE know what values to supply? Recall that at the start of this chapter we discussed the built-in SSE databases, specifically the `model` database. SSE takes the default options for parameters from this database unless they are otherwise specified. Thus, it is important to consider carefully any modifications to the `model` database.

The database name is obviously essential, and in this case, `ApressFinancial` is the chosen name.

The `ON` parameter provides SSE with specifics about the data files to be created, rather than taking the defaults. Admittedly in this instance, there is no need to specify these details, as by taking the defaults, SSE would supply the parameters as listed anyway.

This can also be said for the next set of parameters, which deal with the Transaction Log found with LOG ON. In this instance, there is no need to supply these parameters, as again the listed amounts are the SSE defaults.

Finally, the collation sequence we specify is actually the default for the server.

Taking all this on board, the command could actually be entered as follows, which would then take all the default settings from SSE to build the database:

```
CREATE DATABASE ApressFinancial
```

We can then set the database options as outlined during the discussion of the script earlier in the chapter.

Similarly, if we want to delete the database using T-SQL code, it's a simple case of ensuring that we are not connected within that particular query pane to ApressFinancial via the USE command. Then we use the command DROP followed by the object we want to drop, or delete, and then the name of the object.

```
USE Master
GO
DROP DATABASE ApressFinancial
```

Summary

In this chapter, we looked at designing and creating our example database. Though just a sample database, it still required careful thought.

In the next chapter, we will take a look at security within SSE and start investigating how to ensure your SSE database is safe and secure from within and outside your organization.

CHAPTER 4

■ ■ ■

Security

Security is important—more so, in fact, than design, creation, and performance. If your database had no security measures in place, absolutely anyone could come along and steal or corrupt the data, causing havoc to you and your company. And not just in one database, but in every database in every server.

Security can be enforced in many ways in SSE: by Windows itself through Windows authentication; by restricting users' access to sensitive data through views; or by specifically creating users, logins, and roles that have explicit levels of access.

This chapter covers some parts of security, although it is impossible to talk about every area of security, mainly because we haven't seen much of SQL Server's contents yet! In Chapter 1, we looked at the difference between Windows authentication and SQL Server authentication, so already you know your options with regard to the type of security you might wish to use.

So what's next? You need to understand **logins**, **users**, **roles**, and **schemas**.

Logins

The only way anyone can connect to SQL Server is via a login. As discussed in Chapter 1, this doesn't necessarily mean that every user has to have a specific login within SQL Server itself. With Windows authentication, if a user belongs to a specific Windows group, just by belonging to that group, providing that group is contained within SQL Server, the account will have access to SQL Server.

When a database is created, initially only the database owner has any rights to complete any task on that database, whether that be to add a table, insert any data, or view any data. This was the case when we first created our ApressFinancial database in Chapter 3. It is only when the database owner grants permissions to other users that they gain extra access to complete tasks.

It is common practice on Windows XP Professional, Windows 2003, etc., to create a Windows group and place Windows user accounts into that group. On Windows XP Home, creating groups is not possible. By not having the ability to create groups, each user that is required to access the database needs to be defined individually. Recall from Chapter 1 we created AJMason and VMcGlynn as Limited and Administrator accounts, respectively. Because they were members of BUILTIN/Users, they could connect to SQL Server if they wished. Let's now see what authority they have and how to proceed with adding users that require access to our database but don't currently have it.

Try It Out: Logins and SQL Server Access

1. Select Start ➤ Log Off. From the screen that comes up, select Switch User and then select VMcGlynn.

2. Select Start ➤ All Programs ➤ Microsoft SQL Server 2005 ➤ SQL Server Management Studio Express to start up SSMSE with a connection dialog box as you have seen several times. If you click Connect, you should be able to connect to SSE.

3. It is possible within Object Explorer to navigate to ApressFinancial, and expand the node (see Figure 4-1), and investigate any object within that database (although there is very little in there at the moment). Notice as well on the top line of Object Explorer to the right side the connection has a name of XP-HOME\VMcGlynn.

Figure 4-1. *Administrators can automatically log on to your database.*

4. Log off from VMcGlynn and log on to AJMason, who you should recall has a Limited account. Navigate and once again start SSMSE. This time, within the Object Explorer, if you navigate to ApressFinancial, you will receive an error message (see Figure 4-2).

Figure 4-2. *AJMason login is unable to access database.*

5. If we wished to add AJMason so that this user could log in to our database but did not want to make the account an Administrator account on Windows XP Home, we can add the user within SSMSE.

6. Navigate within Object Explorer to the Security node and expand it along with the Logins node. Right-click and select New Login, as shown in Figure 4-3.

Figure 4-3. *New Login option*

7. In the dialog box that appears, click Search (see Figure 4-4) to display the Select User or Group dialog box where we will begin our search for AJMason, as shown in Figure 4-5.

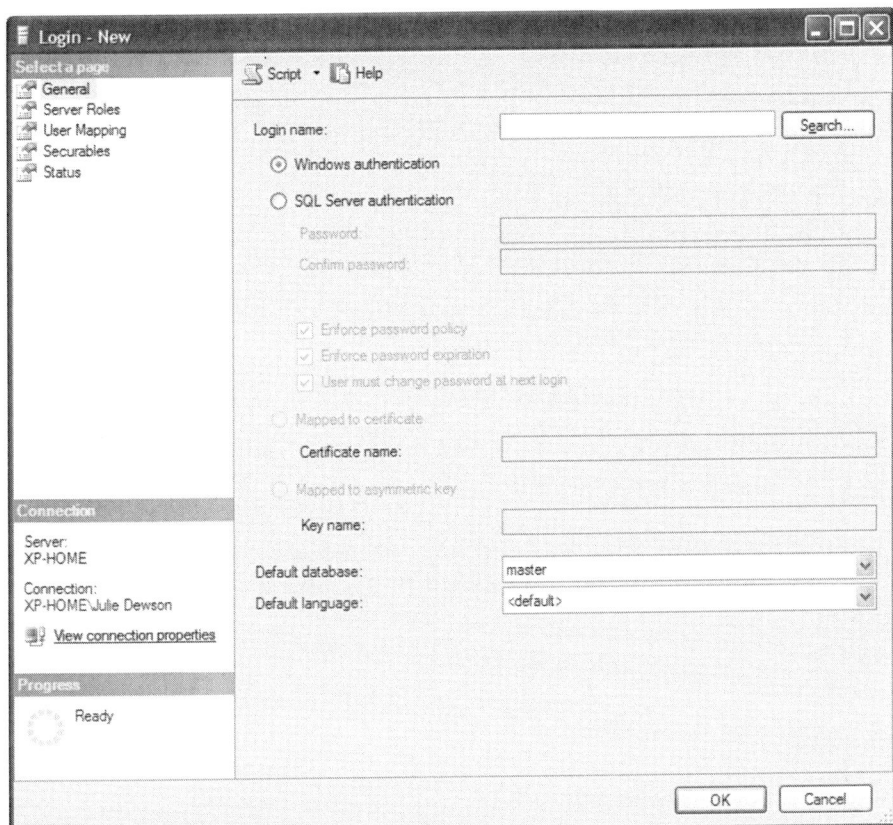

Figure 4-4. *Creating a new login*

Figure 4-5. *Searching for logins*

8. Click Advanced, which will let you complete the search for the user or built-in security principal you want. Click Find Now. After a few moments, the grid below will be populated, and you can find AJMason, which you need to highlight as shown in Figure 4-6.

Figure 4-6. *Finding AJMason*

9. Clicking OK after highlighting AJMason brings us back to the Select User or Group dialog box, where we will see our group has been added, as shown in Figure 4-7. We can then click OK.

Figure 4-7. *AJMason found, ready for adding*

■Note We are now back at the Login - New screen where the group will be populated. If you clicked OK at this point, this would still allow AJMason to connect only to SQL Server and nothing else. AJMason could not do anything else such as create a table or even connect to `ApressFinancial`.

10. We need to give AJMason access to the databases we wish to allow this user to use. It is vital that you only allow users or groups of users access to the resources they need and don't use the "allow everything, it's easier" approach that I have seen on my travels. We only want AJMason to see the `ApressFinancial` database, so move to the User Mapping option page and select the `ApressFinancial` database on the Users Mapped to This Login section of the screen shown in Figure 4-8. For the moment, click the Script button at the top of the screen. (When you click this button, it doesn't matter which of the three options you choose when selecting where to put the T-SQL.) We will come back to logins in the next section when we examine roles.

Figure 4-8. *Giving a login access to a database*

11. The SQL generated from Figure 4-8 follows. We will look at it in more detail in a moment when we examine adding a login more closely.

```
USE [master]
GO
CREATE LOGIN [XP-HOME\AJMason]
FOR LOGIN [XP-HOME\AJMason]
GO
```

12. Going back to SSMSE, you can see in Figure 4-9 that we have moved to the Status page. Here we can grant or deny access to SQL Server for a Windows account, SQL Server login, or in our case Windows group. The second set of options is for enabling or disabling SQL Server logins. The final set of options, specific to SQL Server authentication, allows an account to be unlocked after it has been locked out.

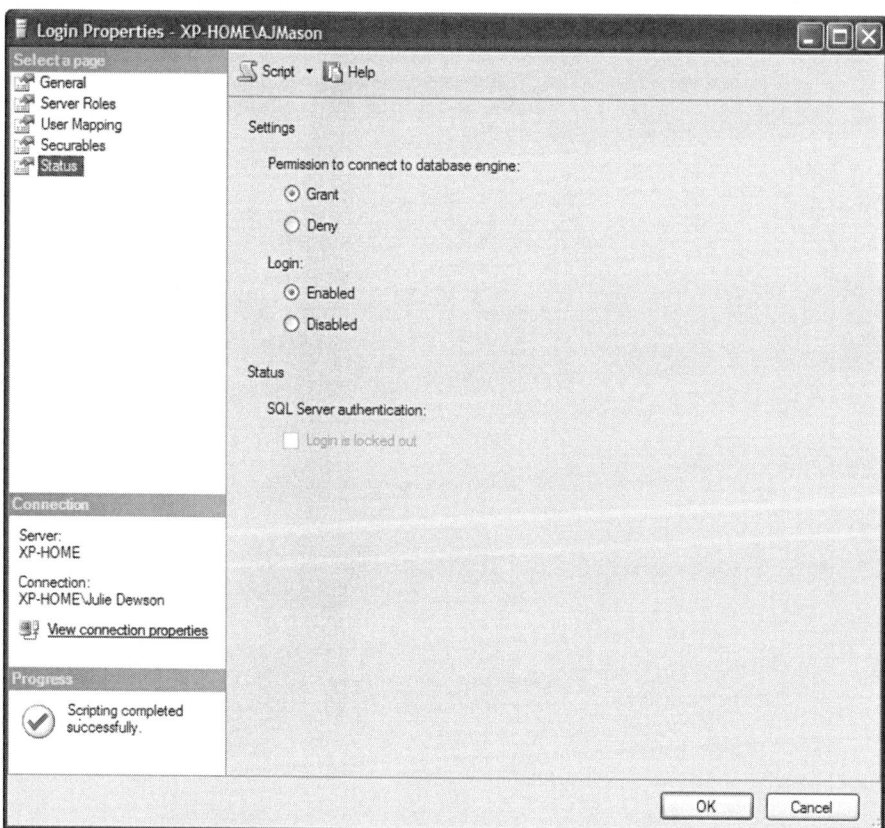

Figure 4-9. *Login status*

13. We could now click OK to add AJMason, but click Cancel instead, which will allow the login creation to be demonstrated using T-SQL.

If we had created the new login and placed it within SQL Server, we could now switch the user account to AJMason and successfully connect. However, as AJMason, we would only be able to explore the ApressFinancial database we created in Chapter 3.

As I mentioned at the start of this discussion, the process would be the same if you wished to add a single user.

For SQL Server authentication, each user needs to be added separately. The process is very similar to that for adding users with Windows authentication, but you must specify a password expiration and enforce password complexity. This will force the Windows password policies for expiration and complexity that exist on this account to apply to this login's password.

Moving back to the T-SQL script for creating AJMason, re-created next, we can investigate what is happening but also see the full T-SQL code required.

Try It Out: T-SQL, Logins, and SQL Server Access

1. The first action that needs to be completed is to add the login to SQL Server.

```
USE [master]
GO
CREATE LOGIN [XP-HOME\AJMason]
FROM WINDOWS
WITH DEFAULT_DATABASE=[master],
DEFAULT_LANGUAGE=[us_english]
GO
```

The format of this syntax is straightforward. In this case, CREATE LOGIN instructs SQL Server that you want to create a new login called XP-HOME\AJMason, where XP-HOME is the name of the computer or network domain in which AJMason can be found. You should change the prefix to match your own setup. Here the definition appears surrounded with optional square brackets in case of spaces in the name.

Next, the keywords FROM WINDOWS inform SQL Server that you are creating a login with Windows authentication. After that you define the name of the database that the login will connect to when a connection is made using WITH DEFAULT_DATABASE. At present this is pointing to a system database, but you could alter it to ApressFinancial or any other database. However, it is safer to initially point to a system database in case the creation to the defined database fails. Finally, the second option specifies the default language the connection will use, although it is possible at any time to alter the language using the Set Language option. This will allow this group to connect to SQL Server.

2. Once you have placed the code in your query pane, you can execute it by pressing either Ctrl+E or F5, or clicking the Execute button on the toolbar. Once it finishes executing, you should see the new login in the Security node within Object Explorer on the left, as shown in Figure 4-10. If you right-click the new login and select Properties, you will see the same screen and details as you saw when you created the login graphically.

Figure 4-10. *AJMason specifically added*

3. We can then give the login access to SQL Server or disable it by using the ALTER LOGIN command. It is also possible to alter the login's default database. In our code example, when creating the login, you will see that the default database was called master. It would be better for the login to connect to the correct database. The following code informs SQL Server that it should connect our login to the ApressFinancial database by default, rather than the master database as defined previously. Remember to change the prefix as appropriate.

```
ALTER LOGIN [XP-HOME\AJMason]
WITH DEFAULT_DATABASE=ApressFinancial
```

Note At this point, you will stop AJMason from being able to connect to SSE. No connection to the server will occur, and an error message will be produced. It is not until the next step that this access is granted again.

4. The final piece in the jigsaw is to grant the Windows account access to the database, which will then allow the login to use the ApressFinancial database. To do this, we need to switch from the master database to the ApressFinancial database with the USE keyword followed by the name of the database. We then need to create the user AJMason in ApressFinancial and define this user as having the ability to logon. You need to enter and run the following code to make this happen:

```
USE ApressFinancial
GO
CREATE USER [XP-HOME\AJMason]
FOR LOGIN [XP-HOME\AJMason]
```

It is not necessary in the CREATE USER section to have the domain or computer prefix there and a CREATE USER AJMason; however, by keeping the domain name there, if you have remote connections, it gives an instant visible representation of the login.

Although AJMason now can log on to the server and connect to your database, there is not much more this user can do; therefore, if you log in to SQL Server with this account, you may be wondering why. We will go into this a little later in the chapter, as there are a couple of points we need to cover first.

Server Logins and Database Users

As you now know, there are two steps to complete, whether you want to create a SQL Server authentication–based login or a Windows authentication–based login. The first is a server login, which was the first part of creating a login that we went through. A server login is one that, when used, can connect only to the server itself. It cannot use any of the user databases within SQL Server. The second step was creating the database user; in the graphical section that we looked at first, this is when we selected the databases we wanted to use.

Within SQL Server, permissions can be granted at multiple levels, including the server and database level. Examples of server-level permissions include creating new logins or managing server properties. Examples of database permissions include being able to read data from a table or being able to create new tables. One server login can be associated with multiple users in different databases. Generally, when using Windows authentication, a database username is the same as the login name, but this does not have to be the case. It does, however, simplify administration. In this book, we will mostly be dealing with database-level permissions, but we will briefly examine server roles in the following section.

Roles

Three different types of roles exist within SQL Server: fixed server roles, database roles (which refers to the general roles included during installation of SQL Server, component-specific roles such as those for Reporting Services that are added when the component is installed, and user-defined roles), and application roles.

Fixed Server Roles

Within SQL Server, specific predefined roles are set up to allow certain tasks and to restrict other tasks. Someone with the right permissions, such as a system administrator, can assign these roles to any user ID or group of user IDs within SQL Server.

If you look at the Server Roles node in Object Explorer, you will see a list of roles as shown in Figure 4-11. But what do they mean? You get a little hint if you move to the Server Roles node within SSMSE.

Figure 4-11. *Fixed server roles*

■**Note** It is not possible to create your own server role.

These roles, available for anyone to use across the server, can perform the following tasks:

- `bulkadmin`: Run BULK INSERT statements.
- `dbcreator`: Create, alter, or drop databases as well as restore them.
- `diskadmin`: Administer disk files.
- `processadmin`: Kill a login running T-SQL code.
- `securityadmin`: Manage logins including passwords for SQL logins and login permissions.
- `serveradmin`: Administrate the server and carry out tasks such as changing options and even starting and shutting down the server.
- `setupadmin`: Work with more than one server with which they are linked and manage the linked server definitions.
- `sysadmin`: Perform any activity.

Server roles are static objects. They contain groups of actions that operate at the server level rather than at the database level. When creating a new login, you could assign these server roles to it if you wanted the login to carry out server actions as well as any database-related actions, if required.

If your Windows account belongs to the `BUILTIN/Administrators` group, it automatically belongs to the `sysadmin` server role. You can check this yourself by highlighting the `sysadmin` server role, right-clicking it, and selecting Properties to bring up the dialog box shown in Figure 4-12. You should see `BUILTIN/Administrators` listed. As more logins are created, they can be added to this role via the Add button.

Figure 4-12. *Members of the sysadmin role*

Although we are not going to alter this for our example database, having Windows XP administrators automatically being administrators for SQL Server can be a bit of a security issue. Many companies batten down their computers so that no user is an administrator of his or her local

machine. By doing this, they stop people adding their own software, shareware, games, or whatever to a machine that is administrated and looked after by a support team.

This helps keep the machine stable, and throughout your organization everyone will know that a piece of software developed on one machine will work on any other. Therefore, users won't have administrator rights on their XP machine and won't have those rights in SQL Server. This is not the case in all organizations. By leaving the Administrators group in the sysadmin role, everyone who has administrator rights on their PC will have system administrator rights within SQL Server. As the owner of the database, you have now lost control of the security and development of your SQL Server database.

> ■**Note** Because this book assumes that you're using either a standalone PC or a secure set of users, it is safe to keep the Administrators group. However, you will find that this group is usually removed from database setups to keep the security of the database intact. It is worth keeping in mind that before removing the login, or removing it from the sysadmin role, you should set up a new group or user as a system administrator to prevent locking yourself out.

Database Roles

Database roles deal with actions that are performed at the database level. Actions within SQL Server can be grouped into different types.

Following are the existing database roles installed with SQL Server and what they can or cannot do:

- dbo/db_owner: Specifies the owner of the database
- db_accessadmin: Can manage access to a database for logins
- db_backupoperator: Can back up the database
- db_datareader: Can read data from all user-defined tables
- db_datawriter: Can perform any write actions to user tables
- db_ddladmin: Can perform Data Definition Language (DDL) actions like creation of tables
- db_denydatareader: Cannot read data from user tables
- db_denydatawriter: Cannot write data from user tables
- db_securityadmin: Can modify database role membership and manage permissions
- public: Can see any database objects that are created with public, or full rights, access (Every user that you create will belong to the public database role.)

Although you will put the existing database roles to use, you'll find it helpful to create new database roles, a common task in SQL Server, when you want to be very specific about permissions particular users have. You do this by creating a specific database role, and then adding the Windows accounts/Windows groups/SQL Server logins to your role. If you wanted to group several groups together, you might create a new role.

Application Roles

Databases are written for applications. However, not all databases exist for just one application. Application roles allow you to define one role for accessing a database based on the application that

is connecting, rather than having security for different groups of users or single users. Let's look at an example.

Consider a central database that holds client data. This database is in turn accessed from the sales order department, which has its own separate database. The client database is also accessed from the debt recovery department, which also has its own database.

As a database administrator, you may set up user groups for each application. Say you have a Debt Recovery group and a Sales Order Processing group. Debt Recovery would want to see information that was hidden from the Sales Order group, such as how in debt a customer is. But what if a user, such as AJMason, worked in both the debt recovery and sales order departments, in two different part-time jobs, for instance? While working as part of the Sales Order group, AJMason could see information that was not pertinent to that group.

You can set up an application role for Sales Order and another for Debt Recovery, thus removing the conflict of having two different sets of security settings for the one user. Also, when users move departments, you are not wasting time revoking one set of roles to give them a new set of roles for their new department.

An application role overrides any user security settings and is created for giving an application access to SQL Server. Therefore, the Sales Order Processing application would define the access for anybody using it.

An application role has no users; it is used when you wish to define what an application can access within your database and what it cannot. We need to create an application role for examples shown later in this book, so let's do this now.

Try It Out: Creating a New Application Role

1. Navigate to the ApressFinancial database, expand the Security node, right-click Roles, and select New Application Role. In the dialog box that appears, enter a useful role name and a password as shown in Figure 4-13. This role will be for the banking application through which users will want to look at checks, cash withdrawals, etc.

Figure 4-13. *Creating a new application role*

2. Click Securables in Object Explorer on the left-hand side, and then click Add. This is how we begin to define what objects we want to assign to this role.

3. In the Add Objects dialog box that appears, leave the options as they are shown in Figure 4-14 and click OK.

Figure 4-14. *Selecting the type of objects to add*

4. This brings up a dialog box where you can select object types. Click Object Types as shown in Figure 4-15.

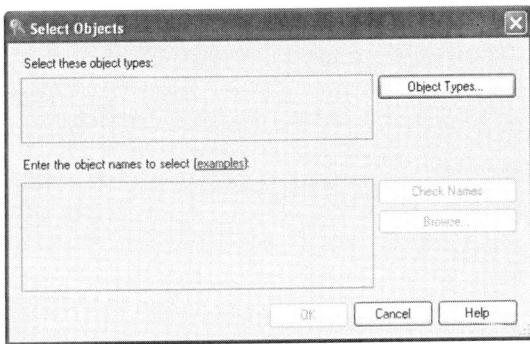

Figure 4-15. *The Select Objects dialog box*

5. As we don't have much within our database that we can give permissions to just now, select Databases, as shown in Figure 4-16, and click OK. We are going to give this application authority to access our database, but not the ability to do anything (mainly because we don't have anything it can do yet).

Figure 4-16. *Selecting the database*

6. Clicking OK in Figure 4-16 brings you back to the Select Objects dialog box that was shown in Figure 4-15 but with the Databases object populated. Click Browse, and we now see a list of all the databases within the server. As shown in Figure 4-17, select `ApressFinancial`, as this is the only database this role will access. Then click OK twice.

Figure 4-17. *ApressFinancial database selected*

7. This brings us back to the Securables screen where we can allow or deny specific actions, as you see in Figure 4-18. Leave everything unchecked for the moment; we will come back to this later in the book when we look at stored procedures in Chapter 10.

Figure 4-18. *Application roles' explicit permission settings*

8. Click OK to finish creating our application role.

Schemas

In the following chapters, we will be creating SQL Server objects to hold and work with our data. We could create these objects so that each could be seen as its own small part of the overall solution. It would make for better organization, though, if objects that could be seen as subsets of the whole solution were grouped together. For instance, in our example, we could group share details and share prices together as share information, or group the financial transactions the customer makes using the transactions and transaction types tables together. These groupings could then be used as the basis of security to the underlying data for when a SQL Server connection tries to access the data. These groupings we have just talked about exist in SSE and are called **schemas**. Therefore, a schema is a method of creating a group and placing objects within that group, which can then be used to grant or revoke permissions as a group to SQL Server connections.

Prior to SQL Server 2005, each object was owned by a user account. Whenever a user left, quite often it would mean moving the ownership of objects for that user's account to a new account. As you can imagine, in a very large system, this could take hours or even days to complete. Now objects are owned by schemas, and the many objects that exist will be contained within one schema in our very large system. Although a schema will still be owned by a SQL Server account, as you will see when we take a look at the syntax in a moment, the number of schemas should be a fraction of the number of objects, even in very large systems, and therefore moving ownership will be easier and faster.

So by having a schema within our solution and assigning objects to that schema, not only are we improving security, but we are also grouping logical units of the database together, making our solution easier to understand and use.

To create a schema is very simple, and the syntax is defined as follows:

```
CREATE SCHEMA schema_name AUTHORIZATION owner_name
```

We can now see this in action.

Try It Out: Creating Schemas and Adding Objects

1. Open up a Query Editor window so we can create our first schema. This schema will be used to keep all our transaction details together. Enter the following code:

   ```
   USE ApressFinancial
   GO
   CREATE SCHEMA TransactionDetails AUTHORIZATION dbo
   ```

2. When we execute this by pressing F5 or Ctrl+E, or clicking the Execute button, we should see it successfully complete. This will have created a new schema where the owner of the schema is the dbo user. This means that any login with sysadmin privileges will automatically have access to this schema because they map to the dbo user in all databases. If you execute the code successfully, you'll see the following message:

   ```
   Command(s) completed successfully.
   ```

3. We can then create further schemas for other groupings, such as one for share details or customer details including products. Enter the following code:

   ```
   CREATE SCHEMA ShareDetails AUTHORIZATION dbo
   GO
   CREATE SCHEMA CustomerDetails AUTHORIZATION dbo
   ```

4. Execute this code, which will add the schemas to the database. If you execute the code successfully, you'll see the following message:

```
Command(s) completed successfully.
```

5. If we move to Object Explorer, we can see our schemas in place, as shown in Figure 4-19.

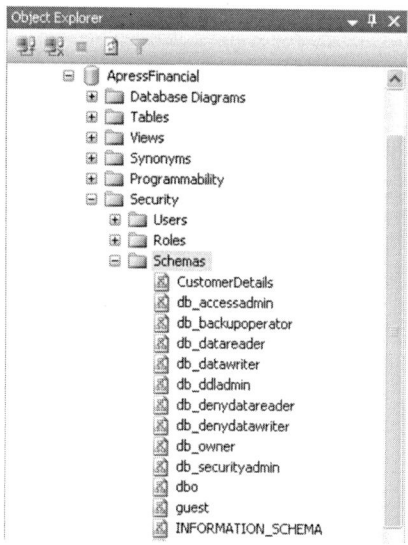

Figure 4-19. *Schemas applied*

Before You Can Proceed with Your Solution

You have now created a database and gained an understanding of the different roles in SQL Server. Before you can proceed and create objects such as tables, you need to clear a couple of obstacles. After the database, the next objects you will create are tables, which you will learn about in the next chapter. So what security considerations do you need to check before you can do this?

First of all, you must be using a database user account that has the authority to add tables to the specific database you are required to, in this case ApressFinancial. This is the next security issue we will tackle, and you should keep in mind what you learned in the previous chapter about accounts and roles. You also need to have access to that specific database. Let's look at the issue of access to the database if you are using a user ID that did not create the database.

The database was re-created at the very end of the previous chapter under user ID XP-HOME\RDewson, which is the user ID that I am connected as. The user who created the database is the database owner, also known as dbo. So how can you check who created the database if you did not? At the end of the last chapter, I asked you to create the database under your own user ID, which, if you followed the instructions so far and you are a local administrator of the machine SQL Server is installed on, you should have the right privileges within SQL Server to do.

If you are working with SQL Server already installed on a Windows XP/Windows Server 2003 machine, you need to ensure that your user ID is set up as an administrator user ID, as demonstrated in Chapter 1, or set up specifically as an administrator within SQL Server.

This next section will demonstrate how you check the identity of the database owner.

Try It Out: Checking the Database Owner

1. Ensure that SSMSE is open.

2. Navigate to the database that you wish to check on, in this case ApressFinancial.

3. Click the ApressFinancial database node in Object Explorer on the left-hand side of the screen once, and then right-click.

4. Select Properties to bring up the Database Properties dialog box shown in Figure 4-20. On the General tab, you will see an item named Owner. This is the fully qualified Windows XP/Windows Server 2003 account preceded by the domain or local machine name.

Figure 4-20. *Database Properties dialog box*

5. Click Cancel to close this dialog box.

Ownership of tables and other database objects is just as important. If you create a table using the same login ID as that which you created the database with, or use a logon ID that is a member of the sysadmin role that is also implicitly mapped to the dbo user in the database, the table will have a default schema of dbo. However, if you logged in with a different user ID, the table would have that user's default schema as the prefix to the table name, replacing the dbo prefix.

Now that we know who the database owner is, it is up to that user, or another user who has system administration rights (in other words, a login that has the sysadmin server role or has the db_owner database role), to allow any other specified user the ability to create tables within the database. We have a user called AJMason who is not a system administrator, but has limited access.

The next section will go through a scenario where, as a developer, AJMason has no rights to create any new items. However, we will rectify this situation in the next section, where we will alter AJMason so that this user can connect to SQL Server and create a table.

Try It Out: Allowing A User to Create a Table

1. Log on to SQL Server as a `sysadmin`, which would be your own login or, as in this book, rdewson. Navigate to the ApressFinancial node, then navigate down through Security and Users and find the XP-HOME\AJMason node we created a few pages earlier (see Figure 4-21).

Figure 4-21. *Finding AJMason*

2. Highlight XP-HOME\AJMason, right-click, and select Properties. We are not going to assign this user any server roles, but we are going to assign this user to the `db_owner` role, as you see in Figure 4-22. This will allow the user to create tables as well as create and work with other objects and data. We could have selected `db_ddladmin`, but this would only have allowed the user to create objects and not create data.

Figure 4-22. *Assigning AJMason to db_owner*

AJMason is now in a position to log in to SQL Server and create tables in the `ApressFinancial` database.

Summary

There is a great deal to cover concerning security and its different aspects. I would like to recap everything that you have seen just one last time to ensure that you understand how everything fits together.

Before you can connect to SQL Server, an administrator of the SQL Server installation must give you permission to connect. In a Windows authentication setup, the administrator would allow either your Windows account or a group that contains your Windows account to connect to SQL Server. He or she can do this by either using the GUI and creating a login via the Security node or using the `CREATE LOGIN ... FROM WINDOWS` T-SQL statement. If you are in a SQL Server authentication setup, a user ID and password would be created within SQL Server, again either via the Security/Logins node or by using the `CREATE LOGIN ... PASSWORD = 'password'` syntax.

Once a connection has been made, you can create a user login within the database using the `CREATE USER ...` syntax. This will allow either the Windows account or the SQL Server login access to the database.

It is then possible to place the user into a role: either a predefined role or, more likely, a custom role that you create. This role can be used to determine what can and cannot be accessed within SQL Server tables, views, stored procedures, and any other object. Therefore, a role allows groups of users in one statement to be granted or revoked access to objects within SQL Server. Without roles, as new people join and as old people leave, or people move between departments, you would need to grant or revoke privileges as required—quite an onerous task.

Finally, when creating objects, as you will see in the next few chapters, these objects are owned by schemas. This allows for groups of objects to belong to a specific schema rather than to a specific user login. This also reduces the overhead of granting privileges and allows the grouping of objects that belong together, making your application easier to understanding.

This chapter continued our coverage of security within SSE. At this point in the book, you now know about SQL Server authentication and Windows authentication, and you have discovered how to control access to databases. Even during the installation process, the `sa` login and password enforcement were discussed on that special account. Our discussions on security are by no means finished because there are still several areas that we need to explore together, which we will do as we go through the book.

Security is the most important part of ensuring that your organization continues to have the ability to work. A security breach could result in lost income and will certainly mean that many people will be unable to do their work. It can also lead to unfulfilled orders, backlogs, or even fraudulent transactions. Regardless of whether you have the most well-designed database or the most poorly performing application ever, if you allow the wrong person into the wrong database, the result will be catastrophic.

Defining Tables

Now that we've created the database, it obviously needs to have the ability to store information. After all, without this, what is the point of a database? The first area that needs to be worked on is the table definitions.

To be functional, a database needs at least one table, but it can have many and, depending on the solution you are building, the number of tables can become quite large. Therefore, it is important that you as a developer know as much about tables, their structures, and their contents as possible. The aim of this chapter is to teach just that, so that you have a sound base to work from regarding tables, which you can then use for the creation of other objects associated with tables.

The design of a table is crucial. Each table needs to contain the correct information for its collection of columns to allow the correct relationships to be established. One of the skills of a database developer or administrator is to ensure that the final design is the correct solution, hence avoiding painful alterations once further development of the system is in progress. For example, if we designed a system where the table definitions had some major problems and required columns to be moved around, any applications using the tables would have to be revisited. This can involve an enormous amount of work. We looked at database design in Chapter 3, where we also created the database in which our tables will reside, so we know what tables we need and what data they will store.

So that we can successfully create a table, this chapter will cover the following:

- How to define a table
- The different types of data that can be stored
- How and where a table is stored
- Creating a table using SQL Server Management Studio Express
- Dealing with more advanced areas of table creation including
 - How to make a row unique
 - Special data states
- Dealing with images and large text data

What Is a Table?

A table is a repository for data, with items of data grouped in one or more columns. Tables contain zero or more rows of information. An Excel spreadsheet can be thought of as a table, albeit a very simple table with few or no rules governing the data. If you look at Figure 5-1, you will see that the first three columns contain data that can be assumed to be first name, last name, and date of birth, but the fourth column is free-format and varies between a hotel room number, a house number, and a flat number. There is no consistency. In fact, in Excel, all the columns could in reality contain any data.

A	B	C	D	E	F
Robin	Dewson	24/03/1964	80	Fat-Belly Dr	UK
Bernie	McGee	15/10/1955	121b	The Crescent	UK
Anthony	Jawad	31/12/1969	Flat 7	Bank Street	UK

Figure 5-1. *Excel showing part address details*

What sets a table inside SSE apart from other kinds of tables is that an SSE table will have specific types of data held in each column, and a predetermined type of data defined for a column can never change without affecting every row of data within that column for that table. If you use Excel, in a specific column you could have a character in one row, a number in the next row, a monetary value in the following row, and so on. That cannot happen in a database table. You can store all of these different values, but they would all have to be stored as a data type that holds strings, which defeats the purpose of using a database in the first place.

At the time a table is created, every column will have a specific data type. Therefore, very careful consideration has to be made when defining a table to ensure that the column data type is the most appropriate. There is no point in selecting a generic data type (a string, for example) to cover all eventualities, as you would have to revisit the design later anyway. Leaving all data types as a string would cause problems with math functions, for example, when you needed to add taxes to a price or when you want to add a number of days to a date for delivery times over order date.

A table's purpose is to hold specific information. The table requires a meaningful name and one or more columns, each given a meaningful name and a data type.

When it comes time to create a table, you do have to be connected to SSE with a login that belongs to the correct server and a database role that can create tables, such as sysadmin or db_ddladmin. When you create a table, it has to be owned within the database, and this is done by assigning the table to a schema. Recall that Chapter 4 discussed a schema for grouping objects and as a basis for object security.

Some data types have fixed storage specifications, whereas with other data types you have to decide for yourself how many characters the maximum will be. If you had a column defined for holding surnames, it would hold character values. There would also be no sense in setting the maximum length of this column at 10 characters, as many surnames are longer than this. Similarly, there would be little sense in saying the maximum should be 1,000 characters. A sensible balance has to be reached.

The rows of data that will be held in a table should be related logically to each other. If a table is defined to hold customer information, this is all it should hold. Under no circumstances should you consider putting information that was not about a customer in the table. It would be illogical to put, for example, details of customer's orders within it.

Defining a Table: SQL Server Management Studio Express

You have learned a great deal about SSE before you have even created your first table. However, it is essential to know all of this information before creating a table and looking at the ramifications of it all going horribly wrong. You also now know why you have to be careful with users to ensure that the user has enough security privileges to build tables.

Defining a table can be completed either in SQL Server Management Studio Express (SSMSE) or Query Editor. We can also create a table through a number of other means using developer tools and languages, but these two methods are the ones this book will focus on. We will create the first

table with SSMSE. This is the `Customers` table, which will hold details about each customer. But before we can do this, it is necessary to look at the different types of data that can be stored.

Different Table Data Types

SSE has many different data types that are available for each column of data. This section will explain the different data types and help you down the path of choosing the right type for each column.

You will find that several data types may look similar, but keep in mind that each data type has a specific use. For example, unless you really need to define characters to be stored as Unicode, don't use the n prefix data types. Unicode characters use up more space than standard characters due to the potentially wide range of characters that SSE has to store. Also, when looking at numbers, if the largest value you will store in a column is 100, then don't go for the data type that will allow the largest number to be stored. This would be a waste of disk space.

Let's take a look at the data types you can use in a table. Afterwards, you'll see data types you can use in a program.

char

The `char` data type is fixed in length. If you define a column to be 20 characters long, 20 characters will be stored. If you enter less than the number of characters defined, the remaining length will be space filled to the right. Therefore, if a column were defined as `char (10)`, "aaa" would be stored as "aaa ". Use this data type when the column data is to be of fixed length, which tends to be the case for customer IDs and bank account IDs.

nchar

The `nchar` type is exactly the same as `char`, but will hold characters in Unicode format rather than ANSI. The Unicode format has a larger character set range than ANSI. ANSI character sets only hold up to 256 characters. However, Unicode character sets hold up to 65,536 different characters. Unicode data types do take up more storage in SSE; in fact, SSE allocates double the space internally, so unless there is a need in your database to hold this type of character, it is easier to stick with ANSI. If you define the column with no value, a length of 1 will be defined. However, this is not good practice, and all lengths should be defined for clarity.

varchar

The `varchar` data type holds alphanumeric data, just like `char`. The difference is that each row can hold a different number of characters up to the maximum length defined. If a column is defined as `varchar(50)`, this means that the data in the column can be up to a maximum of 50 characters long. However, if you only store a string of 3 characters, only three storage spaces are used up. This definition is perfect for scenarios where there is no specific length of data; for example, people's names or descriptions where the length of the stored item does not matter. The maximum size of a `varchar` column is 8,000 characters. However, if you define the column with no size, that is, `varchar()`, the length will default to 1.

You can also use another setting that can exceed the 8,000-character limit, by defining the data type with the constant `max`. You would use this when you believe the data to be below 8,000 characters in length but you want to account for instances when the data may exceed this limit. If you know that you will exceed the 8,000-character limit in at least one row, use this option. Finally, you should use `max` for large blocks of text, because it will eventually supersede the `text` data type.

nvarchar

The nvarchar type is defined in a similar way to varchar, except it uses Unicode and therefore doubles the amount of space required to store the data.

text

■**Caution** This data type will be removed in a future release of SQL Server Express, so you should use varchar(max) instead.

It is useful to understand this data type in case you come across it in any legacy systems that have been upgraded to SQL Server Express 2005. If you need to hold any character data that will always be longer than 8,000 characters, you should not use varchar(max). This is where the text data type comes into play. These data types can hold up to 2GB of data, and could be used to hold notes about customers in a call center, for example. However, text data types are usually different from other data types. Because such a large amount of data can be stored in this data type, it doesn't make sense to store this data within each row of SSE. If you think about it, you would very quickly have a vast database holding very little data. Therefore, if you are storing data within this data type, the data itself is held elsewhere. A pointer is held within SSE in the column defined as a text data type, pointing to where the data is physically held. However, you can store up to 8,000 characters of physical data, if you wish, within this data type within the row; but really, if you have decided to use text as a data type, you are expecting large amounts of data, and therefore it would be best to keep the data outside the database. Microsoft has stated that keeping this data within a row will cease being supported in future versions of SSE.

ntext

This data type is very similar to text, with the exception that the data is stored as Unicode, and only 1GB of characters can be stored because this data type takes double the amount of space to store one character of text. This data type will also be removed in a future version of SSE, and therefore you should use nvarchar(max) instead.

image

image is very much like the text data type, except this is for any type of binary data, which includes images but could also include movies, music, and so on. Like text, it will also be removed in a future version of SSE, so you should use varbinary(max) instead.

int

The int, or integer, data type is used for holding numeric values that do not have a decimal point (whole numbers). There is a range limit to the value of the numbers held: int will hold any number between the values of –2,147,483,648 and 2,147,483,647.

bigint

A bigint, or big integer, data type is very similar to int, except that much larger numbers can be held. A range of –9,223,372,036,854,775,808 through to 9,223,372,036,854,775,807 can be stored.

smallint

The smallint data type, or small integer, holds small integer numbers in the range of –32,768 through to 32,767. Do take care when defining columns with this data type and make sure there really is no possibility of exceeding these limits. There is always a big danger when creating a column with this data type that you have to go back and change the data type, so if in doubt, select int.

tinyint

The tinyint, or tiny integer, data type is even smaller than smallint and holds numbers from 0 through to 255. It could be used to hold a numeric value for each US or Canadian state or perhaps every county in the United Kingdom.

decimal/numeric

Both of these data types hold the same precision and ranges of data. The range is from –10 to the power 38 + 1 through to 10 to the power 38 – 1. These are quite large ranges, from –0.00000000000000000000000000000000000001 through to 10,000,000,000,000,000,000,000,000,000. However, do take care with this, as you cannot store 38 digits to the right and left of the decimal point. You can only store up to and including 38 digits. So, the greater the precision required to the right of the decimal point, the fewer digits are left to represent the whole number.

float

This is used for numbers where the decimal point is not fixed. float data types hold very large numbers in the range of –1.79E+308 through 1.79E+308. There is a warning with this data type: the values cannot always be seen as 100% accurate, as they can be approximate. The approximation arises from the way the number is physically stored as binary code. You will have problems where a number ends in .3, .6, or .7. The value stored has to be approximated, or rounded, as some values can't be stored accurately, for they may have more decimal places than can be catered to. A well-known example is the value of Pi.

real

The real data type is very much like float, except that real can store only numbers in the range of –3.40E+38 through 3.40E+38. This data type also holds an approximate value.

money

The money data type is used for holding numeric values up to four decimal places. If you need to use more than four decimal places, you need to look to another data type, such as decimal. This data type doesn't actually store the currency symbol to signify the monetary type, so you should not use this data type for different currency values, although you can combine a column using this data type with a second column defining the currency type. The money data type has a range of –922,337,203,685,477.5808 through 922,337,203,685,477.5807. If you need to store the currency symbol of the currency that is held here ($ or USD for dollars, £ or GBP for British pounds, etc.), you would need to store this separately, as the money data type does not hold the currency symbol. A column defined as money will hold the money to 1/10,000 of a decimal unit, which is a bit of a waste if you are storing the values as Turkish Lira.

smallmoney

This data type is similar to money with the exception of the range, which lies between –214,748.3648 and 214,748.3647.

datetime

This will hold any date and time from January 1, 1753, through to December 31, 9999. However, it stores not only a date, but also a time alongside it. If you just populate a column defined as datetime with a date, a default time of 12:00:00 will be stored as well.

smalldatetime

This data type is very much like datetime, except the date range is January 1, 1900, through to June 6, 2079. The reason for the strange date at the end of the range lies in the binary storage representation of this datetime.

timestamp

This is an unusual data type, as it is used for a column for which you would not be expected to supply a value. The timestamp data type holds a binary number generated by SSE, which will be unique for each row within a database. Every time a record is modified, the column with this data type in the record will be modified to reflect the time of modification. Therefore, you can use columns with this data type in more advanced techniques where you want to keep a version history of what has been changed.

uniqueidentifier

This data type holds a Globally Unique Identifier, or GUID. This is similar to the timestamp data type, in that the identifier is created by an SSE statement when a record is inserted or modified. The identifier is generated from information from the network card on a machine, processor ID, and the date and time. If you have no network card, the uniqueidentifier is generated from information from your own machine information only. These IDs should be unique throughout the world.

binary

Data held in this data type is in binary format. This data type is mainly used for data held as flags or combinations of flags. For example, perhaps you wanted to hold flags about a customer. You need to know whether the customer is active (value = 1), ordered within the last month (value = 2), last order was for more than $1,000 (value = 4), or meets loyalty criteria (value = 8). This would add up to four columns of data within a database. However, by using binary values, if a client had a value of 13 in binary, the client would have values 1 + 4 + 8, which is active, last order more than $1,000, and meets the loyalty criteria. When you define the column of a set size in binary, all data will be of that size.

varbinary

This data type is very much like binary, except the physical column size per row will differ depending on the value stored. varbinary(max) can hold values more than 8,000 characters in length and should be used for holding data such as images.

bit

This data type holds a value of 0 or 1. Usually, `bit` is used to determine true (1) or false (0) values.

xml

XML data can be held in its own special data type rather than in a `varchar(max)` column. There are special query clauses that can then be used to query and work with this data. Prior to SQL Server Express 2005, XML data was almost an afterthought with no data type, and earlier versions of SSE had extremely limited functionality to work with the XML data that did exist.

Different Program Data Types

There are three more data types that can be used within a program, which we will take a look at now.

cursor

Data can be held in a memory-resident state called a **cursor**. It is like a table, as it has rows and columns of data, but that's where the similarity ends. There are no indexes, for example. A cursor is used to build up a set of data for processing one row at a time.

table

A `table` data type has similarities to both a cursor and a table. It holds rows and columns of data, but the data cannot be indexed. In this case, you deal with the data a "set at a time," like a normal table. We'll look at both the `cursor` and `table` data types later in the book, as they are more advanced topics.

sql_variant

It is possible to have a data type that can hold a few different data types. I will be honest, I don't recommend using this data type as it shows that you are unsure of your data and what type of data to expect. Before putting data into a data type, I feel you need to be sure what type of data you are getting. Although we have `sql_variant` as a program data type, it can also be used as a column data type, but the same arguments apply. We won't look at this data type any further within this book.

Columns Are More Than Simple Data Repositories

Assigning a data type to a column defines what you expect the column to hold. But column definitions have more power than just this. It is possible to fill the column with a seed value, or even with no value whatsoever.

Default Values

As a row is added to a table, rather than requiring developers to add values to columns that could be populated by SSE, such as a column that details using a date and time when a row of data was added, it is possible to place a default value there instead. The default value can be any valid value for that data type. A default value can be overwritten and is not "set in stone."

Generating IDENTITY Values

For those readers who have used MS Access, the IDENTITY keyword option is similar to AutoNumber.

When adding a new row to an SSE table, you may wish to give this row a unique but easily identifiable ID number that can be used to link a row in one table with a row in another. Within the ApressFinancial database, there will be a table holding a list of transactions that needs to be linked to the customers table. Rather than trying to link on values that cannot guarantee a unique link (first name and surname, for example), a unique numeric ID value gives that possibility, providing it is used in conjunction with a unique index. If you have a customer with an ID of 100 in the Customers table and you have linked to the Transactions table via the ID, you could retrieve all the financial transactions for that customer where the foreign key is 100. However, this could mean that when you want to insert a new customer, you have to figure out which ID is next via some T-SQL code or using a table that just held "next number" identities. But fear not, this is where the IDENTITY option within a column definition is invaluable.

By defining a column using the IDENTITY option, what you are informing SSE is that

- The column will have a value generated by SSE.
- There will be a start point (seed).
- An increment value is given, informing SSE by how much each new ID should increase.
- SSE will manage the allocation of IDs.
- Values cannot be modified, as the column is totally controlled by SSE internally.
- Each row will be unique by virtue of the ID being unique.

You would have to perform all of these tasks if SSE did not do so. Therefore, by using this option in a column definition, you can use the value generated to create a solid, reliable, and unique link from one table to another, rather than relying on more imprecise selection criteria.

The Use of NULL Values

Columns can be defined as NULL or NOT NULL. (In the Table Designer, you can check or uncheck the Allow Nulls option.) These two different constraints define whether data must be entered into the column or not. A NULL value means that there is absolutely nothing entered in that column—no data at all. A NULL value is in a special data state, with special meaning. It means that the data value is unknown.

If a value is a NULL value, no data has been inserted into the column for a given row. You have to use the T-SQL IS NULL and IS NOT NULL operators to test for this value. Take the example of a column defined to hold characters, but where one of the rows has a NULL value within it. If you execute a query that carries out string manipulation, the row with the NULL value might cause an error, or the row might not to be included in the result set.

Image and Large Text Storage in SSE

Storing images and large amounts of text is different from storing other kinds of information within SSE. Images can take up large amounts of space. The following also holds true for large amounts of text.

Several scenarios exist where, by holding large amounts of data, SSE and the SSE installation will end up running into problems. I'll explain why in a minute, but first of all you will see what you should do in SSE to handle such data.

If you wish to store large numbers of images or large amounts of text (by large, I mean more than 8KB or 4KB if you want to store the Unicode version of the text), you should store these outside SSE on the hard drive somewhere. SSE then holds a file location in the column to point to where the image or text data is held. From there you can retrieve the information and use it as necessary. This gives you as a developer or database administrator the ability to store large amounts of data on a different volume from the SSE installation, on a different server, or even just in a different directory.

However, if you do wish to hold image data within a table, then if you define a column as varbinary(max), it is possible to hold up to 2^{31} bytes of data, or around 2GB.

Having this control over how you store information can keep your database to a small size. Using file pointers, like c:\temp\mypicture.jpg, allows SSE to run without taking up vast amounts of the database space declared at setup. After all, if you set your database's initial size to 20MB with 5MB extensions, it would not take too many pictures to fill this space up if you held the images in a column. Not only that, SSE has built-in functionality, called **transactions**, where a copy of the data can be taken before and after any modification. If that data included a 5MB graphic or volume of text, SSE would have to keep track of that information while the transaction was in progress.

Note When executing a transaction, SSE takes a copy of the data and then applies your changes. If everything looks OK, you can inform SSE to save them or, if there are problems, you can inform SSE to roll back those changes. Therefore, SSE needs to keep copies of all the data involved in the transaction.

If your application does use images or large amounts of text within a column, keep a close eye on disk space and where the information is stored. By doing so, you can avoid situations where your SSE database stops when the limit of disk space is met on your hard drive or it has no growth options left.

In Chapter 12, there will be discussions about manipulating and inserting images into the database and how this works. However, just keep in mind the information just given so that you can start planning now what solution would be best for your database.

Creating a Table in SQL Server Management Studio Express

This is the first table in our example application. Every organization has to have a set of customers and will need to store these details. Within this table, we will hold information such as each customer's name and an ID to an external system where addresses are held. The only product that our company has where a customer can have an ongoing cash balance with funds that aren't cleared is a bank account. This means our table will also hold the bank account ID, the current balance, and any amount clearing.

Try It Out: Defining a Table

1. Ensure that SSMSE is running.

2. Expand Object Explorer so that you can see the ApressFinancial database, created in Chapter 3.

3. Expand the ApressFinancial database so that you can see the Tables node, as shown in Figure 5-2.

Figure 5-2. *ApressFinancial with no tables*

4. Right-click the Tables node and select New Table. This will take you into the Table Designer. Figure 5-3 shows how the Table Designer looks when you first enter it.

Figure 5-3. *Creating our first table with no columns as yet*

5. From this screen, you need to enter the details for each column within the table. Enter the first column, `CustomerId`, in the Column Name column. When naming columns, try to avoid using spaces. Either keep the column names without spaces, like I have done with `CustomerId`, or use an underscore (_) instead of a space. It is perfectly valid to have column names with spaces. However, to use these columns in SQL code, we have to surround the names by square brackets, [], which is very cumbersome.

At the moment, notice that Column Properties in the middle of Figure 5-3 is empty. This will fill up when you start entering a data type after entering the column name. The Column Properties section is just as crucial as the top half of the screen where you enter the column name and data type.

6. The drop-down combo box that lists the data types is one of the first areas provided by SSE to help us with table creation. This way we don't have to remember every data type there is within SSE. By having all the necessary values listed, it is simple enough to just select the most suitable one. In this instance we want to select `bigint`, as shown in Figure 5-4.

Figure 5-4. *Selecting our data type*

7. The final major item when creating a column within a table is the Allow Nulls check box option. If you don't check the box, some sort of data must be placed in this column. Leaving the check box in the default state will allow NULL values in the column, which is not recommended if the data is required (name, order number, etc.). You can also allow NULLs for numeric columns, so instead of needing to enter a zero, you can just skip over that column when it comes to entering the data. In this instance, we want data to be populated within every row, so remove the check mark.

8. The Column Properties section for our column will now look like the screen shown in Figure 5-5. Take a moment to peruse this section. We can see the name, whether we are allowing NULLs, and the type of data we are storing. There will be changes to what is displayed depending on the data type chosen.

Figure 5-5. *More in-depth properties for columns*

9. We want this column to be an identity column. If you have not already done so, within the Column Properties area expand the Identity Specification node, as we need to set the Is Identity property to Yes. This will set the Identity Increment to 1 and the Identity Seed to 1 as well, as shown in Figure 5-6.

☐ Identity Specification	Yes
(Is Identity)	Yes
Identity Increment	1
Identity Seed	1

Figure 5-6. *Defining a column as having an identity*

10. It is now possible to add in a few more columns before we get to the next interesting item as in Figure 5-7. Go ahead and do so now. Not everybody will have more than a first name and last name, although some people may have initials. Therefore, we will allow NULL values for any initials they may have. We leave the box checked on the CustomerOtherInitials column, as shown in Figure 5-7. We also alter the length of this column to 10 characters, which should be more than enough.

Figure 5-7. *A column that will allow NULL values*

11. We can now define our final columns, which you see in Figure 5-8. The last column will record when the account was opened. This can be done by setting the default value of the DateAdded column. The default value can be a constant value, the value from a function, or a value bound to a formula defined here. For the moment we will use an SSE function that returns the current date and time, GETDATE(), as shown in Figure 5-8. Then every time a row is added, it is not necessary for a value to be entered for this column, as SSE will put the date and time in for you.

Figure 5-8. *The table column definition is now complete.*

■**Note** In Chapter 3, when I discussed normalization, I also covered when data should be denormalized. The Customers table is the one place we do need a small amount of denormalization for speed of access. To speed up the process when a client goes to a cash point, we will have a column that holds his or her account number so that we send a single row of data to the cash point. We can therefore cross-check with his or her card as well as his or her cleared and uncleared balance for display. This data will also be held within the CustomerProducts and Transactions tables. If the account number was stored in the CustomerProducts table only, we would have to send two rows of data to the cash machine: one with the account number and one with the balances.

12. Before we save the table, we need to define some properties for it such as the schema owner. On the right, you should see the Table Properties dialog window, as shown in Figure 5-9. If this is not displayed, you can press F4, or from the menu select View ➤ Properties Window. First of all, give the table a name, Customers, and give the table some sort of description. We then move to the schema owner details. When you click the Schema combo box, it presents you with a list of possible schemas the table can belong to. In Chapter 4, we built the schema we want to use for this table, CustomerDetails.

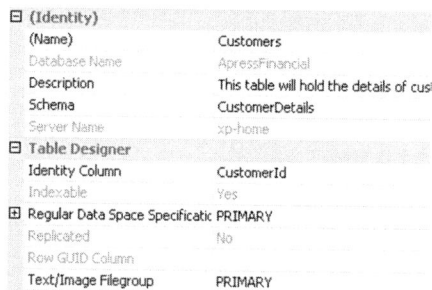

Figure 5-9. *Table properties*

13. Now that we are finished, we can save the table either by clicking the Save toolbar button, which sports a floppy disk icon, or by clicking the X (close) button on the Table Designer to close the window, which should bring up the Save dialog box, asking whether we want to save the changes. By clicking Yes, we get a second box asking for the name of the table if we didn't enter a table name in the Table Properties dialog window as shown in Figure 5-10, or the table is saved using the name specified and we are returned to SSMSE.

Figure 5-10. *Saving a table that was not given a name*

14. If you now right-click the table and select Properties, you can see important details about the table, as shown in Figure 5-11. The first section details who is currently connected. Then we see the date the table was created, its name, and the schema name of the owner of the table.

Figure 5-11. *Table properties*

Now that a table has been created in SSMSE, let's look at creating a table within Query Editor.

Defining a Table Through Query Editor

The next table that needs to be created is the one that will hold the details of the financial transactions for each customer. These transactions will not just be simple money in and money out transactions, but will also be those financial transactions involving shares when a dividend is received or a tax credit if the shares are held in a product that is tax free. We know from our design that details of which product the transaction relates to will be held in a separate table, so we need to create a link between our transaction table and one holding some sort of reference data. It is also necessary to have a link between this table and our CustomerDetails.Customers table. Finally, if the transaction relates to shares and is not recording the finances involved, we need to record that this is the case. To clarify this last point, when a client buys some shares, there will be two records: one for the money leaving the account to buy the shares, and another showing the physical number of shares purchased.

Try It Out: Defining a Table Through Query Editor

1. Ensure that you are pointing to the ApressFinancial database in Query Editor, as shown in Figure 5-12.

Figure 5-12. *Selecting the database from the toolbar*

2. In Query Editor, enter the following code:

```
CREATE TABLE TransactionDetails.Transactions
    (TransactionId bigint IDENTITY(1,1) NOT NULL,
    CustomerId bigint NOT NULL,
    TransactionType int NOT NULL,
    DateEntered datetime NOT NULL,
    Amount numeric(18, 5) NOT NULL,
    ReferenceDetails nvarchar(50) NULL,
    Notes nvarchar(max)  NULL,
    RelatedShareId bigint NULL,
    RelatedProductId bigint NOT NULL)
```

■**Note** Notice that when you type this code into Query Editor, the keywords are colored. For example, CREATE TABLE is in blue and NOT NULL is in gray. This helps you to avoid typing mistakes.

3. Execute the code by either pressing Ctrl+E or F5, or clicking the toolbar Execute button.

4. You should now see the following message in the results pane:

```
The command(s) completed successfully.
```

5. However, you may have received an error message instead. This could be for a number of reasons, from a typing mistake through to not having the authority to create tables. I could list every message that you could receive at this point, but I would be doing so for many pages. Taking one example here, as you can see, the error messages generated are usually self-explanatory. This is informing me that I have a typing error on line 5.

```
Msg 102, Level 15, State 1, Line 5
Incorrect syntax near 'NUL'.
```

6. Now move to Object Explorer. If it is already open, you will have to refresh the Details pane (by right-clicking the Tables node and selecting Refresh). You should then see the TransactionDetails.Transactions table alongside the CustomerDetails.Customers table created previously.

How It Works: Defining a Table Through Query Editor

Using Query Editor to define a table works very much like in SSMSE but without the graphical aids. Recall that SSMSE has prompts for column name, data type, and so on, but here you have to type in every detail. All of the code will be discussed in a moment. However, many people prefer to create a table this way. Having to switch between cursor and keyboard when using the graphical designer can be slower than keying in the details in Query Editor. There is not a lot of time required to create a table this way, and we can build up the table creation as we go along. The query can be saved to a file until it is time to run it and build the table in the database.

Let's now take a look at the T-SQL code that we used to create the table. This code does not include all the options available for creating a table, as there are a large number not used within this book. If you need to use more options or discover what they are, check in Books Online. When it comes to putting a database solution into a production environment, you should consider these options, although some of them will be for larger enterprise production solutions.

The basic syntax for creating a table is as follows:

```
CREATE TABLE [database_name].[schema_name].table_name
    (column_name data_type [length] [IDENTITY(seed, increment)] [NULL/NOT NULL])
```

There are a greater number of possible options, but for the moment let's just concentrate on the ones mentioned previously. You should be able to create most tables using this syntax.

The items listed in square brackets in the CREATE TABLE syntax are optional; however, there are times when we will require them. Let me explain. Take the first option, database_name: if you are in the master database and you wish to create a table in the ApressFinancial database, you would have to either switch to that database with the USE statement or use the database_name option. Usually you will be in the database where you want to create the table, but this option is ideal when creating code that will be executed unattended. It will ensure that the table is built for the correct database rather than trusting that you are in the right area.

The schema_name option allows us to assign the table to the correct and relevant schema, rather than in the default schema of the user connected.

Next we define the columns. Column name and data type are mandatory. However, depending on the data type, the length is optional. You must prefix the first column with an opening parenthesis (and once you have defined the last column, close the list with a closing parenthesis). Each column should be separated from the previous column by a comma. There is a limit of 1,024 columns for a table. If you get anywhere close to that number, you should sit back and reevaluate your table design, because chances are the design needs to be revised.

Defining a Table: Using a Template

SSE has a third method of building tables, although this is my least favored method. A large number of templates are built into SSMSE for everyday tasks. It is also possible to build your own template for repetitive tasks, which is where I can see more power for developers in this area.

Templates can be found in their own explorer window. Selecting View ➤ Template Explorer or pressing Ctrl+Alt+T brings up the Template Explorer window, displayed initially on the right-hand side of SSMSE.

Try It Out: Creating a Table Using a Template

1. Expand the Table node on the Template Explorer. About halfway down you will see a template called Create Table, as shown in Figure 5-13. Double-click this to open up a new Query Editor pane with the template for creating a table.

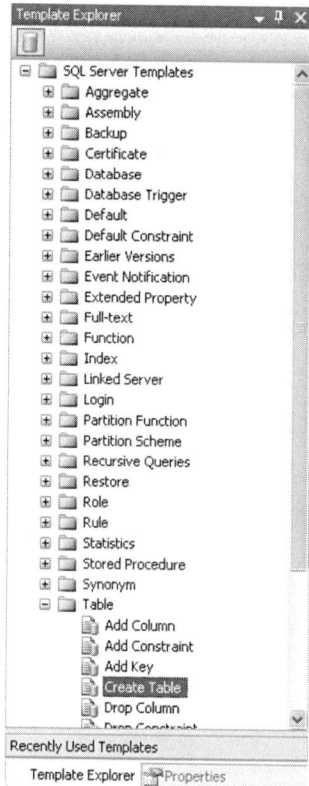

Figure 5-13. *List of templates*

2. Take a close look at the following, which is the listing from the template. A template includes a number of parameters. These are enclosed by angle brackets (<>).

```
-- =========================================
-- Create table template
-- =========================================
USE <database, sysname, AdventureWorks>
GO

IF OBJECT_ID('<schema_name, sysname, dbo>.<table_name,
             sysname, sample_table>', 'U') IS NOT NULL
  DROP TABLE <schema_name, sysname, dbo>.<table_name, sysname,
sample_table>
GO
```

```
CREATE TABLE
    <schema_name, sysname, dbo>.<table_name, sysname, sample_table>(
    <columns_in_primary_key, , c1> <column1_datatype, , int>
    <column1_nullability,, NOT NULL>,
    <column2_name, sysname, c2> <column2_datatype, , char(10)>
    <column2_nullability,, NULL>,
    <column3_name, sysname, c3> <column3_datatype, , datetime>
    <column3_nullability,, NULL>,
    CONSTRAINT <contraint_name, sysname, PK_sample_table>
        PRIMARY KEY (<columns_in_primary_key, , c1>)
)
GO
```

3. By pressing Ctrl+Shift+M, you can alter these parameters to make a set of meaningful code. Do this now, so that the parameters can be altered. Figure 5-14 shows most of our third table, TransactionDetails. TransactionTypes. The reason I say most is that our template code only deals with three columns, and our table has four columns. Before choosing to display this screen, you could have altered the code to include the fourth column, or you could modify the base template if you think that three columns are not enough. When you scroll down, you will see a parameter called CONSTRAINT. You can either leave the details as they are or blank them out; it doesn't matter, as we will be removing that code in a moment.

Figure 5-14. *Template parameters for TransactionTypes*

4. After clicking OK, the code is as follows. The main point of interest is the IF statement after switching to the ApressFinancial database. At the time of writing, there is a bug that should be fixed: the code doesn't create the table name parameter correctly. The following code corrects this bug. However, this code queries SSE's system tables to check for a TransactionTypes table within the dbo schema. If it does exist, the DROP TABLE statement is executed. This statement will delete the table defined from SSE, if possible. An error message may be displayed if the table has links with other tables or if someone has a lock on it, thus preventing the deletion. You'll learn all about locks in Chapter 8.

```
-- =======================================
-- Create table template
-- =======================================
USE ApressFinancial
GO
```

```
IF OBJECT_ID('TransactionDetails.TransactionTypes', 'U')
        IS NOT NULL
  DROP TABLE TransactionDetails.TransactionTypes
GO

CREATE TABLE TransactionDetails.TransactionTypes(
    TransactionTypeId int NOT NULL,
    TransactionDescription nvarchar(30) NOT NULL,
    CreditType bit NOT NULL,
    CONSTRAINT  PRIMARY KEY ()
)
GO
```

5. The full code for the `TransactionTypes` table follows. Once you have entered it, you can execute it. Note that there are two changes here. First, we put in the `IDENTITY` details for the `TransactionTypeId` column, but we are not going to place the fourth column in at this time. We will add it when we take a look at how to alter a table in the section "The ALTER TABLE Statement" later in this chapter. Finally, we remove the `CONSTRAINT` statement, as we are not creating a key at this time. Then execute the code.

```
-- =======================================
-- Create table template
-- =======================================
USE ApressFinancial
GO

IF OBJECT_ID('TransactionDetails.TransactionTypes', 'U') IS NOT NULL
  DROP TABLE TransactionDetails.TransactionTypes
GO

CREATE TABLE TransactionDetails.TransactionTypes(
    TransactionTypeId int IDENTITY(1,1) NOT NULL,
    TransactionDescription nvarchar(30) NOT NULL,
    CreditType bit NOT NULL
)
GO
```

Now that we have our third table, we can look at altering the template of the CREATE TEMPLATE, as it would be better to have the IDENTITY parameter there as well as four or five columns.

Creating and Altering a Template

The processes for creating and altering a template follow the same steps. All templates are stored in a central location and are available for every connection to SSE on that computer, therefore templates are not database or server restricted. The path to where they reside is

```
C:\Program Files\Microsoft SQL Server\
90\Tools\Binn\VSShell\Common7\IDE\SSMSEETemplates\Sql
```

It is also possible to create a new node for templates from within the Template Explorer by right-clicking and selecting New ➤ Folder.

■**Note** Don't create the folder directly in the `Sql` folder, as this is not picked up by SSMSE until you exit and reenter SSMSE.

You could create different formats of templates for slightly different actions on tables. You saw the CREATE TABLE template previously, but what if you wanted a template that included a CREATE TABLE specification with an IDENTITY column? This is possible by taking a current template and upgrading it for a new template.

Try It Out: Creating a Template from an Existing Template

1. From the Template Explorer, find the CREATE TABLE template, right-click it, and select Edit. This will display the template that we saw earlier. Change the comment and then we can start altering the code.

2. The first change is to add that the first column is an IDENTITY column. We know where this is located from our code earlier: it comes directly after the data type. To add a new parameter, input a set of angle brackets, then create the name of the parameter as the first option. The second option is the type of parameter this is, for example, sysname, defining that the parameter is a system name, which is just an alias for nvarchar(256). The third option is the value for the parameter; in this case we will be including the value of IDENTITY(1,1). The final set of code follows, where you can also see a fourth column has been defined with a bit data type.

■**Tip** You can check the alias by running the `sp_help sysname` T-SQL stored procedure.

```
-- =======================================
-- Create table template with IDENTITY
-- =======================================
USE <database, sysname, AdventureWorks>
GO

IF OBJECT_ID('<schema_name, sysname, dbo>.<table_name, sysname,
sample_table>', 'U') IS NOT NULL
  DROP TABLE
     <schema_name, sysname, dbo>.<table_name, sysname, sample_table>
GO

CREATE TABLE
  <schema_name, sysname, dbo>.<table_name, sysname, sample_table>(
  <columns_in_primary_key, , c1> <column1_datatype, , int>
<identity,,IDENTITY (1,1)>
  <column1_nullability,, NOT NULL>,
  <column2_name, sysname, c2> <column2_datatype, , char(10)>
  <column2_nullability,, NULL>,
  <column3_name, sysname, c3> <column3_datatype, , datetime>
  <column3_nullability,, NULL>,
  <column4_name, sysname, c4> <column4_datatype, , bit>
  <column4_nullability,, NOT NULL>,
  CONSTRAINT <contraint_name, sysname, PK_sample_table>
    PRIMARY KEY (<columns_in_primary_key, , c1>)
)
GO
```

3. Now the code is built, but before we test it, we shall save this as a new template called `CREATE TABLE with IDENTITY`. From the menu, select File ➤ Save CREATE TABLE.sql As, and from the Save File As dialog box, save this as CREATE TABLE with IDENTITY.sql. This should update your Template Explorer, but if it doesn't, try exiting and reentering SSMSE, after which it will be available to use.

So, now you know how to create a table. This chapter has covered several options for doing so, but there is one point that you should keep in mind when building a table, whether you are creating or modifying it. When creating a table in SSMSE, you should always save the table first by clicking the Save toolbar button. If you have made a mistake when defining the table and you close the table, and in doing so save in one action, you will get an error message informing you that an error has occurred, and all your changes will be lost. You will then have to go back into the Table Designer and reapply any changes made.

Try also to get used to using both SSMSE and Query Editor, as you may find that Query Editor gives you a more comfortable feel to the way you want to work. Also, you will find that in the Query Editor pane, you can save your work to a file on your hard drive as you go along. You can also do this within SSMSE; however, the changes are saved to a text file as T-SQL, which then need to be run through Query Editor anyway.

The ALTER TABLE Statement

If, when using the original template, we had created the table with only three columns, we would have an error to correct. One solution is to delete the table with `DROP TABLE`, but if we had placed some test data in the table before we realized we had missed the column, this would not be ideal. There is an alternative: the `ALTER TABLE` statement, which allows restrictive alterations to a table layout but keeps the contents. SSMSE uses this statement when altering a table graphically, but here I will show you how to use it to add the missing fourth column for our `TransactionDetails.TransactionTypes` table.

Columns can be added, removed, or modified using the `ALTER TABLE` statement. Removing a column will simply remove the data within that column, but careful thought has to take place before adding, removing, or altering a column.

■**Caution** Removing a column especially could have a major repercussive effect on existing code that you have either in your database or other applications that access your database.

There are two scenarios when adding a new column to a table: should it contain `NULL` values for all the existing rows, or should there be a default value instead? Any new columns created using the `ALTER TABLE` statement where a value is expected (or defined as `NOT NULL`) will take time to implement. This is because any existing data will have `NULL` values for the new column; after all, SSE has no way of knowing what value to enter. When altering a table and using `NOT NULL`, you need to complete a number of complex processes, which include moving data to an interim table and then moving it back. The easiest solution is to alter the table and define the column to allow `NULL`s, add in the default data values using the `UPDATE` T-SQL statement, and alter the column to `NOT NULL`.

■**Note** It is common practice, when adding columns with `ALTER TABLE`, to allow `NULL` values, as the default value may not be valid in some rows.

Try It Out: Adding a Column

1. First of all, open up Query Editor and ensure that you are pointing to the ApressFinancial database. Then write the code to alter the TransactionDetails.TransactionTypes table to add the new column. The format is very simple. We specify the table prefixed by the schema name we want to alter after the ALTER TABLE statement. Next we use a comma-delimited list of the columns we wish to add. We define the name, the data type, the length if required, and finally whether we allow NULLs or not. As we don't want the existing data to have any default values, we will have to define the column to allow NULL values.

```
ALTER TABLE TransactionDetails.TransactionTypes
ADD AffectCashBalance bit NULL
GO
```

2. Once we've altered the data as required, we then want to remove the ability for further rows of data to have a NULL value. This new column will take a value of 0 or 1. Again, we use the ALTER TABLE statement, but this time we'll add the ALTER COLUMN statement with the name of the column we wish to alter. After this statement are the alterations we wish to make. Although we are not altering the data type, it is a mandatory requirement to redefine the data type and data length. After this, we can inform SSE that the column will not allow NULL values.

```
ALTER TABLE TransactionDetails.TransactionTypes
ALTER COLUMN AffectCashBalance bit NOT NULL
GO
```

3. Execute the preceding code to make the TransactionDetails.TransactionTypes table correct.

Creating the Remaining Tables

Now that three of the tables have been created, we need to create the remaining four tables. We will do this as code placed in Query Editor. There is nothing specifically new to cover in this next section, and therefore only the code is listed. Enter the following code and then execute it as before. You can then move into SSMSE and refresh it, after which you should be able to see the new tables.

```
USE ApressFinancial
GO
CREATE TABLE CustomerDetails.CustomerProducts(
        CustomerFinancialProductId bigint NOT NULL,
        CustomerId bigint NOT NULL,
        FinancialProductId bigint NOT NULL,
        AmountToCollect money NOT NULL,
        Frequency smallint NOT NULL,
        LastCollected datetime NOT NULL,
        LastCollection datetime NOT NULL,
        Renewable bit NOT NULL
)
ON [PRIMARY]
GO
CREATE TABLE CustomerDetails.FinancialProducts(
        ProductId bigint NOT NULL,
        ProductName nvarchar(50) NOT NULL
) ON [PRIMARY]
```

```
GO
CREATE TABLE ShareDetails.SharePrices(
          SharePriceId bigint IDENTITY(1,1) NOT NULL,
          ShareId bigint NOT NULL,
          Price numeric(18, 5) NOT NULL,
          PriceDate datetime NOT NULL
) ON [PRIMARY]

GO
CREATE TABLE ShareDetails.Shares(
          ShareId bigint IDENTITY(1,1) NOT NULL,
          ShareDesc nvarchar(50) NOT NULL,
          ShareTickerId nvarchar(50) NULL,
          CurrentPrice numeric(18, 5) NOT NULL
) ON [PRIMARY]

GO
```

Defining a Primary Key

Defining a primary key can be completed in SSMSE with just a couple of mouse clicks. This section will demonstrate how easy this actually is. For more on keys, see Chapter 3.

Try It Out: Setting a Primary Key

1. Ensure that SSMSE is running and that you have navigated to the ApressFinancial database. Find the ShareDetails.Shares table, and right-click and select Modify. Once in the Table Designer, select the ShareId column. This will be the column we are setting the primary key for. Right-click to bring up the pop-up menu shown in Figure 5-15.

Figure 5-15. *Defining a primary key*

2. Select the Set Primary Key option from the pop-up menu. This will then change the display to place a small key in the leftmost column details. Only one column has been defined as the primary key, as you see in Figure 5-16.

Table - ShareDetails.Shares*	XP-HOME.Apress...ctionTyp	
Column Name	Data Type	Allow Nulls
ShareId	bigint	☐
ShareDesc	nvarchar(50)	☐
ShareTickerId	nvarchar(50)	☑
CurrentPrice	numeric(18, 5)	☐
		☐

Figure 5-16. *Primary key defined*

3. However, this is not all that happens, as you will see. Save the table modifications by clicking the Save button. Click the Manage Indexes/Keys button on the toolbar. This brings up the dialog box shown in Figure 5-17. Look at Type, the third option down in the General section. It says Primary Key. Notice that a key definition has been created for you, with a name and the selected column, informing you that the index is unique and clustered (more on indexes and their relation to primary keys in Chapter 6).

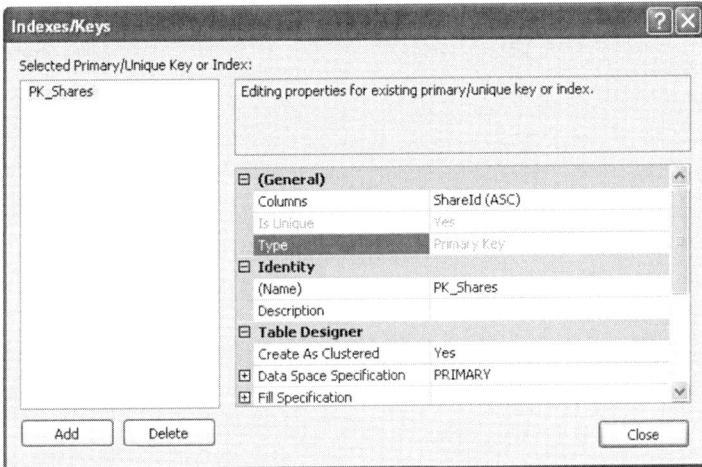

Figure 5-17. *Indexes/Keys dialog box*

That's all there is to creating and setting a primary key. A primary key has now been set up on the `ShareDetails.Shares` table. In this instance, any record added to this table will ensure that the data will be kept in `ShareId` ascending order (this is to do with the index, which you will see in Chapter 6), and it is impossible to insert a duplicate row of data. This key can then be used to link to other tables within the database at a later stage.

Creating a Relationship

We covered relationships in Chapter 3, but we've not created any. Now we will. The first relationship that we create will be between the customer and customer transactions tables. This will be a one-to-many relationship where there is one customer record to many transaction records. Keep in mind that although a customer may have several customer records, one for each product he or she has bought, the relationship is a combination of customer and product to transactions because a new `CustomerId` will be generated for each product the customer buys. We will now build that first relationship.

Try It Out: Building a Relationship

1. Ensure that SSMSE is running, and that `ApressFinancial` database is selected and expanded. We need to add a primary key to `CustomerDetails.Customers`. Enter the code that follows and then execute it:

```
ALTER TABLE CustomerDetails.Customers
ADD CONSTRAINT
    PK_Customers PRIMARY KEY NONCLUSTERED
```

```
(
CustomerId
)
WITH( STATISTICS_NORECOMPUTE = OFF, IGNORE_DUP_KEY = OFF,
ALLOW_ROW_LOCKS = ON, ALLOW_PAGE_LOCKS = ON)
ON [PRIMARY]
GO
```

2. Find and select the `TransactionDetails.Transactions` table, and then right-click. Select Modify to invoke the Table Designer.

3. Once in the Table Designer, right-click and select Relationships from the pop-up menu shown in Figure 5-18. Or click the Relationships button on the Table Designer toolbar.

Figure 5-18. *Building a relationship*

4. This brings up the Relationship Designer. As it's empty, you need to click Add. This will then populate the screen as shown in Figure 5-19.

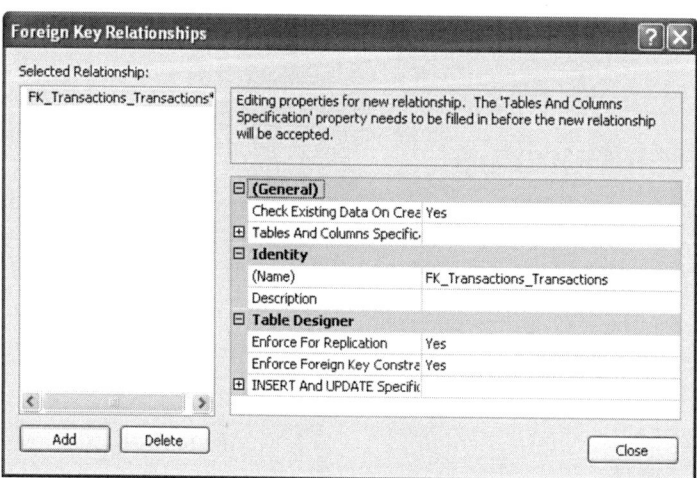

Figure 5-19. *Foreign Key Relationships dialog box*

5. Expand the Tables and Columns Specified node, which will allow the relationship to be built. Notice that there is now an ellipse button on the right, as shown in Figure 5-20. To create the relationship, click the ellipse button.

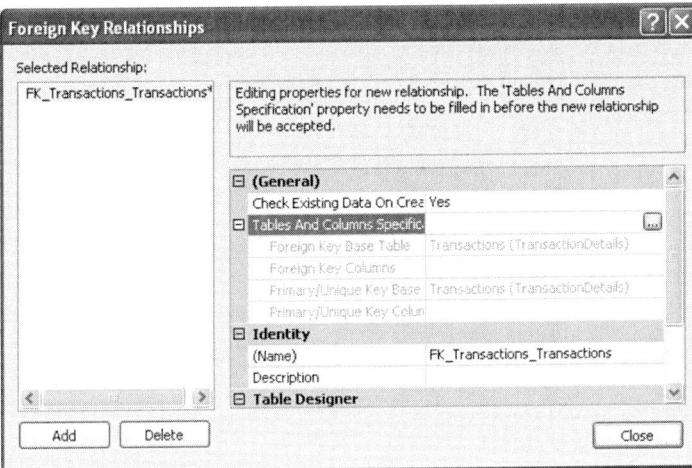

Figure 5-20. *Adding tables and columns*

6. The first requirement is to change the name to make it more meaningful. Quite often you will find that naming the key FK_ParentTable_ChildTable is the best method, so in this case change it to FK_Customers_Transactions as the CustomerDetails.Customers table will be the master table for this foreign key. We also need to define the column in each table that is the link. We are linking every one customer record to many transaction records, and we can do so via the CustomerId. So select that column for both tables, as shown in Figure 5-21. Now click OK.

Figure 5-21. *Columns selection*

Note In this instance, both columns have the same name, but this is not mandatory. The only requirement is that the information in the two columns be the same.

7. This brings us back to the Foreign Key Relationships definition screen, shown in Figure 5-22. Notice that at the top of the list items in the grayed-out area you can see the details of the foreign key we just defined. Within the Identity section there is now also a description of the foreign key. Ignore the option Enforce for Replication, which you'll see is in the Table Designer section if you scroll down in the Foreign Key Relationships screen.

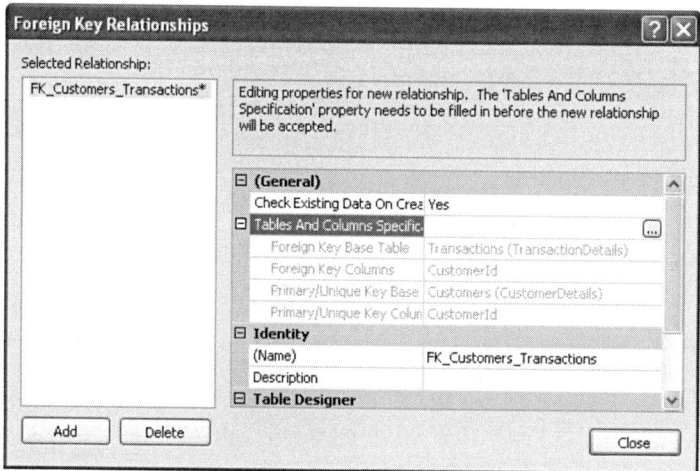

Figure 5-22. *Foreign key with description*

8. There are three other options we are interested in that are displayed at the bottom of the dialog box, as shown in Figure 5-23. Leave the options at the defaults.

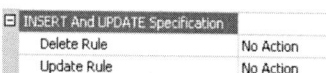

Figure 5-23. *Insert and Update Specification*

9. Closing this dialog box does not save the changes. Not until you close the Table Designer will the changes be applied. When you do so, you should see the dialog box in Figure 5-24 notifying you that two tables are to be changed. Click Yes to save the changes.

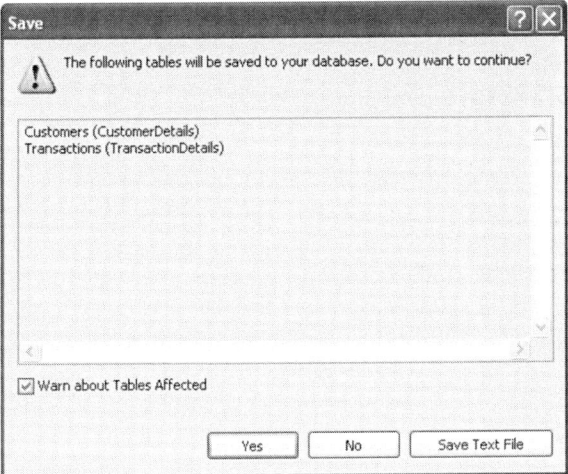

Figure 5-24. *Saving changes*

The relationship is now built, but what about those options we left alone? Let's go through those now.

Check Existing Data on Creation

If there is data in either of the tables, by setting this option to Yes we instruct SSE that when the time comes to physically add the relationship, the data within the tables is to be checked. If the data meets the definition of the relationship, the relationship is successfully inserted into the table. However, if any data fails the relationship test, the relationship is not applied to the database. An example of this would be when it is necessary to ensure that there is a customer record for all transactions, but there are customer transactions records that don't have a corresponding customer record, which would cause the relationship to fail. Obviously, if you come across this, you have a decision to make. Either correct the data by adding master records or altering the old records, and then reapply the relationship, or revisit the relationship to ensure it is what you want.

By creating the relationship, you want the data within the relationship to work, therefore you would select No if you were going to go back and fix the data after the additions. What if you still miss rows? Would this be a problem? In our scenario, there should be no transaction records without customer records. But you may still wish to add the relationship to stop further anomalies going forward.

Enforce Foreign Key Constraints

Once the relationship has been created, it is possible to prevent the relationship from being broken. If you set Check Existing Data on Creation from higher up in the dialog box to Yes, you are more than likely hoping to keep the integrity of the data intact. That option will only check the existing data. It does nothing for further additions, deletions, etc., on the data. However, by setting the Enforce Foreign Key Constraints option to Yes, we will ensure that any addition, modification, or deletion of the data will not break the relationship. It doesn't stop changing or removing data providing that the integrity of the database is kept in sync. For example, it would be possible to change the customer number of transactions, providing that the new customer number also exists with the CustomerDetails.Customers table.

Delete Rule/Update Rule

If a deletion or an update is performed, it is possible for one of four actions to then occur on the related data, based on the following options:

No Action: Nothing happens to any related data.

Cascade: If you delete a customer, all of the transaction rows for that customer will also be deleted.

Set Null: If you delete a customer, then if the `CustomerId` column in the `TransactionDetails.Transactions` table could accept `NULL` as a value, the value would be set to `NULL`. In the customers/transactions scenario, we have specified the column cannot accept `NULL` values. The danger with this is that you are leaving "unlinked" rows behind, a scenario that can be valid, but do take care.

Set Default: When defining the table, the column could be defined so that a default value is placed in it. On setting the option to this value, you are saying that the column will revert to this default value. Again a dangerous setting, but potentially a less dangerous option than Set Null, as at least there is a meaningful value within the column.

■**Note** If at any point you do decide to cascade deletions, please do take the greatest of care, as it can result in deletions that you may regret. If you implemented this on the `CustomerDetails.Customers` table, when you delete a customer, then all the transactions are gone. This is ideal for use if you have an archive database to which all rows are archived. To keep your current and online system lean and fast, you could use delete cascades to quickly and cleanly remove customers who have closed their accounts.

Using the ALTER TABLE SQL Statement

It is also possible to build a relationship, or referential integrity constraint, through a T-SQL statement. This would be done using an `ALTER TABLE` SQL statement. This time, a relationship will be created between the `Transactions` table and the `Shares` table. Let's now take a few moments to check the syntax for building a constraint within T-SQL code.

```
ALTER TABLE child_table_name
WITH NOCHECK|CHECK
ADD CONSTRAINT [Constraint_Name]
FOREIGN KEY (child_column_name, ...,)
REFERENCES [master_table_name]([master_column_name, ...,])
```

We have to use an `ALTER TABLE` statement to achieve the goal of inserting a constraint to build the relationship. After naming the child table in the `ALTER TABLE` statement, we then decide whether we want the foreign key to check the existing data or not when it is being created. This is similar to the Check Existing Data on Creation option you saw earlier.

Now we move on to building the constraint. To do this, we must first of all instruct SSE that this is what we are intending to complete, and so we will need the `ADD CONSTRAINT` clause.

Next, we name the constraint we are building. Again, I tend to use underscores instead of spaces. However, if you do wish to use spaces, which I wholeheartedly do not recommend, you'll have to surround the name of the key using the [] brackets. I know I mentioned this before, but it's crucial to realize the impact of having spaces in a column, table, or constraint name. Every time you wish to deal with an object that has a name separated by spaces, then you will also need to surround it with square brackets. Why make extra work for yourself?

Now that the name of the constraint has been defined, the next stage is to inform SSE that a FOREIGN KEY is being defined next. Recall that a constraint can also be used for other functionality, such as creating a default value to be inserted into a column.

When defining the foreign key, ensure that all column names are separated by a comma and surrounded by parentheses. The final stage of building a relationship in code is to specify the master table of the constraint and the columns involved.

The rule here is that there must be a one-to-one match on columns on the child table and the master table, and that all corresponding columns must match on data type.

It is as simple as that. When building relationships, you may wish to use SSMSE, as there is a lot less typing involved and you can also instantly see the exact correspondence between the columns and whether they match in the same order. However, with T-SQL you can save the code, and it is ready for deployment to production servers when required.

Try It Out: Using SQL to Build a Relationship

1. In a Query Editor pane, enter the following T-SQL and execute it by pressing Ctrl+E or F5, or clicking the Execute button:

```
USE ApressFinancial
GO
ALTER TABLE TransactionDetails.Transactions
WITH NOCHECK
ADD CONSTRAINT FK_Transactions_Shares
FOREIGN KEY(RelatedShareId)
REFERENCES ShareDetails.Shares(ShareId)
```

2. You should then see that the T-SQL has been executed successfully.

```
The command(s) completed successfully.
```

That's it. The relationship is created in the second batch of T-SQL code, the first batch ensuring that we are pointing to the right database. Once the index is built, it is possible to alter the table to add the relationship.

With our code, although we are executing an ALTER TABLE statement, no columns are being altered, but a constraint is being added. A relationship is a special type of constraint, and it is through a constraint that a relationship is built.

A constraint is, in essence, a checking mechanism, checking data modifications within SSE and the table(s) that it is associated with.

Summary

In this chapter, we have looked at our first real work in building our database solution. You have built your first set of tables and applied different data types to columns within tables that match the expected data to hold. You have also seen how to alter a table and place constraints on columns for keys and relationships as well as how to have auto-incrementing numbers when rows of data are added. Templates have also been used for creating tables, and you have seen how to create your own template.

In the next chapter, we will move our design on and take a look at indexing the data as well as diagramming the database that you have built.

CHAPTER 6

■■■

Creating Indexes and Database Diagramming

Now that we've created the tables, we could stop at this point and just work with our data from here. However, this would not be a good decision. As soon as any table contained a reasonable amount of information, and we wished to find a particular record, it would take SQL Server Express a fair amount of time to locate it. Performance would suffer, and our users would soon get annoyed with the slowdown in speed.

In this scenario, the database is like a large filing cabinet in which we have to find one piece of paper, but there's no clear filing system or form of indexing. If we had some sort of cross-reference facility, it would likely be easier to find the information we need. And if that cross-reference facility were in fact an index, this would be even better, as we might be able to find the piece of paper in our filing cabinet almost instantly. It is this theory that we need to put into practice in our SSE database tables. Generally, indexing is a conscious decision by a developer who favors faster conditional selection of records over modification or insertion of records.

In this chapter, you'll learn the basics of indexing and how you can start implementing an indexing solution. This chapter covers the following topics:

- What an index is
- Different types of indexes
- Size restrictions on indexes
- Qualities of a good index and a bad index
- How to build an index in code as well as graphically
- How to alter an index

Let's begin by looking at what an index is and how it stores data.

What Is an Index?

In the previous chapter, you learned about tables, which are, in essence, repositories that hold data and information about data—what it looks like and where it is held. However, a table definition is not a great deal of use in getting to the data quickly. For this, some sort of cross-reference facility is required, where for certain columns of information within a table it should be possible to get to the whole record of information quickly.

If you think of this book, for example, as a table, the cross-reference you would use to find information quickly is the index at the back of the book. You look in the book index for a piece of information, or **key**. When you find the listing for that information in the index, you'll find it's

associated with a page number, or a **pointer**, which directs you to where you can find the data you're looking for. This is where an **index** within your SSE database comes in.

You define an index in SSE so that it can locate the rows it requires to satisfy database queries faster. If an index does not exist to help find the necessary rows, SSE has no other option but to look at every row in a table to see if it contains the information required by the query. This is called a **table scan**, which by its very nature adds considerable overhead to data-retrieval operations.

Note There will be times when a table scan is preferable to using an index. For example, if SSE needs to process a reasonable portion of rows in a table, as a rule of thumb 10 percent or more of them, a table scan is likely to be better than using an index. Table scans aren't always bad.

When searching a table using an index, SSE does not go through all the data; rather, it focuses on a much smaller subset of that data, as it will be looking at the columns defined within the index, which is faster. Once the record is found in the index, a pointer states where the data for that row can be found in the relevant table.

There are different types of indexes you can build onto a table. An index can be created on one column, called a **simple** index, or on more than one column, called a **compound** index. The circumstances of the column or columns you select and the data that will be held within these columns determine which type of index you use.

Types of Indexes

Although SSE has three types of indexes, clustered, nonclustered, and primary and secondary XML indexes, we will concentrate only on clustered and nonclustered ones in this book, as XML and XML indexes are quite an advanced topic.

The index type refers to the way the index and the physical rows of data are stored internally by SSE. The differences between the index types are important to understand, so we'll delve into them in the sections that follow.

Clustered

A clustered index defines the physical order of the data in the table. If you have more than one column defined in a clustered index, the data will be stored in sequential order according to columns: the first column, then the next column, and so on. Only one clustered index can be defined per table. It would be impossible to store the data in two different physical orders.

Going back to our earlier book analogy, if you examine a telephone book, you'll see that the data is presented in alphabetical order with surnames appearing first, then first names, and then any middle-name initial(s). Therefore, when you search the index and find the key, you are already at the point in the data from which you want to retrieve the information, such as the telephone number. In other words, you don't have to turn to another page as indicated by the key, because the data is right there. This is a clustered index of surname, first name, initials.

As data is inserted, SSE will take the data within the index key values you have passed in and insert the row at the appropriate point. It will then move the data along so that it remains in the same order. You can think of this data as being like books on a bookshelf. When a librarian gets a new book, he or she will find the correct alphabetical point and try to insert the book at that point. All the books will then be moved within the shelf. If there is no room as the books are moved, the books at the end of the shelf will be moved to the next shelf down, and so on, until a shelf with enough room is found. Although this analogy puts the process in simple terms, this is exactly what SSE does.

Do not place a clustered index on columns that will have a lot of updates performed on them, as this means SSE will have to constantly alter the physical order of the data and so use up a great deal of processing power.

As a clustered index contains the table data itself, SSE would perform fewer I/O operations to retrieve the data using the clustered index than it would using a nonclustered index. Therefore, if you only have one index on a table, try to make sure it is a clustered index.

Nonclustered

Unlike a clustered index, a nonclustered index does not store the table data itself. Instead, a nonclustered index stores pointers to the table data as part of the index keys; therefore, many nonclustered indexes can exist on a single table at one time.

As a nonclustered index is stored in a separate structure (in fact, it is really held as a table with a clustered index hidden from your view) from the base table, it is possible to create the nonclustered index on a different filegroup from the base table. If the filegroups are located on separate disks, data retrieval can be enhanced for your queries as SSE can use parallel I/O operations to retrieve the data from the index and base tables concurrently.

When you are retrieving information from a table that has a nonclustered index, SSE finds the relevant row in the index. If the information you want doesn't form part of the data in the index, SSE then uses the information in the index pointer to retrieve the relevant row in the data. As you can see, this involves at least two I/O actions—and possibly more, depending on the optimization of the index.

When a nonclustered index is created, the information used to build the index is placed in a separate location from the table and therefore can be stored on a different physical disk if required.

Caution The more indexes you have, the more times SSE has to perform index modifications when inserting or updating data in columns that are indexed.

Primary and Secondary XML

If you wish to index XML data, which I cover only briefly later in the book, it would be best to read Books Online, as this topic is beyond the scope of this book.

Uniqueness

An index can be defined as either **unique** or **nonunique**. A unique index ensures that the values contained within the unique index columns will appear only once within the table, including a value of NULL.

SSE automatically enforces the uniqueness of the columns contained within a unique index. If an attempt is made to insert a value that already exists in the table, an error will be generated, and the attempt to insert or modify the data will fail.

A nonunique index is perfectly valid. However, as there can be duplicated values, a nonunique index has more overhead than a unique index when retrieving data. SSE will need to check whether there are multiple entries to return, compared with a unique index where SSE knows to stop searching after finding the first row.

Unique indexes are commonly implemented to support constraints such as the primary key. Nonunique indexes are commonly implemented to support locating rows using a nonkey column.

Determining What Makes a Good Index

To create an index on a table, you have to specify which columns are contained within the index. Columns in an index do not have to all be of the same data type. You should be aware that there is a limit of 16 columns on an index, and the total amount of data for the index columns within a row cannot be more than 900 bytes. To be honest, if you get to an index that contains more than four or five columns, you should stand back and reevaluate the index definition. Sometimes you'll have more than five columns, but you really should double-check.

It is possible to get around this restriction and have an index that does include columns that are not part of the key: the columns are tagged onto the end of the index. This will mean that the index takes up more space, but if it means that SSE can retrieve all of the data from an index search, then it will be faster. However, to reiterate, if you are going down this route for indexes, perhaps you need to look at your design.

In the sections that follow, we'll examine some of the factors that can determine whether an index is good:

- Using "low-maintenance" columns
- Using primary and foreign keys
- Being able to find a specific record
- Using covering indexes
- Looking for a range of information
- Keeping the data in order

Using Low-Maintenance Columns

As I've indicated, for nonclustered indexes the actual index data is separate from the table data, although both can be stored in the same area or in different areas (e.g., on different hard drives). To reiterate, this means that when you insert a record into a table, the information from the columns included in the index is copied and inserted into the index area. So, if you alter data in a column within a table, and that column has been defined as making up an index, SSE also has to alter the data in the index. Instead of only one update being completed, two will be completed. If the table has more than one index, and in more than one of those indexes is a column that is to be updated a great deal, there may be several disk writes to perform when updating just one record. While this will result in a performance reduction for data-modification operations, appropriate indexing will balance this out by greatly increasing the performance of data-retrieval operations.

Therefore, data that is **low maintenance**—namely, columns that are not heavily updated—could become an index and would make a good index. The fewer disk writes that SSE has to do, the faster the database will be, as well as every other database within that SSE instance. Don't let this statement put you off. If you feel that data within a table is retrieved more often than it is modified, or if the performance of the retrieval is more critical than the performance of the modification, then do look at including the column within the index.

In the example application we're building, each month we need to update a customer's bank balance with any interest gained or charged. However, we have a nightly job that wants to check for clients who have between $10,000 and $50,000, as the bank can get a higher rate of deposit with the Federal Reserve on those sorts of amounts. A client's bank balance will be constantly updated, but an index on this sort of column could speed up the overnight deposit check program. Before the index in this example is created, we need to determine whether the slight performance degradation in the updating of the balances is justified by the improvement of performance of the deposit check program.

Primary and Foreign Keys

One important use of indexes is on referential constraints within a table. A referential constraint is where you've indicated that through the use of a key, certain actions are constrained depending on what data exists. To give a quick example of a referential constraint, say you have a customer who owns banking products. A referential constraint would prevent the customer's record from being deleted while those products existed.

SSE does not automatically create indexes on your foreign keys. However, as the foreign key column values need to be identified by SSE when joining to the parent table, it is almost always recommended that an index be created on the columns of the foreign key.

Finding Specific Records

Ideal candidates for indexes are columns that allow SSE to quickly identify the appropriate rows. In Chapter 8, you'll meet the WHERE clause of a query. This clause lists certain columns in your table and is used to limit the number of rows returned from a query. The columns used in the WHERE clause of your most common queries make excellent choices for an index. So, for example, if you wanted to find a customer's order for a specific order number, an index based on customer_id and order_number would be perfect, as all the information needed to locate a requested row in the table would be contained in the index.

If finding specific records is going to make up part of the way the application works, do look at this scenario as an area for an index to be created.

Using Covering Indexes

As mentioned earlier, when you insert or update a record, any data in a column that is included in an index is stored not only in the table, but also in the indexes for nonclustered indexes. From finding an entry in an index, SSE then moves to the table to locate and retrieve the record. However, if the necessary information is held within the index, there is no need to go to the table and retrieve the record, providing much speedier data access.

For example, consider the ShareDetails.Shares table in the ApressFinancial database. Suppose that you wanted to find out the description, current price, and ticker ID of a share. If an index was placed on the ShareId column, knowing that this is an identifier column and therefore unique, you would ask SSE to find a record using the ID supplied. It would then take the details from the index of where the data is located and move to that data area. If, however, there was an index with all of the columns defined, SSE will be able to retrieve the description ticker and price details in the index action. It will not be necessary to move to the data area. This is called a **covered index**, since the index covers every column in the table for data retrieval.

Looking for a Range of Information

An index can be just as useful for finding one record as it can be for searching for a range of records. For example, say you wish to find a list of cities in Florida with names between Orlando and St. Petersburg in alphabetical order. You could put an index on the city name, and SSE would go to the index location of Orlando and then read forward from there an index row at a time, until it reached the item after St. Petersburg, where it would then stop. Because SSE knows that an index is on this column and that the data will be sorted by city name, this makes it ideal for building an index on a city name column.

It should be noted that SSE indexes are not useful when attempting to search for characters embedded in a body of text. For example, suppose you want to find every author in a publisher's database whose last name contains the letters "ab." This type of query does not provide a means of determining where in the index tree to start and stop searching for appropriate values. The only way

SSE can determine which rows are valid for this query is to examine every row within the table. Depending on the amount of data within the table, this can be a very slow process. If you have a requirement to perform this sort of wildcard text searching, you should take a look at the SSE full-text feature, as this will provide better performance for such queries.

Keeping the Data in Order

As previously stated, a clustered index actually keeps the data in the table in a specific order. When you specify a column (or multiple columns) as a clustered index, on inserting a record SSE will place that record in a physical position to keep the records in the correct ascending or descending order that corresponds to the order defined in the index. To explain this a bit further, if you have a clustered index on customer numbers, and the data currently has customer numbers 10, 6, 4, 7, 2, and 5, then SSE will physically store the data in the following order: 2, 4, 5, 6, 7, 10. If a process then adds in a customer number 9, it will be physically inserted between 7 and 10, which may mean that the record for customer number 10 needs to move physically. Therefore, if you have defined a clustered index on a column or a set of columns where data insertions cause the clustered index to be reordered, this is going to greatly affect your insert performance. SSE does provide a way to reduce the reordering impact by allowing a fill factor to be specified when an index is created.

Determining What Makes a Bad Index

Now that you know what makes a good index, let's investigate what makes a bad index. There are several "gotchas" to be aware of:

- Using unsuitable columns
- Choosing unsuitable data
- Including too many columns
- Including too few records in the table

Using Unsuitable Columns

If a column isn't used by a query to locate a row within a table, there is a good chance that the column does not need to be indexed, unless it is combined with another column to create a covering index, as described earlier. If this is the case, the index will still add overhead to the data-modification operations but will not produce any performance benefit to the data-retrieval operations.

Choosing Unsuitable Data

Indexes work best when the data contained in the index columns is highly selective between rows. The optimal index is one created on a column that has a unique value for every row within a table, such as a primary key. If a query requests a row based on a value within this column, SSE can quickly navigate the index structure and identify the single row that matches the query predicate.

However, if the selectivity of the data in the index columns is poor, the effectiveness of the index is reduced. For example, if an index is created on a column that contains only three distinct values, the index would be able to reduce the number of rows to just a third of the total before applying other methods to identify the exact row. In this instance, SSE would probably ignore the index anyway and find that reading the data table instead would be faster. Therefore, when deciding on appropriate index columns, you should examine the data selectivity to estimate the effectiveness of the index.

Including Too Many Columns

The more columns there are in an index, the more data writing has to take place when a process completes an update or an insertion of data. Although in SSE these updates to the index data take a very short amount of time, it can add up. Therefore, each index that is added to a table will incur extra processing overhead, so it is recommended that you create the minimum number of indexes needed to give your data-retrieval operations acceptable performance.

Including Too Few Records in the Table

There is also absolutely no need to place an index on a table that has only one row. SSE will find the record at the first request, without the need of an index.

This statement also holds true when a table has only a handful of records. Again, there is no reason to place an index on these tables. The reason for this is that SSE would go to the index, use its engine to make several reads of the data to find the correct record, and then move directly to that record using the record pointer from the index to retrieve the information. Several actions are involved in this process, as well as passing data between different components within SSE. When you execute a query, SSE will determine whether it's more efficient to use the indexes defined for the table to locate the necessary rows or to simply perform a table scan and look at every row within the table.

Reviewing Your Indexes for Performance

Every so often, it's necessary for you as an administrator or a developer to review the indexes built on your table to ensure that yesterday's good index is not today's bad index. When a solution is built, what is perceived to be a good index in development may not be so good in production—for example, the users may be performing one task more times than expected. Therefore, it is highly advisable that you set up tasks that constantly review your indexes and how they are performing. This can be completed within SSE via its index-tuning tool, the Database Tuning Advisor (DTA).

The DTA looks at your database and a workload file holding a representative amount of information that will be processed, and uses the information it gleans from these to figure out what indexes to place within the database and where improvements can be made. At this point in the book, I haven't actually covered working with data, so going through the use of this tool will just lead to confusion. This powerful and advanced tool should be used only by experienced SSE developers or database administrators.

Getting the indexes right is crucial to your SSE database running in an optimal fashion. Spend time thinking about the indexes, try to get them right, and then review them at regular intervals. Review clustering, uniqueness, and especially the columns contained within indexes so that you ensure the data is retrieved as fast as possible. Finally, also ensure that the order of the columns within the index will reduce the number of reads that SSE has to do to find the data. An index where the columns defined are `FirstName`, `LastName`, and `Department` might be better defined as `Department`, `FirstName`, and `LastName` if the greatest number of queries is based on finding someone within a specific department or listing employees of a department. The difference between these two indexes is that in the first, SSE would probably need to perform a table scan to find the relevant records. Compare that with the second example, where SSE would search the index until it found the right department, and then just continue to return rows from the index until the department changed. As you can see, the second involves much less work.

Creating an Index

Now that you know what an index is and you have an understanding of the various types of indexes, let's proceed to create some in SSE. There are many different ways to create indexes within SSE, as you might expect. Those various methods are the focus of this section of the chapter, starting with how to use the Table Designer in SQL Server Management Studio Express.

The first index we'll place into the database will be on the `CustomerId` field within the `CustomerDetails.Customers` table.

Creating an Index with the Table Designer

As you may recall from the previous chapter, when the `CustomerId` column was set up, SSE automatically generated the data within this field whenever a new record was inserted into this table. This data will never alter, as it uses the `IDENTITY` function for the column. Thus, the `CustomerId` column will be updated automatically whenever a customer is added. An application written in, for example, C# could be used as the user front end for updating the remaining areas of the customer's data, and it could also display specific customer details, but it would not know that the `CustomerId` requires incrementing for each new record, and it would not know the value to start from.

The first index created will be used to find the record to update with a customer's information. The application will have found the customer using a combination of name and address, but it is still possible to have multiple records with the same details. For example, you may have John J. Doe and his son, John J. Doe, who are both living at the same address. Once you have those details displayed on the screen, how will the computer program know which John J. Doe to use when it comes to completing an update?

Instead of looking for the customer by first name, last name, and address, the application will know the `CustomerId` and use this to find the record within SSE. When completing the initial search, the `CustomerId` is returned as part of the set of values, so when the user selects the appropriate John J. Doe, the application knows the appropriate `CustomerId`. SSE will use this value to specifically locate the record to update. In the following exercise, we'll add this index to the `CustomerDetails.Customers` table.

Try It Out: Creating an Index Graphically

1. Ensure that SSMSE is running and that you have expanded the nodes in the tree view so that you can see the Tables node within the `ApressFinancial` database.

2. Find the first table that the index is to be added to (i.e., the `CustomerDetails.Customers` table). Right-click and select Modify. This will bring you into the Table Designer. Right-click and select Indexes and Keys (see Figure 6-1). It is possible to click a button on the Table Designer toolbar that appears within SSMSE when you enter the Table Designer.

 Figure 6-1. *The Manage Indexes and Keys button*

3. The index creation screen will appear. Recall in Chapter 5 we created a primary key using T-SQL on this table. When the dialog box opens, it will be shown. For our new index, click the Add button to select the index's properties. The screen will look similar to Figure 6-2.

The fields in this dialog box are prepopulated, but you are able to change the necessary fields and options that you might wish to use. However, no matter what indexes have been created already, the initial column chosen for the index will always be the first column defined in the table.

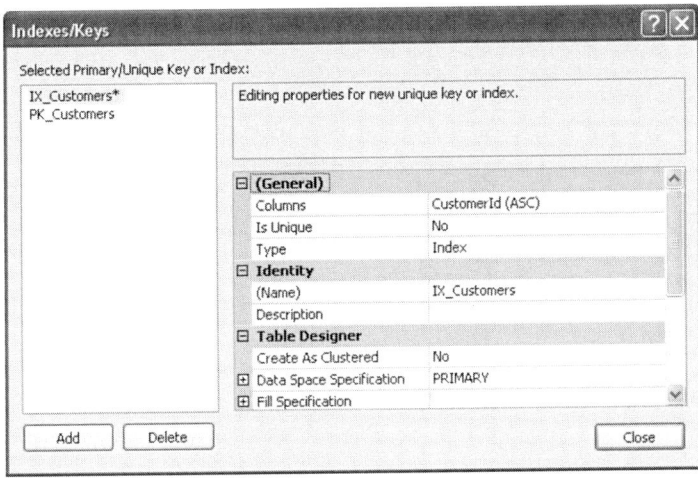

Figure 6-2. *The Indexes/Keys dialog box*

4. The first area to change is the name of the index. Notice that in the (Name) text box, SSE has created a possible value for you. The name is prefixed with `IX_`, which is a good naming system to use. It is also good to keep the name of the table and then a useful suffix, such as the name of the column. In this case, the index will be called `IX_Customers_CustomerId`. It might also be good to place something in the description. However, index names should be self-explanatory, so there really shouldn't be a need for a description.

5. SSE has, in this instance, correctly selected `CustomerId` as the column that will make up the index. Also, it has selected that the index will be ascending. For this example, the default sort order is appropriate. The sort order of the index column is useful when creating an index on the columns that will be used in an `ORDER BY` clause of a query when there are multiple columns with differing sort orders. If the sort order of the columns within the index matches the sort order of those columns specified in the `ORDER BY` clause, SSE may be able to avoid performing an internal sort, resulting in improved query performance.

Tip If an index is only one column, SSE can read the index just as fast in a forward direction as it can backward.

6. As indicated earlier, SSE generates the value of the `CustomerId` column to be the next number in a sequence when a record is added, as this column uses the `IDENTITY` functionality. This value can't be altered within the table, and so taking these two items of information and putting them together, you should be able to deduce that this value will be unique. Therefore, change the Is Unique option to Yes.

7. The final part of creating the index is to change the Create As Clustered option to Yes (see Figure 6-3). The order of the records inserted into SSE won't change. And if you scroll down the screen, you'll see the Re-compute Statistics for This Index option, which should remain No.

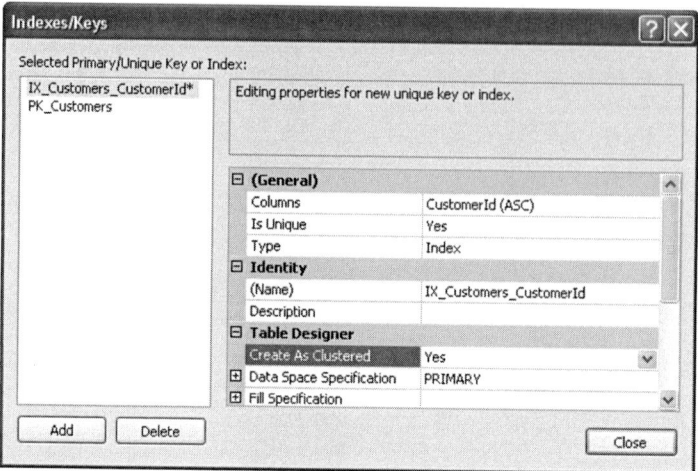

Figure 6-3. *The clustering option in the Indexes/Keys dialog box*

8. Click Close and then close the table modification, answering Yes when you are asked if you wish to save the changes. This will add the index to the database.

Building an index in SSMSE is a straightforward procedure, as you have just seen. Although this is the first index that you have created yourself, it took only a few moments, and there were just a couple of areas where you had to do any reasonable amount of decision making. We will cover those areas now.

Choosing the name of the index and the columns to include is easy and is not worth dwelling on. You should know which columns to include from the discussions at the start of the chapter, where we examined the basics of building indexes.

The first major decision you need to make is determining whether a column carries unique values. The column chosen for our first index is an identity column which, if you recall, is a column that cannot have data entered into it by any SQL command, as the data entered into this column is completed automatically by SSE itself. Also, in an identity column, by default no two rows can have the same value. However, there is no automation to stop any attempt to create duplicate keys. Therefore, there is still a need to inform SSE that the index will be unique.

Moving on to the Create As Clustered setting, the data in this table would be best held in `CustomerId` order. This is because each record that is inserted will have a higher `CustomerId` number than the previous record. Therefore, each time a record is added, it will be added to the end of the table, removing the need for a clustered index. As with the Is Unique option, the Create As Clustered option doesn't need to be selected.

Finally, the Re-compute Statistics for This Index option defines whether SSE automatically recomputes the statistics on the index when data is modified. Take care with this option, as it can cause confusion. For example, setting this option to On actually switches the updating of statistics to Off.

Indexes and Statistics

When retrieving data, SSE obviously has to make some decisions as to the best way to get to that data and return it to the query requesting it. Even if an index has been created on a set of columns, SSE may determine that it is better and faster to use another method to retrieve the data—through a table scan, perhaps. Or maybe there are a couple of indexes that could be chosen to retrieve the same data. No matter what the scenario, SSE has to have some basis of information on which to make sensible and accurate choices. This is where statistics come in.

SSE keeps statistics on each column contained within an index. These statistics are updated over a period of time and over a number of inserts or modifications. The specifics of how all of this works in the background, and how SSE keeps the statistics up to date, is an advanced topic. What you need to know is that if you alter or build an index on a table that has data in it, and you don't let SSE update the statistics on the table, then SSE could be using inaccurate information when it is trying to decide how to retrieve the data. It could even mean that the index change you thought would improve performance has in fact made the performance much slower.

That said, it is not always prudent to let SSE recompute statistics automatically. SSE will do the updates when it feels they are required. This may happen at a busy time of processing; you have no control over when it will happen. However, if SSE does update the statistics, the query that caused the update to start is not impacted, as the statistics are updated asynchronously.

It may be more efficient to manually update the statistics via a scheduled job. This is what you quite often see within production environments that have a number of inserts and modifications to the data.

The CREATE INDEX Syntax

Creating an index using T-SQL is a lot easier than creating a table. In this section, we'll look only at indexes on tables, although there is an object within SSE called a **view** that can also be indexed.

The full syntax for creating an index is not listed here, although you can find it within Books Online once you progress in your SSE knowledge. A reduced version will be sufficient while you are learning SSE. Most of your indexes will use the following version:

```
CREATE [UNIQUE] [CLUSTERED|NONCLUSTERED]
INDEX index_name
ON table (column [ASC|DESC] [ ,...n ] )
[WITH {IGNORE_DUP_KEY|DROP_EXISTING|SORT_IN_TEMPDB}]
[ON filegroup ]
```

Let's go through each point one by one so that the options in this cut-down version are clear:

CREATE: Required. This keyword informs SSE that you will be building a new object.

UNIQUE: Optional. If used, this option will inform SSE that the columns listed in the index will bring back a single unique row. This is enforced by SSE when attempting to insert a duplicate row, as an error message will be returned.

CLUSTERED *or* NONCLUSTERED: Optional. If neither CLUSTERED nor NONCLUSTERED is explicitly listed, the index is created as NONCLUSTERED.

INDEX: Required. This informs SSE that the new object will be an index.

index_name: Required. This is the name of the index being built. This name must be unique for the table, and it is advisable to keep this name unique for the database, using the naming method of IX_table_column discussed earlier.

ON table: Required. This is the name of the table with which the index is associated. Only one table can be named.

column: Required. This is the name of the column(s) in the table that we wish to include in the index. This is a comma-separated list.

ASC: Optional (default). If neither ASC nor DESC is mentioned, then ASC is assumed. ASC informs SSE that it should store the column names in ascending sequence.

DESC: Optional. This informs SSE that the column is to be stored in descending order.

WITH: Optional. It is, however, required if any of the following options have to be used:

- IGNORE_DUP_KEY: This option is only available when the index is defined as UNIQUE. If this option has not been used earlier, then it is not available to you. I'll explain this further in a moment.

- DROP_EXISTING: This option is used if there is an existing index of the same name within the database. It will then drop the index before re-creating it. This is useful for performance if you are not actually changing any columns within the index. More on this in a moment.

- SORT_IN_TEMPDB: When building an index where there is already data within the table, it may be advisable, if the table is a large table, to get the data sorted for the index within the temporary database, tempdb, as mentioned in Chapter 3. Use this option if you have a large table, or if tempdb is on a different hard disk from your database. This option may speed up the building of the index, as SSE can simultaneously read from the disk device where the table is located and write to the disk device where tempdb is located.

ON: Optional. This option is, however, required if you are going to specify a filegroup. It is not required if you wish the index to be built on the PRIMARY filegroup.

filegroup: This is the name of the filegroup on which the index should be stored. At the moment, there is only one filegroup set up, PRIMARY. PRIMARY is a reserved word and is required to be surrounded by square brackets, [], if used.

Two options need further clarification: IGNORE_DUP_KEY and DROP_EXISTING. We'll look at both in the sections that follow.

IGNORE_DUP_KEY

If you have an index defined as UNIQUE, then no matter how hard you try, you cannot add a new row whose values in the index columns match the values of any current existing row. However, there are two actions that you can perform, depending on this setting within an index.

When performing multirow inserts, if the IGNORE_DUP_KEY option is specified, then no error is generated within SSE if some of the rows being inserted violate the unique index. Only a warning message is issued. The rows that violated the unique index are not inserted, although all other rows are inserted successfully.

When performing multirow inserts, if the IGNORE_DUP_KEY option is omitted, an error message is generated within SSE if some of the rows violate the unique index. The batch is rolled back, and no rows are inserted into the table.

■**Caution** The system variable called @@ERROR can be tested after every SSE action to see whether there has been an error in any item of work or through another error-handling command called Try/Catch. If there has been an error, some sort of error handling within the batch will usually be performed. If you have IGNORE_DUP_KEY, no error will be produced when there is an attempt to insert a duplicate row, and the batch will run as if everything had been inserted. So, be warned: it may look like everything has worked, but in fact some rows were not inserted!

DROP_EXISTING

When data is being inserted and modified, there will be times when an index bloats to a state that isn't ideal. Just as an Access database may need to be compacted, indexes within SSE also need to be compacted sometimes. Compacting the index will speed up performance and reclaim disk space by removing fragmentation of the index. To compact an index, you re-create the index without actually modifying the columns or, in fact, starting from scratch and having to rebuild the whole index and visiting every row within the table.

The DROP_EXISTING clause provides enhanced performance when rebuilding a clustered index compared to a DROP INDEX command followed by a CREATE INDEX command. Nonclustered indexes are rebuilt every time the clustered index for a table is rebuilt. So, if you drop a clustered index and then re-create it, the existing nonclustered indexes are rebuilt twice: once from the drop and once from the creation.

DROP_EXISTING also allows an existing index to be rebuilt without explicitly dropping and re-creating the index. This is particularly useful for rebuilding primary key indexes. As other tables may reference a primary key, it may be necessary to drop all foreign keys in these other tables prior to dropping the primary key. By specifying the DROP_EXISTING clause, SSE will rebuild the index without affecting the primary key constraint.

Creating an Index in Query Editor: Template

Not surprisingly, there is a template within Query Editor that you can use as a basis for creating an index. We'll look at this process first, before we build an index natively in Query Editor, as this creates the basis of the SQL syntax for the creation of the index.

Try It Out: Using a Query Editor Template to Build an Index

1. Ensure that Template Explorer is open (press Ctrl+Alt+T or select View ➤ Template Explorer). Navigate to the Index node and expand it. Select the Create Index Basic node and double-click (see Figure 6-4).

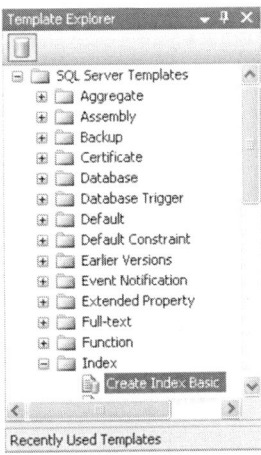

Figure 6-4. *Selecting the Create Index Basic node*

2. A new editor will open with the following code in it. The template that is installed is based on the Adven-tureWorks example. As you saw in the previous chapter, you can create new templates or modify this one.

```
-- =============================================
-- Create index basic template
-- =============================================
USE <database_name, sysname, AdventureWorks>
GO

CREATE INDEX <index_name, sysname, ind_test>
ON <schema_name, sysname, Person>.<table_name, sysname, Address>
(
    <column_name1, sysname, PostalCode>
)
GO
```

3. Alter the template by either changing the code or using the Specify Values for Template Parameters dialog box, which will make the index creating easier. The button to bring up this dialog box should be on the SQL Editor toolbar (see Figure 6-5).

Figure 6-5. *The Specify Values for Template Parameters button*

4. Change the database to the example database, name the index (in this case, it has been named after the table), set schema_name to CustomerDetails, table_name to CustomerProducts, and column_name1 to Cus-tomerId (see Figure 6-6). Then click OK.

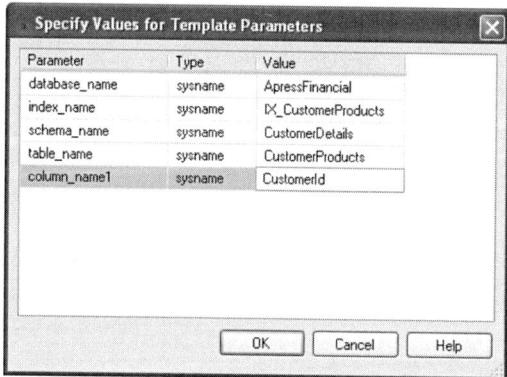

Figure 6-6. *The Specify Values for Template Parameters dialog box*

5. The code now looks as follows:

```
USE ApressFinancial
GO

CREATE INDEX IX_CustomerProducts
ON CustomerDetails.CustomerProducts
(
    CustomerId
)
GO
```

6. Execute the code by pressing F5 or Ctrl+E, or clicking the Execute toolbar button. You should then see the following success message:

```
Command(s) completed successfully.
```

7. Now that you've completed the process, you'll want to check that the index has actually been created as expected. From within Object Explorer, you may need to click the Refresh button on the Object Explorer toolbar or select the Refresh option from the right-click context menu. Navigate to the `CustomerDetails.`
`CustomerProducts` table and expand the Indexes node. This provides you with instant, but limited, information about this index. You can see its name and that it is neither unique nor clustered (see Figure 6-7).

⊟ ▭ Indexes
 ⊞ IX_CustomerProducts (Non-Unique, Non-Clustered)

Figure 6-7. *Index for CustomerProducts*

8. You can see a different perspective of the index if you highlight the index, right-click, and select Properties. Figure 6-8 shows you a layout that offers not only a graphical version of the index, but also a list of many other potential options.

Figure 6-8. *The Index Properties dialog box*

9. The most interesting tab to view once you have data within the table or once you are in production is the Fragmentation tab (see Figure 6-9). As data is modified, indexes are also modified. Similar to a hard drive, an index will also suffer from fragmentation of the data within the index. This will slow down your index, and, as mentioned earlier in this chapter, it is important that you continue to check on your indexes to ensure their best possible speed and performance. A great advancement with SSE is that it is possible to correct the fragmentation while users are still using the system. High availability of SSE has been a priority with this release.

There is no data within the Fragmentation tab at the moment, but if you select the Reorganize Index check box and then click OK, SSE will reduce fragmentation as it reorganizes the index.

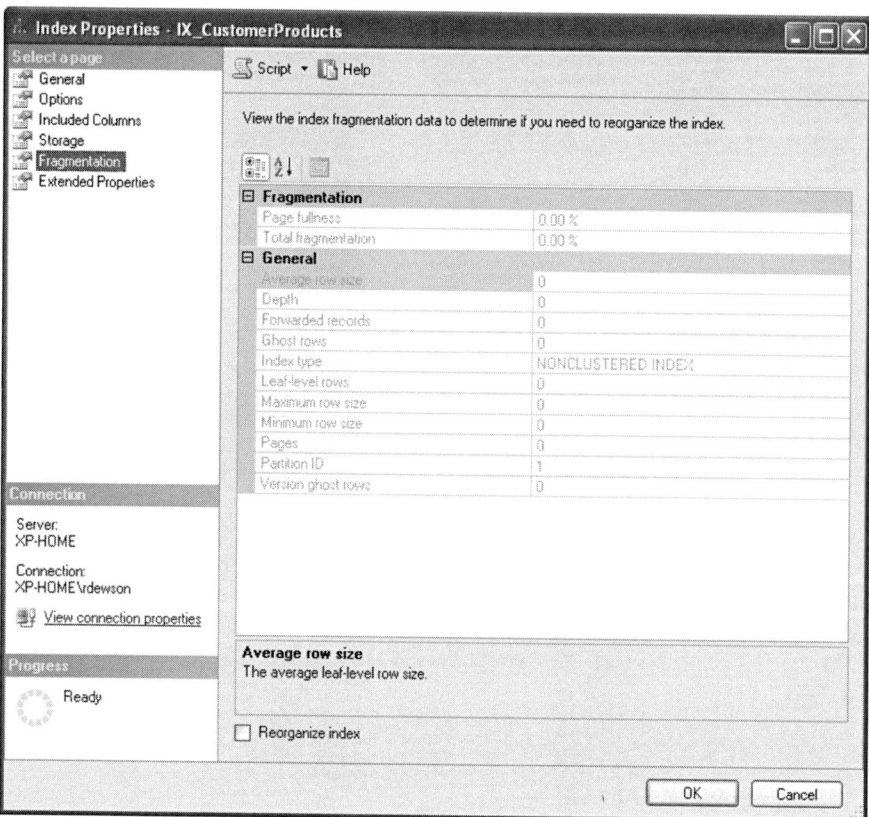

Figure 6-9. *Examining index fragmentation*

The final way to create an index is by coding the whole index by hand in a Query Editor window, which we will look at in the next section.

Creating an Index in Query Editor: SQL Code

In the following exercise, we will create two indexes and a primary key within a Query Editor pane. This will allow us in the next section to build a foreign key between the two tables, TransactionDetails.Transactions and TransactionDetails.TransactionTypes. The code will also demonstrate how to build the index in T-SQL by defining options for the index presented during the CREATE INDEX syntax discussion earlier.

■**Note** The code discussion in the following exercise is broken out into three parts before the code execution, in order to make it easier to follow.

Try It Out: Creating an Index with Query Editor

1. Enter the following code into an empty pane of Query Editor. The first index you will be creating in this section is a uniquely clustered index on the TransactionDetails.TransactionTypes table:

```
USE ApressFinancial
GO
CREATE UNIQUE CLUSTERED INDEX IX_TransactionTypes
ON TransactionDetails.TransactionTypes
(
    TransactionTypeId ASC )
WITH (STATISTICS_NORECOMPUTE  = OFF, SORT_IN_TEMPDB = OFF,
 DROP_EXISTING = OFF, IGNORE_DUP_KEY = OFF, ONLINE = OFF,
 ALLOW_ROW_LOCKS  = ON, ALLOW_PAGE_LOCKS  = OFF)
ON [PRIMARY]
GO
```

2. The second index you'll create is a nonclustered index on the TransactionDetails.Transactions table based on the TransactionType column. You won't make this index clustered, as it would be better to consider either CustomerId or DateEntered as clustered columns.

```
CREATE NONCLUSTERED INDEX IX_Transactions_TType
ON TransactionDetails.Transactions
(
    TransactionType ASC)
WITH (STATISTICS_NORECOMPUTE  = OFF, SORT_IN_TEMPDB = OFF,
DROP_EXISTING = OFF, IGNORE_DUP_KEY = OFF, ONLINE = OFF,
ALLOW_ROW_LOCKS  = ON, ALLOW_PAGE_LOCKS  = OFF)
ON [PRIMARY]
GO
```

3. The final action is to add a primary key to the `TransactionDetails.TransactionTypes` table. You do this through an `ALTER TABLE` statement:

```
ALTER TABLE TransactionDetails.TransactionTypes
ADD CONSTRAINT
    PK_TransactionTypes PRIMARY KEY NONCLUSTERED
    (
    TransactionTypeId
    )
WITH (STATISTICS_NORECOMPUTE = OFF, IGNORE_DUP_KEY = OFF,
ALLOW_ROW_LOCKS = ON, ALLOW_PAGE_LOCKS = ON)
ON [PRIMARY]
GO
```

4. You can now execute the preceding code by pressing F5 or Ctrl+E, or clicking the Execute toolbar button. You should then see the following success message:

```
The command(s) completed successfully.
```

As noted, two different indexes are created in this example. The first one is a unique clustered index, based on the identity column of the `TransactionDetails.TransactionTypes` table. This column was chosen because we will be linking to this table using the `TransactionType` column. Rarely, if ever, will we link on any other column within this table. The overhead is microscopic, though, due to the few records we will be entering, and it is therefore not really a concern. It also allows us to see where to place the keyword within the example.

The second index, built on the `TransactionDetails.Transactions` table, cannot be a unique index; there will be multiple entries of the same value because there are multiple transactions for the same type. However, it is still possible to make this index clustered. Changing the transaction type on a transaction will be rare, or, if we had a full audit trail built within our system, we may "ban" such an action. The only way to change a transaction type around this ban would be to cancel the entry, record the cancel action, and create a new entry. However, a clustered index on transaction types will not give us much of a gain in performance, as there will be few queries of data based on transaction type alone. As mentioned earlier, there are better choices for clustering.

What is interesting about this example is that two indexes are created in one execution—albeit in two batch transactions—whereas in the previous examples, only one index was created at a time. Notice the keyword GO between the two `CREATE` statements creating the index; each index creation has to be completed on its own, without any other SQL statements included. If you need to create more than one index, but you would prefer to build them at the same time, this may be the solution you need. (Please see Chapter 8 for details on transactions.)

An area we have not yet covered is what happens if you try to create an index twice using the same index name. The preceding indexes have already been created, but if you run the query again, SSE will produce error messages informing you that the index already exists. You should see messages like the following:

```
Msg 1913, Level 16, State 1, Line 1
The operation failed because an index or statistics with name
'IX_TransactionTypes'
already exists on table 'TransactionDetails.TransactionTypes'.
Msg 1913, Level 16, State 1, Line 1
The operation failed because an index or statistics with name
'IX_Transactions_TType' already exists on table
'TransactionDetails.Transactions'.
```

```
Msg 1779, Level 16, State 0, Line 1
Table 'TransactionDetails.TransactionTypes' already has a primary key
defined on it.
Msg 1750, Level 16, State 0, Line 1
Could not create constraint. See previous errors.
```

Even if you alter the contents of the index and include different columns, but still use the same name, it is not possible to create another index with the same name as an existing one.

In the last part of the example, we altered the table so that we could add a primary key. There are different types of CONSTRAINTS that can be defined for a table: column constraints are used for default values, as you saw in the previous chapter, but constraints are also used for primary and foreign keys.

Once again, a couple of new areas were covered in this section, but you now have the information you need to be able to create the most common indexes. Indexes need a lot less coding than tables and can be created quickly and easily. However, if you are adding a new index to an existing table that has a substantial amount of information, adding this new index could take a few minutes to complete, depending on the scenario. It is possible to add indexes while the system is being used and the table or clustered indexes are being updated. This is new to SSE and is available in SQL Server 2005 Enterprise Edition, by specifying the index action with REBUILD WITH (ONLINE = ON) option. Take care when doing this, because if anybody tries to access the relevant table while the index is being built, SSE will not recognize the index until it has been built, and when working out the best way to access the data, it will ignore this index. If you are creating the index after removing it for rebuilding statistics, this is when problems may arise if you don't use the ONLINE = ON option. Therefore, if you do need to rebuild an index while keeping the database online for user access, use the REBUILD option.

Dropping an Index

There will be times when an index is redundant and should be removed (i.e., dropped) from a table. Dropping an index is simply a case of executing the DROP INDEX statement, followed by the table name and the index name.

Note If the index is used by a primary key or unique constraint, you cannot drop it directly. In this case, you must use the DROP CONSTRAINT command. The removal of this constraint will also remove the index from the table.

Try It Out: Dropping an Index in Query Editor

1. If you want to drop the index created in the last section, all you need to do is execute the following code. This will remove the index from SSE and also remove any statistics associated with it.

```
USE ApressFinancial
GO
DROP INDEX IX_TransactionTypes ON TransactionDetails.TransactionTypes
```

2. After execution, you should see that everything executed correctly:

```
Command(s) completed successfully.
```

3. Don't forget to re-create the index by running the following code:

```
CREATE UNIQUE CLUSTERED INDEX IX_TransactionTypes
ON TransactionDetails.TransactionTypes
(
    TransactionTypeId ASC
) WITH (STATISTICS_NORECOMPUTE  = OFF, SORT_IN_TEMPDB = OFF,
DROP_EXISTING = OFF, IGNORE_DUP_KEY = OFF, ONLINE = OFF,
ALLOW_ROW_LOCKS  = ON, ALLOW_PAGE_LOCKS  = OFF)
ON [PRIMARY]
GO
```

In the next section, we'll examine what's needed to alter an index.

Altering an Index in Query Editor

Unlike with a table, it is not possible to use an ALTER command to change the columns contained in an index. To do this, you first have to drop the index and then re-create it. The DROP command will physically remove the index from the table; therefore, you should ensure that you know what the contents of the index are before you drop the index, if you want to re-create a similar index.

■**Note** In SSMSE, you can add and remove columns from an index's definition without dropping and re-creating the index, as this is all done for you behind the scenes.

This next exercise demonstrates the steps you need to take to remove an index and then re-create it. You'll learn how to do all of this in two steps, rather than the expected three steps.

Try It Out: Altering an Index in Query Editor

1. First, you want to create an index to retrieve the price of a specific share at a set point in time. The following index will do just that, as you are querying the data using the share ID and the date you want the price for:

```
USE ApressFinancial
GO
CREATE UNIQUE CLUSTERED INDEX IX_SharePrices
ON ShareDetails.SharePrices
(
    ShareId ASC,
    PriceDate ASC
) WITH (STATISTICS_NORECOMPUTE  = OFF, SORT_IN_TEMPDB = OFF,
DROP_EXISTING = OFF, IGNORE_DUP_KEY = OFF, ONLINE = OFF,
ALLOW_ROW_LOCKS  = ON, ALLOW_PAGE_LOCKS  = OFF)
ON [PRIMARY]
GO
```

2. However, it would be better to have the `PriceDate` descending, so that the latest price is at the top, because asking for this information is a query. By including this column, SSE would read only one row rather than an increasing number as more prices were created. It would also be advantageous to include the `Price` itself to avoid a second read to retrieve that column of information from the clustered index.

Note Remember, clustered indexes hold the data, and not pointers to the data. However, in this instance, without the `Price` column, a second read would be performed.

```
CREATE UNIQUE CLUSTERED INDEX IX_SharePrices
ON ShareDetails.SharePrices
(
    ShareId ASC,
    PriceDate DESC,
    Price
) WITH (STATISTICS_NORECOMPUTE  = OFF, SORT_IN_TEMPDB = OFF,
DROP_EXISTING = OFF, IGNORE_DUP_KEY = OFF, ONLINE = OFF,
ALLOW_ROW_LOCKS  = ON, ALLOW_PAGE_LOCKS  =
OFF, DROP_EXISTING = OFF) ON [PRIMARY]
```

Note If you did accidentally run the first set of code, change to `DROP_EXISTING = ON`.

3. Now execute the code using your chosen method, and you should see the following results:

```
The command(s) completed successfully.
```

By using the `DROP_EXISTING` clause of the `CREATE INDEX` command, you can then perform the modification in one execution rather than two. This will drop the index and re-create it.

Note Take care when building indexes. It is possible to use the same columns in the same order more than once, thus creating the same index twice, but under two different index names. This is a waste of time and will place unnecessary overhead on SSE. However, when including key column(s) from a clustered index in a nonclustered index, because the row pointer is actually the clustered index key, SQL is smart enough not to store the information twice in the nonclustered index, and you can explicitly define the order of the clustered index keys as they are used in the nonclustered index.

There are more indexes to build, but we'll take a look at these later.

Diagramming the Database

Now that the database has been built, the tables have been created, the indexes have been inserted, and relationships link some of the tables, it's time to start documenting. To help with this, SSE offers us the database diagram tool, which is the topic of this section.

One of the worst things about database documentation is documenting tables and showing how they relate to one another in a diagram. The database diagram tool can do all of this very quickly and simply, with one caveat: if more than one person is using the database diagram tool on the same database, and there are two sets of changes to be applied to the same table, the person who saves his or her changes last will be the person who creates the final table layout. In other words, the people who save before the last person will lose their changes.

As you developed tables within your database, hopefully you will have commented the columns and tables as you have gone along to say what each column and table is. This is a major part of documentation anyway, and providing you comment columns and tables at the start, it will be less of a chore to add in further comments when you add new columns. If you do have comments on each of your columns within a table, this will help overall with the documentation shown within the diagram.

This said, SSE's database diagram feature is more than just a documentation aid. This tool provides us with the ability to develop and maintain database solutions. It is perhaps not always the quickest method of building a solution, but it is one that allows the entire solution to be completed in one place. Alternatively, you can use it to build up sections of a database into separate diagrams, breaking the whole solution into more manageable parts, rather than switching between nodes in SSMSE.

Database Diagramming Basics

In the book so far, with the creation of databases, tables, indexes, and relationships, as much documentation as SSE will allow has so far been maintained. However, there is no documentation demonstrating how the tables relate to one another within the database. This is where the database diagram comes in.

A database diagram is a useful and easy tool to build simple but effective documentation on these aspects. You build the diagram yourself, and you control what you want to see within the diagram. When you get to a large database solution, you may want diagrams for sections of the database that deal with specific aspects of the system, or perhaps you want to build a diagram showing information about process flows. Although there are other external tools to do this, none is built into SSE that can allow diagrams to be kept instantly up to date.

A diagram will only show tables, columns within those tables, and the relationships between tables in a bare form. You will also see a yellow "key," which denotes a primary key on the table where one has been defined, but that is all the information displayed. It is possible to define the information that is to be displayed about the columns in the table, whether it is just the column name or more in-depth information, such as a column's data type and length, comments, and so on. However, to display more than just the bare essentials, a little bit of work is required.

Although the diagram shows the physical and logical attributes of the database that is being built or has been built, it also allows developers and administrators to see exactly what is included with the database at a glance and how the database fits together.

In the next section, we'll delve a bit deeper into what the SSE database diagram is all about.

The SQL Server Express Database Diagram Tool

SSMSE's database diagram tool aids in the building of diagrams that detail aspects of the database that a developer wishes to see. Although it is a simple and straightforward tool, and it's not as powerful as some other tools on the market for building database diagrams, it is perfect for SSE.

For example, one of the market leaders in database design tools is a product called ERwin, currently owned by Computer Associates. ERwin is a powerful database utility that not only builds diagrams of databases, but also provides data dictionary language output, which can be used to build database solutions. Through links such as OLE DB data providers, these tools can interact directly with databases and so can be used as a front end for creating databases. They can also, at

the same time, keep the created source in alignment and under control from a change control perspective, not only ensuring that the code exists within the database, but also issuing a command to create a new database quickly, if necessary. An example of where this might be useful is when you're creating a new test database. If you want to go further than the SSE database diagram tool provides (you'll learn the tool's boundaries in this chapter), you should be looking at more powerful tools, which can cost a great deal of money.

SSMSE's database diagram utility offers more than just the ability to create diagrams. As mentioned earlier, it can also be used as a front end for building database solutions. Through this utility, SSE allows you to add and modify tables, build relationships, add indexes, and do much more. Any changes built in the tool are held in memory until they are committed using a save command within the tool. However, there are limitations to its overall usefulness.

First of all, the biggest restriction of any diagram-based database tool comes down to the amount of screen space available to view the diagram. As soon as your database solution consists of more than a handful of tables, you will find yourself scrolling around the diagram, trying to find the table you are looking for.

Second, you cannot add stored procedures, schemas, users, views, or any object that is not a table. Other products allow you to include these objects, or they may even build some of them for you.

Finally, for the moment, when altering any of the information you can change within this tool, you are usually using the same dialog boxes and screens as you would in SSMSE.

As you will see as you go through the chapter, the database diagram tool is quite powerful in what it can achieve, but there are some areas of concern that you have to be aware of when working with diagrams. Keep in mind that the database diagram tool is holding all the changes in memory until you actually save the diagram.

For example, if you have a database diagram open, and a table within that diagram is deleted outside of the diagram, perhaps in Query Editor or SSMSE by yourself or another valid user ID, one of two things can happen. First, if you have unsaved changes to the deleted table, saving your diagram will re-create the table, but don't forget that through the earlier deletion all the data will be removed. If, however, you have no changes pending to that table, the table will not be re-created. When you come to reopen the diagram, the table will have been removed.

With several developers working on the database at once, any changes made from the diagramming tool of your SSMSE will not be reflected in any other developer's diagram until their changes are saved and their diagrams are refreshed. If you have multiple diagrams open, and you alter a table and insert or remove a column, this will reflect immediately in all the open diagrams within your own SSMSE only. Don't forget this is an in-memory process, so this process can't reflect on anyone else's diagrams until the changes are saved and the diagrams are refreshed.

Also, if you remove an object in your diagram, when you then save the diagram, the object will be removed and any changes completed by others will be lost. Effectively, the last person who closes his or her diagram wins!

To summarize, if you use the database diagram tool, use it with care. Because many of the processes are in memory, you could inadvertently cause problems.

The Default Database Diagram

Although it's not mandatory, I do feel every SSE database solution should have a default database diagram built into it so that any developer—new or experienced—can instantly see how the database being inspected fits together.

A default database diagram should include every table and every relationship that is held for that database. Unlike other diagrams that may take a more sectionalized view of things, the default database diagram should be all-encompassing.

As mentioned earlier, it is imperative that you keep this diagram up to date. You will notice this statement repeated a few times in this chapter. Don't use the default diagram as the source of development for your database solution. The default diagram includes all the tables, which means

that if you're using the database diagram tool for development, you are potentially logically locking out all other users from touching any other table as part of their development, in case their changes are lost. Only update the diagram with tables and relationships once they have been inserted in the database. We'll look at this in more detail later when we discuss the dangers of using the database diagram tool as a source of development.

Now that you know what diagrams are and what the tool is, it's time to create the first diagram for this database.

Try It Out: Creating a Database Diagram

1. Ensure that SSMSE is running and that the ApressFinancial database is expanded so that you see the Database Diagrams and Tables nodes. Select the Database Diagrams node. If this is the first diagram you are creating for the database, you'll need to install support objects. Without them, you cannot create the diagram, so click Yes at the next dialog box prompt (see Figure 6-10).

Figure 6-10. *Creating a new database diagram*

2. Now right-click and select New Database Diagram (see Figure 6-11).

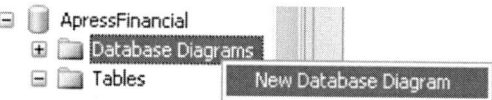

Figure 6-11. *Installing diagramming support*

3. The first screen you'll see when creating the diagram is the Add Table dialog box (see Figure 6-12). Select all of the tables listed, as you want to use all the tables in your diagram, and then click Add. This will "empty" the screen. Click Close.

Figure 6-12. *Selecting tables*

4. After a few moments, you will be returned to SSMSE, but with the database diagram now built. The diagram will not show all the tables at this point and will be very large. You can reduce the size through the Size combo box in the diagram toolbar, as shown in Figure 6-13.

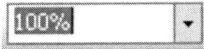

Figure 6-13. *The Size combo box*

5. You'll then see a diagram similar to that shown in Figure 6-14. (Don't be surprised if the layout is different, though.)

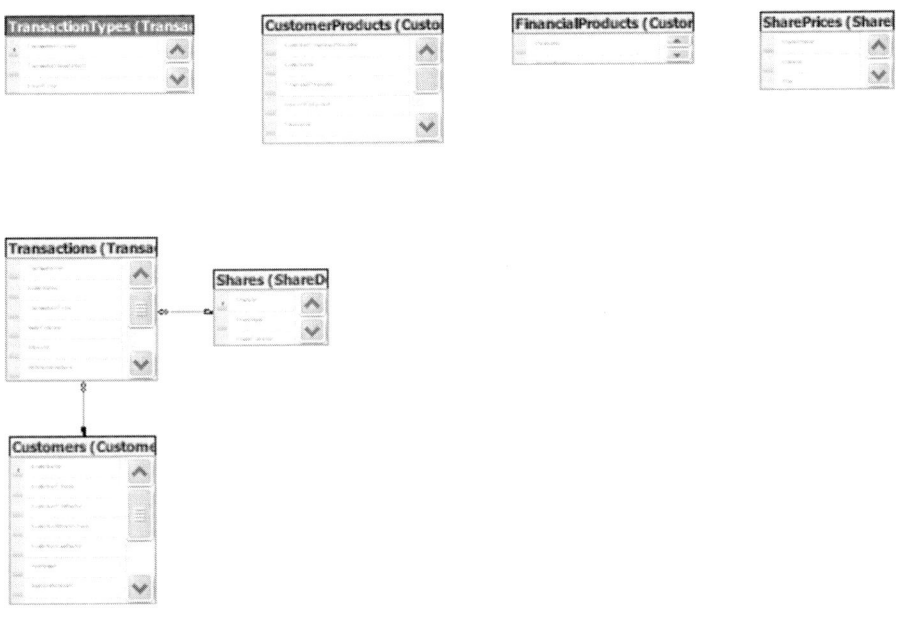

Figure 6-14. *Tables with relationships built so far*

That's all there is to building a basic diagram.

The Database Diagram Toolbar

Let's next take a look at the toolbar and how each of the buttons works within the diagram. The whole toolbar is shown in Figure 6-15.

Figure 6-15. *Database diagram toolbar*

The first button is the New Table button, as shown in the following image. You click this button to create a new table within the Database Designer, which is a similar process to that shown in Chapter 5.

The difference is that you need to use the Properties window for each column rather than having the properties at the bottom of the Table Designer.

When building the diagram, you selected every table. If you hadn't done so and you wanted to include a table added since you created the diagram, clicking the Add Table button (see the following image) would bring up the Add Table dialog box shown earlier to add in any missing tables.

The Add Related Tables button, shown next, adds tables related to the selected table in the designer.

It is also possible to delete a table from a database from the designer using the following button.

If you just wish to remove the database from the diagram rather than the database, you can use the next button to accomplish this. You would use this button, for example, if a table no longer formed part of the "view" the database diagram was built for.

Any changes made to the database within the designer can be saved as a script. Use the following Generate Change Script button to accomplish this.

If you wish to set a column to be the primary key, select the relevant column within a table and click the Set Primary Key button (shown next).

It is possible to create a place within the diagram to put ad hoc text. Use the following New Text Annotation button to do this.

Each table displayed is set to a standard layout. It is possible to change this to a different predefined layout or to create your own custom view. The following Table View drop-down list enables you to change the layout or create your own custom version.

Relationships that exist between tables will, by default, show as a line. However, it is possible to show the name of the relationship as a label by clicking the following button.

Diagrams are ideal methods of documenting the database. Diagrams can be printed for meetings, discussions about future development, and so on. The following button shows the line breaks in pages that will be printed.

Page breaks in diagrams remain as they were first set up until they are recalculated. You are able to view the page breaks, arrange your tables accordingly, and then recalculate the page breaks based on the new layout. Clicking the following button will do this recalculation for you.

Tables can be expanded or shrunk manually, but when you select one or more tables using the Ctrl button, click the relevant tables, and then click the following button, you can resize the tables to a uniform size.

It is possible, by clicking the following button, to rearrange tables that have been selected and let SSE make the arrangement choices.

It is possible to rearrange the tables shown in the diagram. SSE will do its best to get the best layout for the tables and relationships when you click the following button.

In Chapter 5, you saw how to build a relationship between two tables. The button you use to do this appears next. This button will bring up the same dialog box as you saw in that chapter.

You can also manage the indexes and keys using the dialog box you saw in Figure 6-2 by clicking the following button.

It is possible to create an index on a table or column called a **full-text index**. This index allows for searching on text data a bit like Google. For example, Google would hold the web pages within a text data type, and using a full-text index would allow for full searching on that page. Although I don't cover this type of index within this book, clicking the following button will manage these indexes.

If you have indexes placed on any XML data types, clicking the next button will allow you to manage these indexes.

In Chapter 5, you learned how to build constraints for tables. Clicking the following Manage Check Constraints button brings up the same dialog box you saw then.

Summary

We've covered yet another major building block in creating a SQL Server Express solution. The last few chapters have shown you how to store data, and in this chapter you've learned about indexes and how to use them to quickly and efficiently retrieve the data stored in the table.

There are many types of indexes, and choosing the right one at the right time to complete the right job is quite an art. This chapter has taken you through the steps to decide which columns will make an efficient index, and then build those columns in the right type of index to make the most of the information.

This chapter also covered database diagramming. Database diagrams should initially be thought of as a form of documentation. Keep in mind, though, that the database diagram tool may expand in future versions of SQL Server to become much more sophisticated and powerful than it is now—although even now it is quite a powerful utility.

Don't be caught out by the fact that changes in the diagram are not applied until the diagram is saved, and that your changes could overwrite another's changes. If you're using the database diagram tool for development in any sort of multiuser environment, take the greatest of care when completing updates (in fact, try to avoid them altogether). Unless you split your database solution into multiple diagrams, with any table being found in at most one diagram, don't use the database designer as a development tool.

So we have made a fair amount of progress with our solution, and it's now time to safeguard what we've done so far, so in the next chapter we look at how to back up and restore our solution.

■■■

Database Backup, Recovery, and Maintenance

Now that we have created a major part of the database in the previous chapters, and before moving on to inserting and manipulating the data, this is a good point to take a moment to back up the database, just so that if things go wrong, it is possible to recover back to a stable point.

What is abundantly clear when working with any sort of system where data is held is that there must be a comprehensible and workable backup and recovery strategy in place for when things go wrong. The recovery may also be required to cater to problems from a hardware failure up to an act of God. In any of these instances, we may move to an **offsite location**, which is a building a safe distance away from our current building housing the computing equipment. That is quite a dramatic step and is a decision that would be taken at a higher level of authority than we probably have; however, we must create a backup of our system and store it according to the recommendations of our board of directors, whether they are for in-house or offsite storage. Companies have gone bust because a good and secure backup storage wasn't in place when their building burned down, for example. This is of course a worst-case scenario, and there will be times that moving out of the current building to a second secure location is not necessary.

This chapter looks at different backup strategies that can be implemented by you as a developer or an administrator, and how they would be implemented. I also show you scenarios where the database is in use 24 hours a day, 7 days a week, and how a backup strategy needs to be formed around such scenarios. From there, you will see how to perform an ad hoc backup of the database as well as scheduled transaction log backups. It will be made clear in this chapter when you would perform both of these types of backups and when they would be useful. Of course, after the backup, you will have to test that the backup can be restored. Generally, this backup will be restored onto a nonproduction system. Some companies have complete environments established to test their disaster-recovery scenarios.

What you have to realize, and what will be demonstrated, is that there are different methods of taking backups depending on what you are trying to achieve. The most common scenarios are discussed and demonstrated in this chapter, but you will also get to look at database maintenance plans.

It is imperative that you get the correct backup strategy in place, and that it works. This point will be repeated throughout the chapter.

So, in this chapter you will learn

- Backup strategies
- When a problem might occur during the backup and restore process
- How to take a database offline, and then bring it back online
- How to create a backup
- Different media for building a backup: what needs to be considered

- About the transaction log and how to back it up
- When to back up the data, and when to back up the transaction log
- Scheduling backups, and what happens if the server isn't running at the scheduled time
- Restoring a database
- Detaching and attaching a database
- How to work with users still attached to the database when you need them not to be connected
- How to build SQL statements for backing up the database structure, and when it is useful to have them
- Building a maintenance plan and knowing when to use it

Transaction Logs

Data within the database is stored, of course, on your hard drive, or on a hard drive on a server. Imagine the work involved on the disk controller, and imagine the work SSE has to do every time data is added, amended, or removed. Writing data is a slow process so, inevitably, every time data is written, SSE slows down. A good comparison is to think how long it takes you to insert, modify, or erase a sentence, even using MS Word, compared to how long it takes you to read a sentence. What if part of the way through writing the data, there was a power outage and you had no uninterruptible power supply (UPS) in place? What a mess you would be in, not knowing what had been written to disk and therefore your tables within your database, and what hadn't!

It is also possible in SSE to update several tables at once. How would you work around the fact that some of the tables had been updated, but when it came to updating a specific table, the update failed? Well, this is where transaction logs come into play. Transactions themselves are covered in Chapter 8, but very simply, a **transaction** is a single unit of work that can contain one or more table modifications that are either all successful and committed to the database or, if some are unsuccessful, all the modifications are discarded. It is also possible to roll back a transaction so that no changes are made to the database, which can either be invoked by SSE or by issuing a specific T-SQL command. But you must be wondering what all this has to do with a transaction log and even wondering what a transaction log is. Before we move on, there is one last area of background information we need to discuss first.

The first item to note is that every database within an SSE instance has its own transaction log. Every time SSE is requested to do any data modifications—whether these are additions, deletions, or modifications—a record is kept of the action. These recorded actions are kept in a place called a **transaction log**. There are several reasons for this.

First of all, a piece of code could in fact do several different updates at once, either to different rows of data or rows of data in different tables or even databases. If one of the updates fails, for example, you are attempting to place ASCII characters into a column that only allows numerics, you may wish to return the values in all the updated fields to their original value. This is called **rolling back** a transaction. SSE achieves this, in part, by looking at the data held in the transaction log. However, any successful action where all the updates are valid could be permanently stored on file, a process called **committing** a transaction.

As more and more actions are placed in the transaction log, it will become full. Some of these actions will still be within a transaction, and others may form part of a completed transaction ready to be committed to the database. At certain points, SSE will want to remove all the actions it can to relieve some space in the transaction log for further actions. One point could be when the transaction log reached 70 percent full. SSE would then issue a **checkpoint**. The use of a checkpoint ensures that any data modifications are actually committed to the database and not held in any buffers, so

that if a problem occurs, such as a power failure, there is a specific point that you can start from. Therefore, at the end of a checkpoint transaction, you know the database is in a consistent and valid state. As SSE knows that at a checkpoint all is well within the database, there is no need to keep the completed transactions recorded in the transaction log stored up to the checkpoint. SSE will therefore issue a truncation of the transaction log to remove these records, minimizing the size of the log on the computer. This is known as **truncating** the transaction log. It is thus necessary to ensure that you have a large enough transaction log defined to hold the largest valid uncommitted transaction, as these transactions obviously will not be truncated. A transaction log can become full with a rogue query as well, one that is incorrectly coded and just keeps adding more and more uncommitted transactions. When the transaction log reaches 70 percent, there is nothing to checkpoint, and eventually the transaction log will fill up and SSE will stop. This is where you will need the help of an experienced database administrator.

If you had a power failure, you might have to "replay" all the work completed since the last backup, and the transaction log could also be used to do this, in certain scenarios. When a data modification is completed via a T-SQL command, the details are recorded first of all in the transaction log. These changes are flushed to disk, and therefore no longer in memory, before SSE starts to make changes to the tables that you are affecting. SSE doesn't write data immediately to disk. It is kept in a **buffer cache** until this cache is full or SSE issues a checkpoint, and then the data is written out. If a power failure occurs while the cache is still filling up, that data is lost. Once the power comes back, though, SSE would start from its last checkpoint state, and from the transaction log, any updates after the last checkpoint that were logged as successful transactions will be performed.

■**Note** A **disk cache** is a space in the system where changes to the tables within the database are held. By doing so, a whole block of data can be written at once, saving on the slow process of disk head movement.

Transaction logs are best kept, if at all possible, on a hard drive separate from that holding the data. The reason for this is that, when data is written to a transaction log, it is written serially. Therefore, if there is nothing else on the hard drive except the transaction log, the disk heads will be in the right place to continue writing each time. A minor overhead, but if performance is an issue, this is worth considering.

Backup Strategies

Backing up a database, no matter how large or small, should form part of your database solution. Even if a backup is only taken once a week, or even once a month, it is crucial that you sit down and decide which backup strategy is right for you. Much of this decision lies in the hands of the product owners for your company, since they must weigh the risk they're willing to take against the cost of minimizing that risk. There are also many different strategies that can be adopted within your overall main backup strategy, depending on days of the week, or perhaps a period within the month.

Based on the strategy that you choose, you will have to decide what type of backup you will need. **Full database backups** take a complete snapshot of a database at any given point. A **differential backup** will back up only the data that has changed since the last full backup. Finally, a **transaction log backup** only backs up the data in the transaction log, which consists of transactions that were recently committed to the database. All of these types of backups can be done while SSE is online and while users are actively hitting your database. To be able to restore a database to any point in time, you will have to use a combination of these backup types.

The first place to start with the backup strategy is to look at your application and ask yourself the following questions:

- How much of the data can be lost, if any, at any point of failure? In other words, how crucial is it that no data is lost?

- How often is the data updated? Do you need regular backups from a performance viewpoint as well as a recovery viewpoint? For historical databases that only periodically have their data modified, you would, at most, complete a backup postpopulation.

- Do you need to back up all the data all of the time, or can you do this periodically, and then only back up the data that has altered in the meanwhile?

- How much data needs to be backed up and how long do you need to keep the copies of the backups?

- In the event of catastrophic failure, how long will it take to completely rebuild the database, if it's even possible?

There are many more questions that can be asked, but for the moment, these are the most crucial questions that need answers. If you can afford to allow data updates to be lost, this is a straightforward periodic database backup; for example, back up the whole database once a week. This is simple and easy to complete and requires little thought. This is a rare scenario and found usually in data warehousing systems or those with a large amount of static data that can be rebuilt.

Looking at the next question, if a large number of updates are taking place, a more complex solution is required. For every data modification, a record is kept in the transaction log file, which has a limited amount of space. This amount of space was defined when you set up the database as a fixed maximum size or, if you are allowing it to grow unrestrictedly, equals the amount of hard drive there is. If you backed up and cleared the transaction log file, this would free up the space logically and also aid performance. The smaller the active part of the transaction log file, the better. The more transactions there are in the transaction log file, the longer it will take to recover from a corrupt database. This is due to the fact that a restore will have to restore the data, and then every transaction log backup to the point of failure. That is, each transaction log will have to be restored to update the database, not just the latest log file. If you have multiple small files and they are held on media that has to be mounted each time, such as a tape, you will have to take mounting time into consideration as well.

The third question, though, covers the real crux of the problems. If you need to back up all the data each time, how often does that need to take place? This could well be every night, and many production systems do just this. By completing a full data backup every night, you are allowing yourself to be in a state where only one or two restores may need to occur to get back to a working state in a disaster scenario. This would be the data backup, followed by the single transaction log backup, if one was taken in the meantime, to be restored. Much better than having one data backup to be restored, and then a log file for every day since the data file backup. What happens if the failure is on a Friday noon and you completed your last whole database backup on a Saturday evening? That would take one data file and six transaction log file restores to complete.

Therefore sit down and take stock. As often as you can, take a full database backup, then from there take a differential backup, followed by transaction log backups. However, you have to weigh this against the time that a full backup takes over a differential backup or a transaction log; how much processing time you have to complete these backups; and the risk level on having to complete, for example, six transaction log restores.

The problem is, there is no universally right answer. Each situation is different, and it is only through experience, which comes from testing your solution thoroughly, that you find out what is best for that situation.

Whatever your backup strategy, the best possible scenario is to complete a backup when nobody is working within the database. If there are times when you can make the database unavailable, this is an ideal opportunity to take the backup. Although SSE can perform full backups while the database is online and active, you will gain performance benefits by having an inactive

database. The first example, discussed shortly in the section "Taking a Database Offline," demonstrates one method of doing this.

When Problems May Occur

Obviously, when taking a backup, it must work; otherwise, you have wasted your time, but, crucially, you are leaving your database and organization in a vulnerable position if the backup has failed. If you have time within your backup window to verify that a backup has been successful, you must do it. As you will see, SSE gives you different options on how to do this. It cannot be stressed strongly enough that verifying a backup is just as crucial as taking the backup in the first place. There have been situations where a backup has been taken every night; however, no one has noticed that the backup has failed, and then when a hardware failure occurs, there's no backup to use as a restore on a new machine. In one case I know of, almost a week's worth of data was lost. Luckily, the weekend backups had succeeded; otherwise, the company would have been in a major data loss situation. The cause was that the tapes being inserted for the backup were not large enough to hold the backup being performed. Therefore the tape became full, and so the backup failed. Obviously, this was a case of a company that failed not only to verify the backup, but also to have processes in place to check that its backup strategy was still working after a period of implementation. The only sure and positive way of ensuring a backup has succeeded is to restore the database to a specific restore test location and check the data. Although you will see SSE does have a method of checking a backup, this still isn't a guarantee that the backup worked. Do take time to complete regular restores to a location to test that everything is okay.

You should always review your backup strategy on a regular basis. Even better, put in place jobs that run each day, giving some sort of space report so that it is possible to instantly see that a potential problem is looming. SQL Server Reporting Services is a new tool that would be ideal for producing and distributing space reports to database administrators and developer owners alike.

Taking a Database Offline

SSE does not have to be offline to perform a backup, as you will see as we go through the book and work through creating SSE-defined backups using wizards and T-SQL. In most environments, you will not have the luxury of taking a database offline before backing it up, because users are constantly making data changes. Backing up a database can take a long time, and the longer it takes, the longer users cannot be working with the data while it is offline.

By taking your database offline, you do not have to use SSE to perform the backup. This strategy is one where you take a disk backup, which means the hard drive is backed up, rather than a specific database within a server. However, don't forget that by taking your database offline, it means you will have to take a backup of the directory using some sort of drive backup.

If you have your database on a server, no doubt some sort of server backup strategy is in place, and so your database would be backed up fully and successfully through this method; if you can take your database "out of service" in time for those backups, you should do so. This does allow you to think about your SSE deployment strategy. If you have several databases that can be taken offline as part of the backup, it is worth considering whether they can all reside on the same physical server and set your server backup times accordingly. However, this is a rare scenario, and even rarer within a production environment. Taking the database offline means taking your database out of service. Nobody can update or access the data, and nobody can modify table structures, etc. In this next section, we will take `ApressFinancial` offline, allowing a physical backup to be taken. Just to reiterate and clarify: this is being demonstrated only to complete your knowledge of backups, and it will be rare that you perform this action in a live scenario.

Try It Out: Taking a Database Offline

1. Open SQL Server Management Studio Express (SSMSE) and open a Query Editor pane. Enter and execute the following code:

```
USE master
GO
ALTER DATABASE ApressFinancial
SET OFFLINE
```

2. Try to click some of the nodes for the `ApressFinancial` database, for example, the Tables node. We will be reminded that the database is offline, and therefore cannot be viewed or modified, as shown in Figure 7-1. We will also not be able to access the database through any application such as Query Editor.

Figure 7-1. *Database is offline and therefore unable to be opened*

To take a database offline, SSE must be able to gain exclusive access to the database. This means that no user can be in the database when we issue the command. If users are connected, the query will continue to execute until all users are disconnected.

As I said earlier, that's all there is to it. Our database is now no longer available for any updates of data, or modifications, and so can be backed up using any backup utility that takes files from a hard drive.

If you have to restore from a backup completed this way, don't forget to take the database offline first, then restore from the backup, and then bring the database back online, ready for use:

```
USE master
go
ALTER DATABASE ApressFinancial
SET ONLINE
```

There is one area to note when using backup strategies that employ these methods. If you have a server backup that runs, for example, at 0200 hours, do you fancy getting up every night, just before 2 a.m., taking the database offline, and then bringing the server back up once the backup is complete? No—not many people would. Of course, there are installations where people are working through the night, so this is less of a problem, but what if they are busy? Or forget? Then your whole backup will fail because the files are in use, and therefore the server will not back up these files.

So let's now look at a more friendly method of backing up the data by using SSE instead.

Backing the Data Up

Using SSE to back up the database will be the method employed by the majority of readers. By using SSE, we are keeping the backup of the database under the control of an automated process that can control its own destiny, and as you will find out later, it can also control the system when things go wrong.

The backup will be split into two parts. The first part, which will be covered here, will be when we perform the backup manually each time. Obviously, this means we have to be available to perform the backup, but this can be rectified quite easily. Once this has been covered, the next section will schedule a backup to run at a specific time, which will relieve us of needing to be available to complete a backup at the specified time.

Let's start by looking at the manual backup.

Try It Out: Backing Up the Data

1. Ensure SSMSE is running. Find our database, right-click, select Tasks, and then click Back Up.

2. This then brings up the Back Up Database dialog box. Take a moment to peruse this dialog box. As Figure 7-2 shows, a lot appears on this screen, which will be dealt with a section at a time.

Figure 7-2. *Backing up a database (full recovery model)*

Although we have chosen our own database to back up, we could alter our choice of database by changing the value in the combo box. Next is the backup type, and we have up to three options to choose from: Full, Differential, and Transaction Log. This all depends on the recovery model we have in place. During the creation of the database, we specified the Simple recovery model, which means that only the first two options are available. As I mentioned at the time, this should not be used in production. It is the default for SSE when creating databases, however.

The first possibility, **full backup**, is straightforward. Selecting the Full option will ensure that the whole database will be backed up. All the tables, the data, and so on are backed up if we use this option. We also back up enough of the transaction log (transactions committed via code, but not physically applied to the relevant tables at the point of backup).

The second possibility is the **differential backup**, also known as an **incremental backup**. Use the Differential backup option when the circumstances are such that a full backup is either not required or not possible. This just performs a backup on data changed since the last backup. For example, we may take a full backup at the weekend, and then a differential backup every night. Then when it comes to restoring, we would restore the weekend full backup, and then add to that the latest differential backup taken from that point. A word of warning here: if you do take a week's worth of differential backups, and you back up to magnetic tape rather than a hard drive, do keep at least enough different tapes for each differential backup. Therefore, use a complete backup when necessary and then in the interim use a differential backup.

The last possibility, the **transaction log backup**, should be used as the most frequent method of backup in a production environment when using the Full recovery model. As data is modified, details of the modifications are stored in the transaction log. These remain in place until an action "truncates" the transaction log, which means that the transaction log will increase constantly in size. When you issue a transaction log backup, you are just backing up the transaction log, which will issue a checkpoint, and all committed transactions will be stored onto the backup. This means that if a system failure occurs, you would restore from a full backup, then from your differential backups for the week, and finally from any transaction log backups after that point. This is not the case though in Simple recovery, which obviously takes a more simplistic method of backing up and restoring.

So you are probably wondering why not just use differential backups? Transaction logs can fill up during the working day, or perhaps you have set differential backups to happen weekly because there is so little data modification. However, you do need to account for when a transaction log may fill up before you reach the next differential backup. By taking a backup of the transaction log, this is a great deal faster than the other two methods. Certainly in heavily used databases you may have several transaction log backups in the day. You see how to do this using T-SQL after we take our first full backup. At least one backup must exist before we can take a transaction log backup, as we need a point from which the transaction log can roll committed transactions forward.

SSE is not meant for production environments, hence why it defaults to the Simple recovery method when creating a database; however, you can change this by changing the Recovery Model property shown earlier in this book in Figure 3-11.

■**Note** If we were backing up the `master` database, the only option that would be available to us would be a complete database backup via the Full option.

A name and description of your backup are always useful. You will create different backups through time, so a good description is always something that will help at a later date. I recommend that you use some sort of date and time as part of the description, as this will make it easier to find, and which mode of backup you have chosen.

Different types of backups will have different expiry dates. This means that after the defined date, the media you have stored your backup on will allow the data to be overwritten if using SSE (you can't delete the file manually!). For example, you might have a weekly full backup that you want to keep three instances of, and then the first full backup of the month you may wish to keep for six months, or even longer if it is a database that you must keep for government legislation. In this option, you can retain the backup for a set number of days (for example, 21 days) or for a set period of time (a specific date covers for uneven days in a month, or a year, for example).

A default destination is defined that might be more than acceptable. It will be on our hard drive, in a location below where our data is. It is best to have a directory set aside for backups so that they are easy to find, perhaps with a name such as

SQL Server Backups. However, in production this is not recommended. What if the hard drive fails? We can gain a substantial performance improvement by backing up the database to a separate disk or to a RAID 1 drive if one is available. This is because data is written to the backup file in a sequential manner. It is also advisable to give the backup file a meaningful name. In this instance, it has been given the name of the database, `ApressFinancial`.

3. Move to the Options tab, as shown in Figure 7-3, where we can define what options we want to happen as part of this backup.

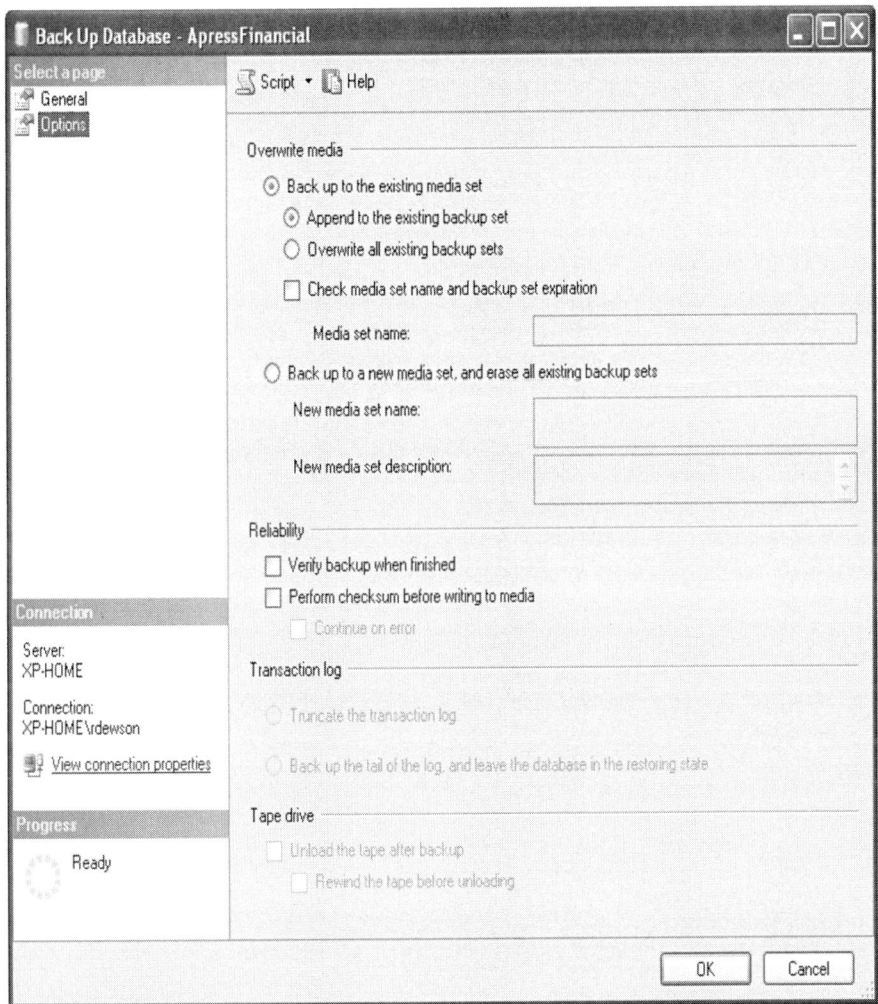

Figure 7-3. *Database backup options*

The first section of this dialog box deals with what you want to happen on second and subsequent backups. The first time the backup is run, it will create the backup files, but when you run subsequent backups, do you want to append to the current data or overwrite it? If this is a full backup, you may overwrite, as you should be placing this full backup over an old unrequired backup. However, if this is a differential backup, where it is perhaps the second or third of the week, you would append to the existing backup set. This would be after the previous differential backups and would mean that if you needed to do a restore, all the backups would be one after another and therefore would provide the fastest retrieval.

The option Check Media Set Name forces the backup to check that where the data is going to be backed up to is still a valid name and, if appending, that the data set has not expired.

You would use the option Back Up to a New Media Set, and Erase All Existing Backup Sets when any previous backups were no longer required. This is ideal when moving the database from development to either user testing or even production and you do not want to be able to restore from an incorrect backup. There is no point in wishing to restore a production server from a development backup after all.

The second section deals with the reliability of the backup. It is possible to simply back up the data and trust that everything worked perfectly, meaning no data transmission errors occurred between your SSE and the backup device, or that no errors occurred when writing the data. A situation such as this is unusual, but there will be times that it does happen. Do you trust that those times will occur when you will not need a backup? I suggest this is something you just cannot and should not rely on. Therefore, although it will increase the amount of time the backup takes, it is good to choose one of the two options in this section. The two options are to allow a verification of the backup where SSE compares what has been backed up with what it expects to have been backed up, and the second option allows for checksum processing whereby SSE performs a mathematical calculation on the data to back up, which will generate a checksum that can then be compared once the data has been transmitted from SSE to the backup device. If you select the second option, you can also specify whether to continue if you get a checksum error.

If you are doing a transaction log backup, the next area of the dialog box would be enabled. You can logically shrink the transaction log by removing all entries that have just been backed up by selecting the first option, Truncate the Transaction Log. To save processing time, the physical transaction log is not shrunk. The second option, Back Up the Tail of the Log, is used when there has been some sort of database corruption. If you wish to back up transaction log records that have not been backed up prior to performing a restore to correct the corruption, you would use this option. To clarify, a database becomes corrupt, and you need to be able to restore up to the last backup, then add all the transactions that have occurred since the last backup. By executing a backup of the tail of the log, you can restore the database and then use this tail log backup to add the missing transactions.

The final area of the dialog box is available if you are using tapes as your backup medium. You can eject the tape once the backup has finished. This is a useful option as the computer operators would know to remove the tape for dispatch to the safe backup area. The second option, which specifies a rewind, is useful for full backups. On differential backups, however, SSE would be halted when running the next backup while the tape device got to the right place on the tape to continue the backup.

4. Clicking OK will then start the backup.Once the backup is finished, you should see the dialog box shown in Figure 7-4.

Figure 7-4. *A successful backup*

The first backup of the ApressFinancial database has now taken place and should have been successful. If we now move to the directory on the hard drive where the backup took place, we will see the ApressFinancial file.

Recall that it was mentioned earlier that a company lost a week's worth of data. It had specified the option to append to media, the tape had become full, and the administrator had not set up the proper scenario to alert someone when a problem occurred. So there was not just one failure in the system, but two; however, it still highlights that if you are appending to the existing backup set, you must check that enough room is available on the medium that you are appending to for the backup to succeed.

Creating a backup of your database and the data is the most crucial action in building a database solution. Get it wrong, and you may as well go home. Well not quite, but if (or when) things go wrong, and you don't have a valid or recent enough backup that is acceptable to the users of your database, it will take a long time for you as a developer to recover from that situation and get back to the excellent working relationship you had beforehand.

The backup taken in the preceding example is the simplest backup to perform. It is a complete backup of our particular SSE database, which happens while we are watching. If it goes wrong, we will instantly see and be able to deal with it. However, most backups in production systems especially, do not happen when you are there; they usually happen through the night. Creating a database backup with a schedule to run at the optimal time is the ideal solution and certainly an area to consider; however, it is not possible to do this with SQL Server Express Edition. To enable job scheduling, which would allow backups to occur automatically, it is necessary to move to a different edition. Therefore this example demonstrates how to complete a backup manually rather than as an automated process.

Before moving on, there are a couple more points concerning backups that you must keep in mind, and it is recommended strongly that these directions be followed. First of all, keep a regular and up-to-date backup of the master and msdb system databases. SSE very rarely becomes corrupted, but it can happen for any number of reasons, from a hard drive failure to a developer altering the database in error. It really doesn't matter, but if you don't have a backup of the master database, you could find yourself in trouble. However, be warned. Restoring the master database should not be performed unless you really have to, and only if you are experienced with SSE. Restoring the master database is not like restoring other databases, and has to be completed outside SSMSE. This book quite deliberately does not cover having to restore the master database, since it is a very advanced topic. If you wish to know more, take a look at Books Online for more details.

When it comes to the msdb database and when to back it up, it could be that a daily backup is required. If you recall, this database holds job schedules and other information pertinent to the SSE Agent for scheduling. If you have jobs that run each day, and you need to keep information about when jobs were run, a daily backup may be required. However, if you only wish to keep a backup of jobs, etc., that are set up and there is no need to know when certain jobs ran and whether they were successful or not, perhaps look at backing up this database weekly.

The model database should be backed up if any details within the model database have been altered. This should be pretty infrequent, and therefore backing up this database need not be as regular as any other database; once a week is probably frequent enough.

Backing up tempdb is not necessary, as this should be seen as a transient database, which has no set state.

Note When SSE is restarted, tempdb is dropped and is re-created as part of the startup process.

As you can see, it is not just your own databases that need to be considered and remembered when it comes to dealing with your backup strategy. A database within SSE is not an insular arrangement and affects the system databases just as much.

If in doubt, back it up more frequently than is required!

Backing Up the Database Using T-SQL

Now that we have backed up the database using the wizard, it is useful to demonstrate performing a backup with T-SQL. These commands and statements can be used within a stored procedure that can be scheduled to run at required intervals as part of an overnight task.

There are two different types of backups. It is possible to back up either the database or specific filegroups or files that are part of the database. The code for the database backup follows. The highlighted code will demonstrate which of the two possible options is the optional default used when neither option is specified.

```
BACKUP DATABASE { database_name | @database_name_var }
TO < backup_device > [ ,...n ]
[ [ MIRROR TO < backup_device > [ ,...n ] ] [ ...next-mirror ] ]
[ WITH
     [ BLOCKSIZE = { blocksize | @blocksize_variable } ]
     [ [ , ] { CHECKSUM | NO_CHECKSUM } ]
     [ [ , ] { STOP_ON_ERROR | CONTINUE_AFTER_ERROR } ]
     [ [ , ] DESCRIPTION = { 'text' | @text_variable } ]
     [ [ , ] DIFFERENTIAL ]
     [ [ , ] EXPIREDATE = { date | @date_var }
     | RETAINDAYS = { days | @days_var } ]
     [ [ , ] PASSWORD = { password | @password_variable } ]
     [ [ , ] { FORMAT | NOFORMAT } ]
     [ [ , ] { INIT | NOINIT } ]
     [ [ , ] { NOSKIP | SKIP } ]
     [ [ , ] MEDIADESCRIPTION = { 'text' | @text_variable } ]
     [ [ , ] MEDIANAME = { media_name | @media_name_variable } ]
     [ [ , ] MEDIAPASSWORD = { mediapassword | @mediapassword_variable } ]
     [ [ , ] NAME = { backup_set_name | @backup_set_name_var } ]
     [ [ , ] { NOREWIND | REWIND } ]
     [ [ , ] { NOUNLOAD | UNLOAD } ]
     [ [ , ] STATS [ = percentage ] ]
     [ [ , ] COPY_ONLY ]
]
```

If instead you just wish to back up specific files or filegroups, the difference in the code is highlighted in the BACKUP DATABASE statement shown here:

```
BACKUP DATABASE { database_name | @database_name_var }
     <file_or_filegroup> [ ,...f ]
TO <backup_device> [ ,...n ]
[ [ MIRROR TO <backup_device> [ ,...n ] ] [ ...next-mirror ] ]
[ WITH
     [ BLOCKSIZE = { blocksize | @blocksize_variable } ]
     [ [ , ] { CHECKSUM | NO_CHECKSUM } ]
     [ [ , ] { STOP_ON_ERROR | CONTINUE_AFTER_ERROR } ]
     [ [ , ] DESCRIPTION = { 'text' | @text_variable } ]
     [ [ , ] DIFFERENTIAL ]
     [ [ , ] EXPIREDATE = { date | @date_var }
     | RETAINDAYS = { days | @days_var } ]
     [ [ , ] PASSWORD = { password | @password_variable } ]
     [ [ , ] { FORMAT | NOFORMAT } ]
     [ [ , ] { INIT | NOINIT } ]
     [ [ , ] { NOSKIP | SKIP } ]
     [ [ , ] MEDIADESCRIPTION = { 'text' | @text_variable } ]
     [ [ , ] MEDIANAME = { media_name | @media_name_variable } ]
     [ [ , ] MEDIAPASSWORD = { mediapassword | @mediapassword_variable } ]
     [ [ , ] NAME = { backup_set_name | @backup_set_name_var } ]
     [ [ , ] { NOREWIND | REWIND } ]
     [ [ , ] { NOUNLOAD | UNLOAD } ]
     [ [ , ] STATS [ = percentage ] ]
     [ [ , ] COPY_ONLY ]
```

I can now give a brief description of all the options that are available. We looked at some of these options previously with the Back Up Database dialog box. This will allow you to compare options within T-SQL and within the backup dialog boxes.

database_name | @database_name_var: Either the name of a database or a local variable that gives the name of the database to back up.

file_or_filegroup: The name of the file or filegroup to back up.

backup_device: The name of the logical or physical backup device to use.

MIRROR TO: The backup file is mirrored to two to four different file locations.

BLOCKSIZE: The block size to use, for example, if backing up to CD-ROM, you would set a block size of 2048.

CHECKSUM | NO_CHECKSUM: Specifies whether to perform checksum calculations to ensure the transmission of data or not.

STOP_ON_ERROR | CONTINUE_AFTER_ERROR: Specifies whether to stop on a checksum error or not.

DESCRIPTION: A description of the backup.

DIFFERENTIAL: If this is a differential backup, specify this option. Without this option, a full backup is taken.

EXPIREDATE: The date the backup expires and is therefore available to be overwritten.

RETAINDAYS: The number of days the backup will be kept before the system will allow it to be overwritten.

PASSWORD: The password associated with the backup. This must be supplied when interrogating the backup for any restore operation. There is no strong encryption on this option, so there is the potential that it could be easily broken.

FORMAT | NOFORMAT: Specifies whether to format the storage medium or not.

INIT | NOINIT: Keeps the media header created with the format but erases the contents.

NOSKIP | SKIP: If you want to skip the checking of expiredate or retaindays when using the media set, select the SKIP option. Otherwise, expiredate and retaindays will be checked.

Note A **media set** is an ordered set of backups on the same disk or tape.

MEDIADESCRIPTION: Gives a description to the media set.

MEDIANAME: Names the media set.

MEDIAPASSWORD: Gives the media set its password.

NAME: Names the backup set.

NOREWIND | REWIND: Specifies whether to rewind a tape or not.

NOUNLOAD | UNLOAD: Specifies whether the tape is unloaded or kept on the tape drive.

STATS [= percentage]: SSE will provide a message at this percentage interval telling you how much of the approximate backup has completed. Useful for gauging progress of long-running backups.

COPY_ONLY: Tells SSE that this is a copy of the data. It cannot be used as a full backup point for differential backups, as the differential backups will be in line with the last "pure" full backup. This option is ideal if you take weekly backups for dumping the data to a user test region, as it will not affect the production backup process.

The only remaining option is for files or filegroups where you can name the file or filegroup that the backup is for. The preceding options do not change for files or filegroups.

Try It Out: Backing Up the Database Using T-SQL for a Full Backup and a Differential Backup

1. Open up a fresh Query Editor window. It doesn't matter which database it is pointing to as the BACKUP DATABASE statement defines the database we will be working with.

2. The T-SQL that we need for our full backup follows. Enter the code (keeping the name of where the backup is located via the TO DISK option and the WITH NAME option all on one line). Notice that there are no options defined for several of the options as we are taking the default.

```
BACKUP DATABASE ApressFinancial
TO DISK = 'C:\Program Files\Microsoft SQL Server\MSSQL.1\MSSQL\Backup\
ApressFinancial.bak'
WITH NAME = 'ApressFinancial-Full Database Backup',
SKIP,
NOUNLOAD,
STATS = 10
```

3. Execute the code and you will see results similar to those that follow. The main points to notice are the stats messages that come out in approximations of 10 percentage points. It then lists the number of data pages backed up and the number of log pages backed up. The on file part of the message details which file within the media set the backup now is. In this case, this is the third backup. You will possibly see on file 2 unless, like I did, a second backup has been taken in the meantime. The final message is the one of greatest interest, as it shows that the backup was successful and the amount of time taken.

```
12 percent processed.
21 percent processed.
30 percent processed.
43 percent processed.
51 percent processed.
60 percent processed.
73 percent processed.
82 percent processed.
90 percent processed.
Processed 184 pages for database 'ApressFinancial',
file 'ApressFinancial' on file 2.
100 percent processed.
Processed 1 pages for database 'ApressFinancial',
file 'ApressFinancial_log' on file 2.
BACKUP DATABASE successfully processed 185 pages
in 1.113 seconds (1.361 MB/sec).
```

4. Although useful to see, not many of the options were used. However, Figures 7-5 and 7-6 show the next backup of the database to be taken, which is a differential backup. We will not allow this backup to expire until 60 days have elapsed, as shown in Figure 7-5. We will also be adding this differential backup to the full backup.

Figure 7-5. *Backing up a database (differential)*

5. Figure 7-6 shows that we are appending to the same media set as the full backup and that we have included some reliability checking. Make sure your version matches the figure.

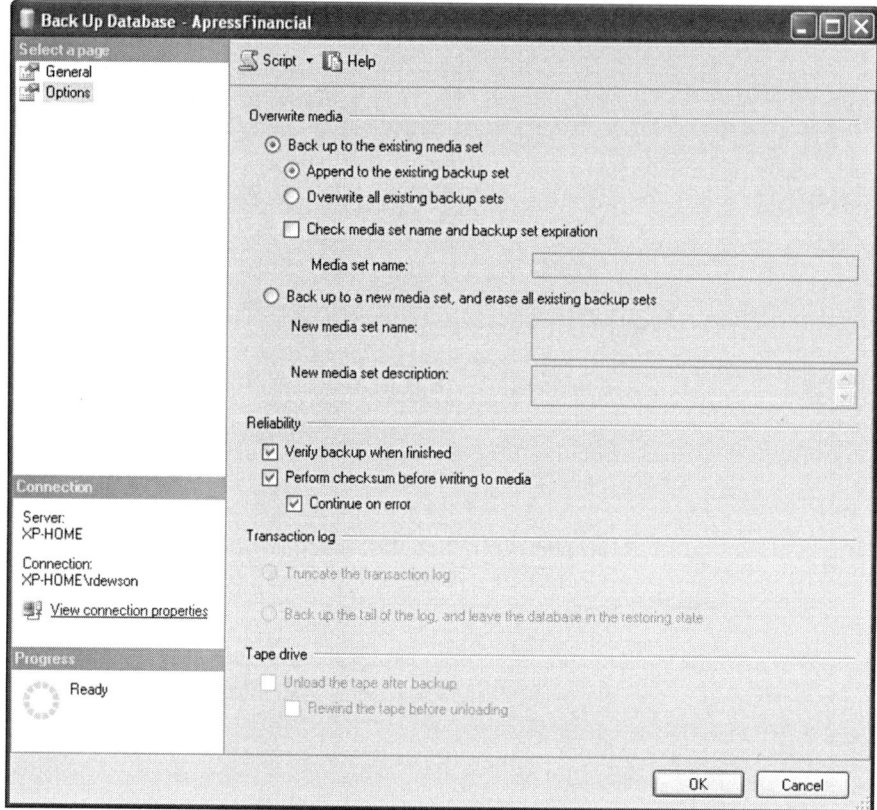

Figure 7-6. *Backing up a database (differential) options*

6. The code that would be the equivalent of these two figures has been split in two. The first part is the differential backup. Again, ensure that the TO DISK, DESCRIPTION, and NAME options are all on the same line of the Query Editor pane.

```
BACKUP DATABASE ApressFinancial
TO DISK = 'C:\Program Files\Microsoft SQL Server\MSSQL.1\MSSQL\Backup\
ApressFinancial.bak'
WITH  DIFFERENTIAL ,
DESCRIPTION = 'This is a differential backup',
RETAINDAYS = 60,
NAME = 'ApressFinancial-Differential Database Backup',
STATS = 10,
CHECKSUM,
CONTINUE_AFTER_ERROR
GO
```

7. The second part is where the reliability checking takes place. This is more complex T-SQL than we have covered, so for the moment just trust that it works and that it does what it is supposed to. You will encounter this code once more when looking at more complex T-SQL later in the book in Chapter 11. However, the basis of the code is that a check is made in the msdb database to retrieve the last backup set that was taken, and we do a "restore" of the database as verification only, without actually restoring any data, and that the restore can complete successfully. If it can't verify the backup set or the restore is okay, you will get an error message.

```
DECLARE @BackupSet AS INT
SELECT @BackupSet = position
  FROM msdb..backupset
 WHERE database_name=N'ApressFinancial'
   AND backup_set_id=
                        (SELECT MAX(backup_set_id)
                         FROM msdb..backupset
                         WHERE database_name=N'ApressFinancial' )
IF @BackupSet IS NULL
BEGIN
    RAISERROR(N'Verify failed. Backup information for database
''ApressFinancial'' not found.', 16, 1)
END
RESTORE VERIFYONLY
FROM DISK = N'C:\Program Files\Microsoft SQL Server\MSSQL.1\MSSQL\Backup\
ApressFinancial.bak'
WITH FILE = @BackupSet,
NOUNLOAD,
NOREWIND
```

8. When the code is executed, you will see something like the results that follow. Again, they contain details of the amount of data backed up as well as which file number on the media set the backup is.

```
19 percent processed.
39 percent processed.
58 percent processed.
78 percent processed.
97 percent processed.
Processed 40 pages for database 'ApressFinancial',
file 'ApressFinancial' on file 4.
100 percent processed.
Processed 1 pages for database 'ApressFinancial',
file 'ApressFinancial_log' on file 4.
BACKUP DATABASE WITH DIFFERENTIAL successfully processed 41 pages
in 0.433 seconds (0.774 MB/sec).
The backup set on file 3 is valid.
```

Transaction Log Backup Using T-SQL

You can back up not only the data, but also, and just as importantly, the transaction log for the database. Just to recap, the transaction log is a file used by databases to log every transaction, including DML actions such as rebuilding indexes. In other words, every data modification that has taken place on any table within the database will be recorded within the transaction log. The transaction log is then used in many different scenarios within a database solution, but where it is most useful, from a database recovery point of view, is when a database crashes. In this case, the transaction log can be used to move forward from the last data backup, using the transactions listed within the transaction log.

If a database crash occurs, the full and differential backups will only take you to the last valid backup. For data entered since that point, the only way to restore the information is to then "replay" the transactions that were committed and recorded as committed in the transaction log. Any

actions that were in progress at the time of the failure that were within a transaction that was still in progress would have to be rerun from the start.

So, to clarify, if you were in the process of deleting data within a table and the power was switched off, you would use your full and differential backups to restore the data. You would then use the information within the transaction log to replay all successful transactions, but because the delete had not been successful, the table would have all the data still within it.

Backing up the transaction log is a good strategy to employ when a large number of updates occur to the data through the day. A transaction log backup should take place at set times throughout the day depending on how large the transaction log has grown and how crucial it was to get your system back up and running after any unexpected outage. When a transaction log is backed up, the transaction log itself is logically shrunk in size so that the transaction log is kept small. It also gives you point-in-time recoverability; this means that you can quickly restore to any time in the past where the transaction was backed up.

Backing up a transaction log is similar to backing up a database. The full syntax is as follows and really only differs from a database backup by using the LOG keyword instead of DATABASE and the options NO_TRUNCATE and NORECOVERY | STANDBY:

```
BACKUP LOG { database_name | @database_name_var }
{
      TO <backup_device> [ ,...n ]
[ [ MIRROR TO <backup_device> [ ,...n ] ] [ ...next-mirror ] ]
      [ WITH
      [ BLOCKSIZE = { blocksize | @blocksize_variable } ]
      [ [ , ] { CHECKSUM | NO_CHECKSUM } ]
      [ [ , ] { STOP_ON_ERROR | CONTINUE_AFTER_ERROR } ]
      [ [ , ] DESCRIPTION = { 'text' | @text_variable } ]
      [ [ ,] EXPIREDATE = { date | @date_var }
      | RETAINDAYS = { days | @days_var } ]
      [ [ , ] PASSWORD = { password | @password_variable } ]
      [ [ , ] { FORMAT | NOFORMAT } ]
      [ [ , ] { INIT | NOINIT } ]
      [ [ , ] { NOSKIP | SKIP } ]
      [ [ , ] MEDIADESCRIPTION = { 'text' | @text_variable } ]
      [ [ , ] MEDIANAME = { media_name | @media_name_variable } ]
      [ [ , ] MEDIAPASSWORD = { mediapassword | @mediapassword_variable } ]
      [ [ , ] NAME = { backup_set_name | @backup_set_name_var } ]
      [ [ , ] NO_TRUNCATE ]
      [ [ , ] { NORECOVERY | STANDBY = undo_file_name } ]
      [ [ , ] { NOREWIND | REWIND } ]
      [ [ , ] { NOUNLOAD | UNLOAD } ]
      [ [ , ] STATS [ = percentage ] ]
      [ [ , ] COPY_ONLY ]
      ]
}
```

To detail the options not covered when looking at backing up the database earlier, let's look at them now:

LOG: Determines that we wish to produce a backup of the transaction log rather than a backup of a database or files/filegroups.

NO_TRUNCATE: Will not truncate the log after the backup. If the database is corrupt, using this option will allow the backup to be attempted at least. Without this option, you will get an error message.

NORECOVERY | STANDBY: Means that after the backup, the database will be in a state whereby it looks to anyone trying to connect as if it were still being restored.

Note The LOG options NO_TRUNCATE and NORECOVERY | STANDBY are used when the database is corrupt and you wish to back up the transaction log prior to performing a restore.

Try It Out: Backing Up the Transaction Log Using T-SQL

1. In a Query Editor pane, enter the following T-SQL code. This should back up the transaction log to the same media set as the full and differential backups. While developing and learning SSE, this is a valid scenario, and in some production setups you may want to back up to the same place as your daily full backup. However, the downside is that if you take several transaction log backups between each differential backup and full backup, SSE will have to "skip" these if they were not required as part of the restore operation. On a tape drive, this could cause significant overhead. In this scenario, you would be better to save the transaction log files to a different media set. Before you run this code, ask yourself, why this will fail.

```
BACKUP LOG ApressFinancial
TO DISK = 'C:\Program Files\Microsoft SQL Server\MSSQL.1\MSSQL\Backup\
ApressFinancial.bak'
WITH NAME = 'ApressFinancial-Transaction Log Backup',
SKIP,
NOUNLOAD,
STATS = 10
```

2. The reason the code will fail is actually twofold. The first is that you have ApressFinancial in Simple recovery mode and therefore there is no Transaction Log backup option available. Run the preceding code; you will get the following error message:

```
Msg 4208, Level 16, State 1, Line 1
The statement BACKUP LOG is not allowed while the recovery model is SIMPLE.
Use BACKUP DATABASE or change the recovery model using ALTER DATABASE.
Msg 3013, Level 16, State 1, Line 1
BACKUP LOG is terminating abnormally.
```

The second reason is that the transaction log backup must follow a full or differential backup and cannot form the first backup. The reason this is the first backup is that we have switched recovery modes, and therefore the other database backups, although they could be used as to restore the system, are discounted.

```
Msg 4214, Level 16, State 1, Line 1
BACKUP LOG cannot be performed because there is no current database backup.
Msg 3013, Level 16, State 1, Line 1
BACKUP LOG is terminating abnormally.
```

3. The following code will alter the database to the correct recovery model, which we will keep for our example from this point onwards, take a full backup, and then do a transaction log backup.

```
ALTER DATABASE ApressFinancial
SET RECOVERY FULL
GO
BACKUP DATABASE ApressFinancial
TO DISK = 'C:\Program Files\Microsoft SQL Server\MSSQL.1\MSSQL\Backup\
ApressFinancial.bak'
WITH NAME = 'ApressFinancial-Full Database Backup' ,
```

```
SKIP,
NOUNLOAD,
STATS = 10
GO
BACKUP LOG ApressFinancial
TO DISK = 'C:\Program Files\Microsoft SQL Server\MSSQL.1\MSSQL\Backup\
ApressFinancial.bak'
WITH NAME = 'ApressFinancial-Transaction Log Backup' ,
SKIP,
NOUNLOAD,
STATS = 10
```

4. This code replicates the Truncate the Transaction Log option, as shown in Figure 7-7. Execute the code.

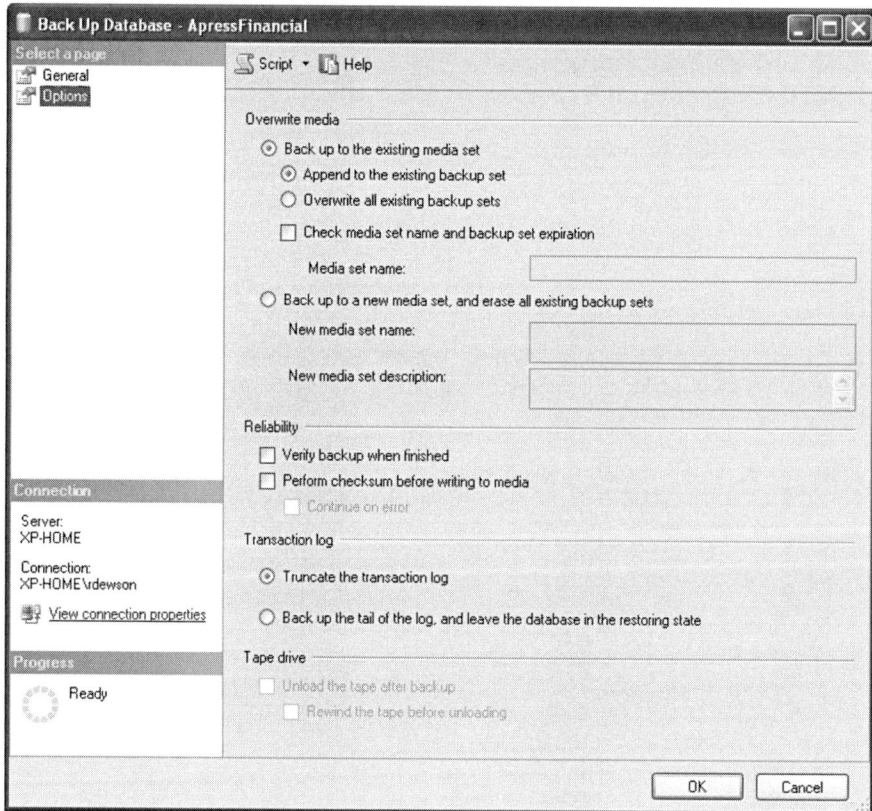

Figure 7-7. *Backing up a transaction log*

5. After execution, you should see output similar to the following where the transaction log has been success-fully backed up and placed in a file. The numbers will differ depending on any other backup actions you have performed. Notice that the first backup, the full backup, is in file 8, and the transaction log is in file 9. Remember that these may differ from your numbers. Keep these file numbers in mind for when we restore our database in the next section.

```
13 percent processed.
22 percent processed.
31 percent processed.
40 percent processed.
54 percent processed.
63 percent processed.
72 percent processed.
81 percent processed.
90 percent processed.
Processed 176 pages for database 'ApressFinancial', file 'ApressFinancial'
on file 8.
100 percent processed.

Processed 1 pages for database 'ApressFinancial', file 'ApressFinancial_log' on
file 8.

BACKUP DATABASE successfully processed 177 pages in 0.722 seconds (2.008
MB/sec).
100 percent processed.
Processed 2 pages for database 'ApressFinancial', file 'ApressFinancial_log'
on file 9.
BACKUP LOG successfully processed 2 pages in 0.203 seconds (0.055 MB/sec).
```

Restoring a Database

Now that the data has been backed up, what if you needed to complete a restore? As has been mentioned, there are two scenarios where a restore could be required: restoring from a backup or restoring when there is a media failure. The second type of restore is not one you wish to perform, but could be set up by creating a long-running transaction, and then simply switching your computer off—not one of life's greatest ideas to do deliberately! This book therefore will not be demonstrating this option, and it is not really for a beginner to attempt. However, I will discuss the concept within this section of the chapter. The first option, a simple restore, is easy to replicate and perform, and this will be the option we will be looking at.

You can choose between two means to restore the database: SSMSE and T-SQL. This is a scenario that you hope you will never perform in a production environment, but it will happen. If you just need a restore within the development environment to remove test data and get back to a stable predefined set of data to complete the testing, this next section should help you. It might also be that you do a weekly refresh of your user test region. Before completing the restore, let's first modify the ApressFinancial database to prove that the restore works, as there is no data within the database yet to prove the restore has worked by that method. Keep in mind, however, that a restore will restore not only the data structures, but also the data, the permissions, and other areas of the database not yet covered in the book, for example, views, stored procedures, and so on.

Restoring Using SQL Server Management Studio Express

The restore demonstrated in the following example will be a complete database restore of our ApressFinancial database. In other words, it will restore all the full and differential backups taken.

Try It Out: Restoring a Database

1. Add a new column to the `ShareDetails.Shares` table using the following code in a Query Editor pane:

```
USE ApressFinancial
GO
ALTER TABLE ShareDetails.Shares
ADD DummyColumn varchar(30)
```

2. We can now restore the database, which will remove the changes we have just completed, using the backup we finished earlier. From the Object Explorer window, select the `ApressFinancial` database, right-click, and select Tasks ➤ Restore ➤ Database. This will bring up the dialog box shown in Figure 7-8.

Figure 7-8. *Restoring a database—General tab*

It is possible to change the database you wish to restore by changing the name in the To Database combo box or by simply overwriting the name that is there. The second option, To a Point in Time, is used if you are restoring the database as well as rolling forward changes recorded in the transaction log. This situation is similar to the scenario mentioned earlier about a power failure or hard drive failure. As we are not doing this here, leave this option as it is.

When taking a backup, details are stored in msdb, but it is possible to restore a database from a backup that is not in msdb. For example, if you are rebuilding a server due to corruption, and msdb was one of the databases corrupted, it is necessary to have the option of finding a backup file and restoring from that instead. Or perhaps the last full backup taken is not the backup you wish to restore. This might occur in a development scenario where you wish to restore to a backup before major changes were done that you wish to remove. There would be no transaction log involved or required to be involved, therefore restoring to a point in time would not be a valid scenario. This is where the From Device option could be used. By selecting this option and clicking the ellipse to the right, you can navigate to any old backup files. Finally, you can click which of the items in the backup you wish to restore. The default is all files to be selected, as you can see in Figure 7-8. The settings shown will give us a backup that is as fresh as possible (the Most Recent Possible value for the To a Point in Time setting).

3. Moving to the Options page, shown in Figure 7-9, there are a number of points to consider:

- *Overwrite the Existing Database*: This is the most likely option to be enabled for a normal restore. You would disable it if you wished to create a restore on the same server but where the restore would alter the name of the database.

- *Preserve the Replication Settings*: A more advanced option for when a database is sending changes to another database. For the time being, you will be leaving this option disabled.

- *Prompt Before Restoring Each Backup*: If you wish a prompt message to appear before each restore file is activated, select this. Ideal if you need to swap media over.

- *Restrict Access to the Restored Database*: After a restore, you may wish to check out the database to ensure the restore is what you wish, or in a production environment to run further checks on the database integrity.

- *Restore the Database Files As*: If you wish to move or rename the MDF and LDF files, this grid will allow you to do this.

- *Leave the Database Ready to Use*: This option will define whether users can immediately connect and work with the data after the restore. If there is a transaction in progress, such as deleting rows within a table, connection could occur once the deletion had been rolled back and the table is back in its "original" state.

- *Leave the Database Non-operational*: With this option you are indicating that the database has been partially restored and you are unsure whether you need to perform additional actions. If there is a transaction in progress, like deleting a table, whatever had been deleted will still be deleted and will not be rolled back.

- *Leave the Database in Read-Only Mode*: A combination of the first two options. If there is a transaction in progress, such as deleting rows in a table, connection could occur once the deletion has been rolled back. However, the changes are also kept in a separate file so that any of these actions that have been rolled back can be reapplied. This might happen if there are several actions within a transaction and some can be reapplied.

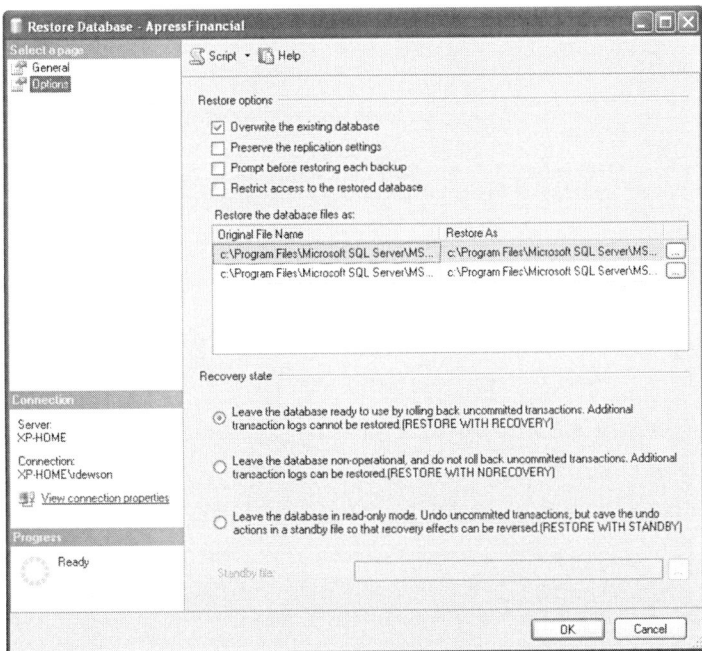

Figure 7-9. *Restoring a database—Options tab*

4. Once you have the option settings you require, ensure that there are no connections to ApressFinancial, which you might find in the query window that was used to alter the ShareDetails.Shares table. Then click OK to perform the restore, and you should see the message in Figure 7-10. If you then move back to the database after clicking OK, you will see that the column we just added has been "removed."

Figure 7-10. *Our restore is successful.*

Restoring Using T-SQL

Using the wizard is a pretty fast way to restore a database, and when under pressure it may even be the best way forward. However, it is not the most flexible way of performing a restore, as some options that are available via T-SQL are not in this wizard. Some of these options were covered when we performed a backup, such as performing checksums when transferring data from the media device back to the database or unloading media at the end of the restore. If there is also a password on the backup medium, this option is not available within the wizard, but with T-SQL you can use passwords. So being comfortable building a restore via T-SQL is important in becoming a more proficient and professional developer or administrator.

The syntax for restoring a database is similar to that for database backups. After looking at the syntax, we will then go through the options you are not familiar with.

```
RESTORE DATABASE { database_name | @database_name_var }
[ FROM <backup_device> [ ,...n ] ]
[ WITH
    [ { CHECKSUM | NO_CHECKSUM } ]
    [ [ , ] { CONTINUE_AFTER_ERROR | STOP_ON_ERROR } ]
    [ [ , ] FILE = { file_number | @file_number } ]
    [ [ , ] KEEP_REPLICATION ]
    [ [ , ] MEDIANAME = { media_name | @media_name_variable } ]
    [ [ , ] MEDIAPASSWORD = { mediapassword |
                    @mediapassword_variable } ]
    [ [ , ] MOVE 'logical_file_name' TO 'operating_system_file_name' ]
            [ ,...n ]
    [ [ , ] PASSWORD = { password | @password_variable } ]
    [ [ , ] { RECOVERY | NORECOVERY | STANDBY =
        {standby_file_name | @standby_file_name_var }
} ]
    [ [ , ] REPLACE ]
    [ [ , ] RESTART ]
    [ [ , ] RESTRICTED_USER ]
    [ [ , ] { REWIND | NOREWIND } ]
    [ [ , ] STATS [ = percentage ] ]
    [ [ , ] { STOPAT = { date_time | @date_time_var }
    | STOPATMARK = { 'mark_name' | 'lsn:lsn_number' }
            [ AFTER datetime ]
```

```
    | STOPBEFOREMARK = { 'mark_name' | 'lsn:lsn_number' }
           [ AFTER datetime ]
  } ]
  [ [ , ] { UNLOAD | NOUNLOAD } ]
]
```

The options not yet covered are as follows:

KEEP_REPLICATION: When working with replication, consider using this option.

MOVE: When completing a restore, the MDF and LDF files that are being restored have to be placed where they were backed up from. However, by using this option, you can change that location.

RECOVERY | NORECOVERY | STANDBY: These three options are the same, and in the same order, as their counterparts (in parentheses) in the wizard:

- RECOVERY (*Leave the Database Ready to Use*): This option will define that after the restore is finished, users can immediately connect and work with the data. If there is a transaction in progress, such as updating rows in a table, not until the updates have been rolled back and therefore the table is back in its "original" state are connections to the database allowed.

- NORECOVERY (*Leave the Database Nonoperational*): With this option, you are indicating that the database has been partially restored, and you are unsure whether you need to perform additional actions. If there is a transaction in progress, like inserting rows in a table, the insertions are *not* rolled back. This allows additional restores to get to a specific point in time.

- STANDBY (*Leave the Database in Read-Only Mode*): A combination of the first two options. If there is a transaction in progress, like deleting rows in a table, the deletion is rolled back. However, the changes are also in a separate file, so that any of these actions that have been rolled back can be reapplied. This might happen if several actions occurred within a transaction and some can be reapplied.

REPLACE: This works the same as the wizard option Overwrite the Existing Database.

RESTART: If a restore is stopped partway through, using this option will restart the restore at the point it was stopped.

RESTRICTED_USER: Use this with the RECOVERY option to only allow users in specific restricted groups to access the database. This would be used to allow further checking by a database owner, or by the dbowner, dbcreator, or sysadmin roles.

STOPAT | STOPATMARK | STOPBEFOREMARK: Used to specify a date and time at which to stop the restore.

The syntax for restoring the transaction log is exactly the same, with the only difference being the definition: you are completing a LOG rather than a DATABASE restore.

```
RESTORE LOG { database_name | @database_name_var }
     <file_or_filegroup_or_pages> [ ,...f ]
[ FROM <backup_device> [ ,...n ] ]
[ WITH
  [ { CHECKSUM | NO_CHECKSUM } ]
  [ [ , ] { CONTINUE_AFTER_ERROR | STOP_ON_ERROR } ]
  [ [ , ] FILE = { file_number | @file_number } ]
  [ [ , ] KEEP_REPLICATION ]
  [ [ , ] MEDIANAME = { media_name | @media_name_variable } ]
  [ [ , ] MEDIAPASSWORD = { mediapassword | @mediapassword_variable }      ]
  [ [ , ] MOVE 'logical_file_name' TO 'operating_system_file_name' ]
           [ ,...n ]
```

```
[ [ , ] PASSWORD = { password | @password_variable } ]
[ [ , ] { RECOVERY | NORECOVERY | STANDBY =
        {standby_file_name | @standby_file_name_var } }
]
[ [ , ] REPLACE ]
[ [ , ] RESTART ]
[ [ , ] RESTRICTED_USER ]
[ [ , ] { REWIND | NOREWIND } ]
[ [ , ] STATS [=percentage ] ]
[ [ , ] { STOPAT = { date_time | @date_time_var }
 | STOPATMARK = { 'mark_name' | 'lsn:lsn_number' }
          [ AFTER datetime ]
 | STOPBEFOREMARK = { 'mark_name' | 'lsn:lsn_number' }
            [ AFTER datetime ]
} ]
[ [ , ] { UNLOAD | NOUNLOAD } ]
]
```

Try It Out: Restoring Using T-SQL

1. Open up an empty Query Editor pane and enter the following code. This will add the column that we want to see "removed" after a restore.

   ```
   USE ApressFinancial
   GO
   ALTER TABLE ShareDetails.Shares
   ADD DummyColumn varchar(30)
   ```

2. Now replace this code with the restore code that follows. Don't execute any of the code just yet, as this piece of code is the first part only. The last full backup we took was after we had moved the recovery model from Simple to Full. Recall that when performing the backups, file 8 was the full backup taken. This is what the first restore will use.

■**Note** Ensure that the FROM DISK option is all on one line. Also recall that FILE = 8 may be FILE = 2 or FILE = 104 depending on the backups taken, and you may see different file numbers as you progress.

```
USE Master
GO
RESTORE DATABASE [ApressFinancial]
FROM  DISK = 'C:\Program Files\Microsoft SQL Server\MSSQL.1\MSSQL\Backup\
ApressFinancial.bak' WITH  FILE = 8,
NORECOVERY,  NOUNLOAD,  REPLACE,  STATS = 10
GO
```

■**Note** We have switched from Simple recovery to Full recovery and took a backup in between. If you restore to a file number prior to the backup taken after the switch, the database will also alter back to Simple recovery mode. Keep this in mind when upgrading from SSE to a full edition that requires a different recovery mode.

You would then include any differential backups that you had done at this point with their file number.

3. The final part of the restore operation is to restore the transaction log file. Once you have entered the following code, you can run all of the code.

```
RESTORE LOG [ApressFinancial]
FROM  DISK = 'C:\Program Files\Microsoft SQL Server\MSSQL.1\MSSQL\Backup\
ApressFinancial.bak' WITH  FILE = 9,
RECOVERY, NOUNLOAD,  STATS = 10
```

4. Once the code has fully executed, the results you should see are similar to those listed here:

```
13 percent processed.
22 percent processed.
31 percent processed.
40 percent processed.
54 percent processed.
63 percent processed.
72 percent processed.
81 percent processed.
90 percent processed.
100 percent processed.
Processed 176 pages for database 'ApressFinancial', file 'ApressFinancial' on
file 8.
Processed 1 pages for database 'ApressFinancial', file 'ApressFinancial_log' on
file 8.
RESTORE DATABASE successfully processed 177 pages in 0.481 seconds (3.014 MB/sec).
100 percent processed.
Processed 0 pages for database 'ApressFinancial', file 'ApressFinancial' on
file 9.
Processed 2 pages for database 'ApressFinancial', file 'ApressFinancial_log' on
file 9.
RESTORE LOG successfully processed 2 pages in 0.029 seconds (0.388 MB/sec).
```

We can now move back to the `ShareDetails.Shares` table and check that the column added has now been removed. You may have to perform a refresh within Object Explorer first to see the changes.

Restoring a database in production will in most instances take place under pressure, as the database will have become corrupt, or been inadvertently damaged. The production system is obviously down and not working, and we have irate users wanting to know how long before the system will be up. This is hopefully the worst-case scenario, but it is that sort of level of pressure that we will be working under when we have to restore the database. Therefore, having the correct backup strategy for your organization based on full, differential, and transaction log backups is crucial. Full database backups for a system that requires high availability so that the restore takes the least amount of time may be what you need.

Detaching and Attaching a Database

Now that we can back up and restore a database, we have available other methods of dealing with the database. There may be a time in the life of our SSE database when we have to move it from one server to another, or in fact just from one hard drive to another. For example, we currently have ApressFinancial on our C drive, and this is getting full, so we would like to move our database to another hard drive. Or perhaps we are moving from an old slower server to a new faster server or a server on a better network. By detaching and reattaching the database, we can do this simply and easily.

I would like to make a couple of points here; they may seem straightforward and really obvious, but better to mention them than cause problems at a later stage. First of all, no updates can be occurring, no jobs can be running, and no users can be attached. Second, just in case, take a full backup before moving the database. This may add time to the process, but it is better to be safe than sorry. Ensure that where you are moving the database to has enough disk space, not only for the move, but also for expected future growth; otherwise, you will be moving your database twice. You should not attach your database to a server without immediately completing a backup on the new server afterwards; this way, you can ensure that the databases are protected in their new state.

Detaching a database physically removes the details from the SSE `master` and `msdb` databases, but does not remove the files from the disk that it resides on. However, detaching the database from SSE will then allow you to safely move, copy, or delete the files that make up the database, if you so desire. This is the only way that a database should be physically removed from a server before moving it to another location.

Detaching and Attaching Using SQL Server Management Studio Express

We'll start by using SSMSE to detach and attach a database.

Try It Out: Detaching a Database

1. First of all, it is necessary to ensure that nobody is logged in to the database, and even if someone is, that the user is not doing any updates. For the moment, I want you to ignore this and to have a connection. Ensure that SSMSE is running and that there is a Query Editor pane with a connection to the `ApressFinancial` database. Find the `ApressFinancial` database in Object Explorer and ensure it is selected. Right-click and select Tasks ➤ Detach.

2. This then brings up the Detach Database dialog box for the `ApressFinancial` database, as shown in Figure 7-11. We haven't removed all the users connected, so you can do this by selecting the Drop Connections check box. The second option, Update Statistics, means that the statistics for indexes, etc., will be updated before the database is detached. The third option, Keep Full Text Catalogs, is when you have set up specialized indexing on text data columns known as full-text indexing. The information is stored separately from the other data files in SSE, so selecting this option will ensure that when the database is detached that the full-text data files are not lost and therefore would need re-creating. The status is Not Ready due to the message indicating that there is still "1 Active connection(s)."

3. Click the message, and the dialog box in Figure 7-12 is displayed. This is a powerful tool within SSE that shows all the processes with connections to your server. This list has already been filtered by SSE because of the message we saw (using the Filter option on the top line) for the `ApressFinancial` database. Only one row is listed, which is the one connection in the Query Editor pane we opened a moment ago.

Figure 7-11. *Detaching a database*

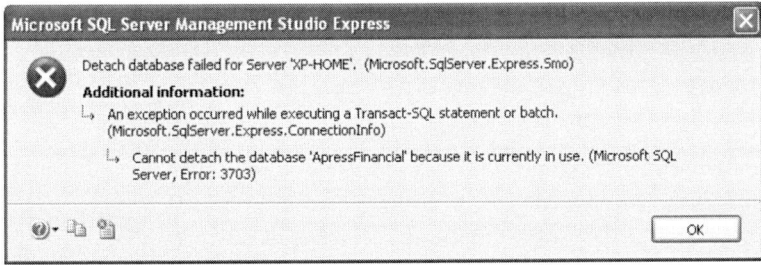

Figure 7-12. *Detaching a database*

4. It is possible to see more information about what the connection is doing by scrolling right and left. To kill the process, right-click the item and select Kill Process. If we were running T-SQL code at the time, it would stop that process immediately. Therefore, if the process is running any data modifications, you have to be sure that this is what you want to do.

Note If you are running updates that are within a transaction, when you issue a kill, SSE will roll back the updates. Therefore, it may still take a long time to remove the connection. Kill is a powerful command with large ramifications in this scenario, so do use it as a last resort.

5. Once the process has been killed or finished (this can be done by changing the database the Query Editor window is pointing to or by enabling the Drop Existing Connections option), you can close the Activity Monitor, which takes you back to the Detach Database dialog box. You can then click OK to finish detaching the database.

That's it. The database is detached, is no longer part of SSE, and is ready to be removed or even deleted from the hard drive. If you check Object Explorer in SSMSE, you will see that the database is no longer listed.

Detaching a database, although seemingly a simple and innocuous operation, has the potential to be fraught with problems and worries. As the example demonstrates, ensuring that there are no users attached to the database at the time of detaching is not as easy as it first may seem. Setting up the database options to eliminate connections or to stop updates is only possible once everyone has been removed from connections to the database. There is no easy way of removing connections safely, as you never know what an application with a connection to the database is doing. You could remove a connection that is in the middle of processing. If you are going down the route of detaching the server, though, there is an obvious reason to do this, such as moving servers or removing the database, and therefore you would have a plan of action to do this. Users would have been notified days or weeks in advance, and the database owner would have coordinated a date and time when nobody should be connected. Also, the database owner would be around if there were any problems, and he or she could make the decision to kill any connections left hanging around.

Detaching the database is a process that removes entries within the SSE system tables to inform SSE that this database is no longer within this instance of SSE and therefore cannot be used. It is as simple as that. If you are removing the database completely, you will need to delete the files from the directory they were created in.

It is possible to detach the database using a system stored procedure, although this does not let you kill the connections. This has to be done via a T-SQL command.

We need to reattach the database before being able to demonstrate this, so let's do that now. This would occur on our new SSE instances after physically moving the files.

Try It Out: Attaching a Database

1. Within Object Explorer, highlight the Databases node, right-click, and select Attach.
2. This brings up the Attach Databases dialog box, shown in Figure 7-13. To add a database, click Add.

Figure 7-13. *Options for attaching a database*

3. This brings up the Locate Database Files Explorer, shown in Figure 7-14. You can use this like other explorers to locate where your database MDF files are. Once you find the database you want to reattach, highlight it and then click OK.

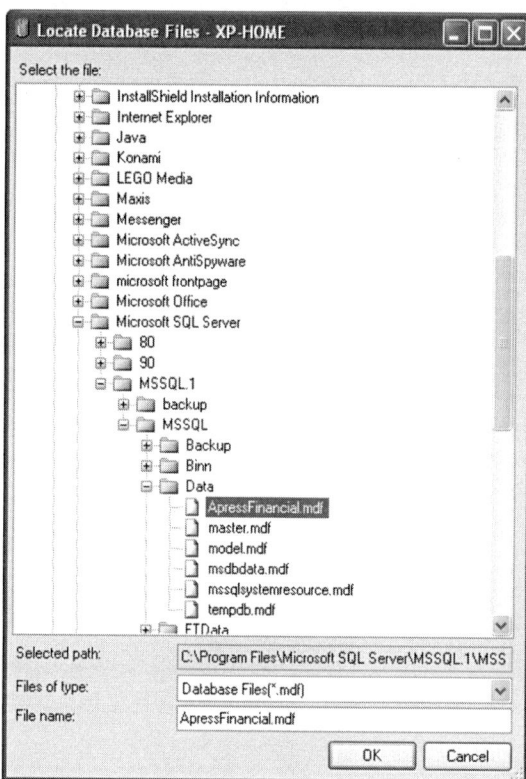

Figure 7-14. *Locating the database to attach*

4. This will bring you back to the Attach Databases dialog box with the details filled in, as you see in Figure 7-15. Take a moment to look over the information in this dialog box. Any problems will be detailed in the Message column. It is possible to attach more than one database, but it is best to do databases one at a time.

5. This then leaves us to click OK to reattach the database. Moving to Object Explorer, you should see your database at the bottom of the list, where it will remain until the explorer is refreshed.

Figure 7-15. *Database located, preparing to attach*

Attaching a database involves informing SSE of the name and the location of the data files and the transaction log files. This data can be placed anywhere on a computer, but it is recommended you place the data in a sensible location. For example, the folders `tempfiles` or `tobedeleted` sport extreme names, but do demonstrate the unsuitability that should be avoided.

When moving the data from one physical server to another, the data does not need to be in a subdirectory of the Microsoft SQL Server installation found under Program Files. In fact, in production environments, this is the last place you would locate the data. You would generally want to keep these files away from any program files or the `pagefile.sys` file, because SSE's performance can be maximized when these files are separated. However, for the purpose of this book, placing the data in the `DATA` directory under the instance of SSE is perfectly valid and acceptable.

Once the two data files have been copied, it is a simple process of a couple of mouse clicks to attach these files to the instance. What happens in the background, very basically, is that SSE takes the name of the database and the location of the data files and places them into internal tables that are used to store information about databases. It then scans the data files to retrieve information, such as the names of the tables, to populate the system tables where necessary.

The main point to keep in mind is the database owner (see Chapters 1, 2, and 5). It is just as important to use a valid database owner and not the `sa` login when attaching a database as it is when creating a database. The database, when it is attached, will be given the owner of the login attaching the database.

Detaching and Attaching Using T-SQL

Detaching and attaching a database is an ideal way to move a database from one server to another as part of an overall solution. It's clean and simple and ideal if you are rolling out a "base" database to many client sites, but it's not the only way of doing it. Detaching a database is simply removing it logically from a server, but keeping the physical files. This allows these files to be moved to anywhere, from another hard drive to a DVD, for further copying to a client computer if need be, and then the database can be reattached at the other end.

Detaching a database removes entries from the master and msdb databases, and therefore the backup history will also be lost. The physical backup files will still be there, so if you do need to complete a restore after a detach and reattach, you can use the From Device option in the Restore Wizard to define the full location in the RESTORE T-SQL command to get to those files.

Note Detaching a database can only be done by a member of the db_owner role.

```
sp_detach_db [ @dbname= ] 'dbname'
    [ , [ @skipchecks= ] 'skipchecks' ]
    [ , [ @KeepFulltextIndexFile= ] 'KeepFulltextIndexFile' ]
```

The options are straightforward, with each being optional. If they are not supplied, the default value, mentioned in the following list, is used:

dbname: The name of the database to detach. If this option is missed, no database will be detached.

skipchecks: NULL (the default) will update statistics. true will skip the updating of statistics.

KeepFulltextIndexFile: true (the default) will keep all the full-text index files that have been generated with this database.

Note **Full-text index files** are special files that hold information about data set up for full-text searching, which is an area outside of the scope of this book. But basically, full-text searching gives the ability to search on all text in a column or multiple columns of data, and is also functionality used by search engines.

You might be expecting that to reattach the database you would use a stored procedure called sp_attach_db. This command does exist, but it will be made obsolete in future versions of SSE. The correct syntax is a "specialized" CREATE DATABASE command:

```
CREATE DATABASE database_name
    ON <filespec> [ ,...n ]
    FOR { ATTACH [ WITH <service_broker_option> ]
        | ATTACH_REBUILD_LOG }
```

The syntax is easy to follow. The first option, ON, will specify the name of the primary database file to attach, which will have the mdf suffix. The second option, <service_broker_option>, we will be ignoring, as this is for a more advanced database.

The third option, ATTACH_REBUILD_LOG, is for situations where you wish to attach a database but at least one transaction log file is missing. Specifying this option will rebuild the transaction log. No database can be attached when SSE believes that there are missing files. If you do use this option, you will lose the full, differential, and transaction log backup chains that exist on SSE, so complete a full backup after attaching to reestablish the backup baseline. This option will tend to be used when

you deliberately wish to lose the transaction log file, such as a read-only version of the database for reporting purposes.

Note If you receive any error messages, reattach *all* files associated with the database, not just the main primary file.

We can now detach and reattach `ApressFinancial`.

Try It Out: Detaching and Reattaching a Database

1. The first test we will do is to try and detach `ApressFinancial` while there are still active connections so that you can see what happens. Open up a Query Editor pane and point it to the `ApressFinancial` database. Then open a second pane and enter the `sp_detach_db` code as follows. Once you have done so, execute the code. Take note that we are explicitly moving this connection to a "safe" system database, away from the database we wish to detach.

```
USE master
GO
sp_detach_db 'ApressFinancial'
```

2. The results you will see will be similar to the following:

```
Msg 3703, Level 16, State 2, Line 1
Cannot detach the database 'ApressFinancial' because it is currently in use
```

3. Close the Query Editor pane opened earlier and any other Query Editor panes that have connections pointing to `ApressFinancial` and then try running the code again. This time you should see the following message:

```
Command(s) completed successfully.
```

4. Now that the database has been detached, we need to reattach it, simulating a move to a new server. Enter in the same Query Editor pane the following code. Replace the `FILENAME` parameters with the path to where your database is located and ensure that the path is all on one line.

```
CREATE DATABASE ApressFinancial
ON (FILENAME='C:\Program Files\Microsoft SQL Server\MSSQL.1\MSSQL\
Data\ApressFinancial.MDF')
FOR ATTACH
```

5. After executing the code, you should see the following message:

```
Command(s) completed successfully.
```

You have now successfully detached and reattached the database.

Producing SQL Script for the Database

The next section demonstrates a different method of backing up the structure of the database and the tables and indexes contained within it by using T-SQL commands to complete this.

Note It is only the structure that will be generated as T-SQL commands; no data will be backed up—only the schema that is needed to re-create the actual database can be scripted here.

The usefulness of this procedure is limited and is really only helpful for keeping structure backups or producing an empty database, but it is useful to know rather than going through the process of copying the database with all the data when the data is not required.

This method tends to be used to keep the structure within a source repository like Visual SourceSafe. It is also very useful for setting up empty databases when moving from development to test, or into production.

Try It Out: Producing the Database SQL

1. Ensure that SSMSE is running and that you have expanded the server so that you can see the ApressFinancial database. Right-click, select Tasks ➤ Generate Scripts. This brings up the Script Wizard, which allows the database to be scripted (the first screen of this wizard after the welcome screen is shown in Figure 7-16). Every attached database will be listed. Select ApressFinancial and click Next.

Figure 7-16. *Scripting—selecting the database*

▓**Note** You can select the check box at the bottom of the screen, which will script all the objects if you wish. This will enable the Finish button so that you can go straight to the end.

2. On the second screen are a number of options about the scripting as well as what objects you wish to script. Take a moment to look it over. Most of these options should be clear to you from the setup options covered in setting up the database so far. (However, at the end of the example is a bulleted list clarifying the options for you.) Figure 7-17 shows the default settings.

Figure 7-17. *Options for the script*

3. In Figure 7-18, a number of options have been altered so that much of the database is scripted, even if there is an error producing the script file. Errors could be as varied as the login not having permissions to see all the objects, for example. Once you have the options you wish to script, then click Next.

Figure 7-18. *Options selected for scripting*

■**Note** The Script Statistics option will significantly increase the time taken to generate the script. Leave this option off in most cases; it is really only useful when moving from a user test environment that is very similar to how the system will work in production.

4. You are now presented with the screen shown in Figure 7-19 that will allow you to specify how the script should be saved. There are three possibilities. Choose to script to a new Query Editor pane and then select Next.

Figure 7-19. *Where to store the script*

5. This will bring you to a summary screen, shown in Figure 7-20, where you can expand what has been selected. You may find that this screen is not of much use, as there are so few screens within this wizard. However, you can use it for categorizing what objects are to be scripted. Take a moment to investigate this screen.

Figure 7-20. *Script summary*

6. Click Finish. The wizard will start to generate the script. At the end you will see a summary of how the script production went. Any errors will be within the Message column on the right, as shown in Figure 7-21.

Figure 7-21. *Generating the script*

The options that were available to you within the wizard are detailed here:

- *Append to File*: If you set this to true, SSE will append the script to the file selected instead of overwriting it.

- *Continue Scripting on Error*: If there are any problems in producing the script, you can decide whether you wish to continue scripting or not.

- *Convert UDDTs to Base Types*: As part of SSE, you can change the base data types, such as int, to your own named type, so you could name a "copy" of int as "myint". This is a bit more advanced, but if you do this, selecting true will convert "myint" back to "int".

- *Generate Script for Dependent Objects*: A very useful option. If there are any dependencies on what you are wanting to script and you haven't selected that object to script, there will be problems rebuilding the object later. Selecting true means that these dependent objects will also be scripted.

- *Include Descriptive Headers*: This will include a date-time stamp as well as a short descriptive header of each object as it is reached within the script.

- *Include If NOT EXISTS*: If you select all the objects to be scripted and set this to true, SSE will put a test around each object so that if that object already is in the database when the script is run, it won't be created. There will be no test for specific columns when scripting a table, but there will be a test for the table itself.

- *Script Behavior*: You can generate a script for creating items or dropping items.

- *Script Collation*: If you wish the SSE collation to be scripted, enable this option. Useful if you are unsure of the collation the script will then be run against.

- *Script Database Create*: This specifies whether you wish a CREATE DATABASE statement to be scripted or not.

- *Script Defaults:* We have some default values that will be set on columns when rows are added. Setting this to true will set these defaults.

- *Script Extended Properties:* Extra properties can be placed on every SSE object. These will be scripted if you select true.

- *Script Logins:* This scripts all Windows and SSE authentication logins.

- *Script Object-Level Permissions:* Each object will have permissions on who can do what. For example, on a table, permissions on who can add, delete, or select the data can be set up. This option will include these choices.

- *Script Owner:* This scripts the owner of the database if specified.

- *Script Statistics:* This specifies whether to script the SSE column and index statistics. It avoids rebuilding them when re-creating the database using the script; however, it will increase the time taken to build the script as well as the size of the script.

- *Script USE DATABASE:* Between each object, this specifies whether to script a USE database statement or not. Ideal if used with scripting-dependent objects.

- *Script Check Constraints:* This will script check constraints.

- *Script Foreign Keys:* Any foreign keys will be scripted.

- *Script Full-Text Indexes:* If you have any full-text indexes, this indicates whether you want to script them or not.

- *Script Indexes:* This specifies whether to script table and view indexes.

- *Script Primary Keys:* This dictates whether to script primary keys or not.

- *Script Triggers:* For any trigger, this specifies whether you wish these to be within the script.

- *Script Unique Keys:* Any unique keys will be scripted.

This concludes our look at the different methods of backing up, restoring, moving, and scripting databases. While this covers every way of ensuring your database structure and data should never be lost, you still need to maintain the database on a regular basis. This is what we will take a look at in the next section of this chapter.

Maintaining Your Database

At this point, we have now created a backup and performed a restore of the example database. We have also covered the different methods to back up and restore the database. However, we have no real plan for regular maintenance and detection of problems in our database strategy. Any jobs for backup of the database or transaction log that we have demonstrated so far are held as single units of work called **steps**. Not only that, there is nothing in place that will look after the data and indexes held within the database to ensure that they are still functioning correctly and that the data is still stored in the optimal fashion. Without a process that runs regularly, we would need to perform all of this by hand, regularly. What a waste of time, and boring to boot!

Unfortunately, with SSE this is what you have to do. When you progress to other editions of SQL Server, though, you will find that you can create **maintenance plans**, which are methods of running a whole set of processes including backups to maintain your database's integrity.

■**Note** Maintenance plans are covered in another book of mine, *Beginning SQL Server 2005 For Developers: From Novice to Professional*, also published by Apress.

Summary

You have seen a great deal in this chapter that is crucial to ensuring that your database is always secure. There are different methods you can use to create a backup of your data and options that allow a database to be moved from one server to another. One of the main points to take away from this chapter is that backing up your data as well as how to complete a restore of the data depends on the model employed. You will also now be aware of the options available in both a backup and a restore and their relevance to areas such as the backup device used, how long a backup will be retained for, and so on. Even when developing your solution, regular backups will ensure that if something goes wrong during the development and your database becomes corrupt, you have a safe point to go back to.

As a manager drummed into me, the unexpected will always happen, but you must always be able to recover from it, no matter what. Therefore, regular backups that are known to work and even the occasional "disaster recovery test" should be done to ensure that you can restore when something unexpected happens. No matter what your managing director says, it is the data of a company that is its most important asset, not the people. Without the data, a company cannot function. If you cannot ensure that the data will be there, the company is in a very dangerous position.

In the next chapter, we will start to put data into our database and then manipulate and retrieve that data, further building up our example.

■ ■ ■

Working with the Data

We have now built our tables, set up the relationships, and backed up our solution, so we are ready to start inserting data. The many tables within the database cover a number of different types of data that can be stored, ranging from characters and numbers to images and XML. This chapter will show you how to insert data into columns defined with all of these data types.

Not all the tables will be populated with data at this point. We will insert data in other tables later on in the book when different functionality of SSE is being demonstrated. Although data is being inserted, the database is still at the stage of being set up, as we are inserting static information at this point in the examples we are building together. To clarify, static data is data that will not change once it has been set up, although there may be further additions to this data at periodic intervals such as when a new share is created.

Not everyone who is allowed to access our database may, or should, be allowed to insert data directly into all of the tables. Therefore, you need to know how to set up the security to grant permission to specific user logins for inserting the data. The only people who really ought to be adding data directly to tables rather than using controlled methods such as stored procedures in production, for example, are special accounts like dbo accounts. In development, any developer should be able to insert data, but any login who would be testing out the application would not have that authority. You will see the reasons for this when we look at the security of adding data later in this chapter, and you will learn about alternative and more secure methods when we look at stored procedures and views.

Once we have set up users correctly, it is time to demonstrate inserting data into SSE. It is possible to insert data using T-SQL through Query Editor or through SSMSE. Although both of these tools will have the same final effect on the database, each works in its own unique way.

When inserting data, you don't necessarily have to populate every column. We take a look at when it is mandatory and when it is not. There are many different ways to avoid inserting data into every column. This chapter will demonstrate the different methods you can use to avoid having to use NULL values and default values. By using these methods, you are reducing the amount of information it is necessary to include with a row insertion. This method of inserting data uses **constraints**. You will see how to define a column constraint through T-SQL in Query Editor as well as in SSMSE.

The T-SQL INSERT Statement Syntax

Before it is possible to insert data using T-SQL code, you need to be familiar with the INSERT statement and its structure.

The INSERT statement is very simple and straightforward in its most minimal form, which is all that is required to insert a row.

```
INSERT [INTO]
    {table_name|view_name}
    [{(column_name,column_name,...)}]
    {VALUES (expression, expression, ...)}
```

Obviously, we are required to start the statement with the type of action we are trying to perform, for example, inserting data. The next part of the statement, INTO, is optional. It serves no purpose, but you will find some do use it to ensure their statement is more readable. The next part of the statement deals with naming the table or the view that the insertion has to place the data into. If the name of the table or view is the same as that of a reserved word or contains spaces, we have to surround that name with square brackets or double quotation marks. However, it is better to use square brackets, because there will be times you wish to set a value such as Acme's Rockets to a column data, which can be added easily by surrounding it with double quotation marks, as covered in the discussion of quoted identifiers earlier in the book.

I cannot stress enough that really, there is nothing to be gained by using reserved words for table, views, or column names. Deciding on easy-to-use and unambiguous object names is part of a good design.

Column names are optional, but it is best practice to list them to help have reliable code, as this ensures that data is only inserted into the columns into which you want it to be inserted. Therefore, it will be necessary to place the column names in a comma-delimited list. The list of column names must be surrounded by parentheses, (). The only time that column names are not required is when the INSERT statement is inserting data into every column that is within the table in the same order as they are laid out in the table. However, this is a potentially dangerous scenario. If you build an INSERT statement, which you then save and use later, you expect the columns to be in a specific order because that is the way they have always been. If someone then comes along and adds a new column, or perhaps alters the order, your query or stored procedure will either not work or give erroneous results, as values will be added to the wrong columns. Therefore, I recommend that you always name every column in anything but a query, which is built, run once, and thrown away.

The VALUES keyword, which precedes the actual values to be entered, is mandatory. SSE needs to know that the following list is a list of values, and not a list of columns. Therefore, you have to use the VALUES keyword, especially if you omit the list of columns as explained previously.

Finally, you will have a comma-separated list surrounded by parentheses covering the values of data to insert. There has to be a column name for every value to be entered. To clarify, if there are ten columns listed for data to be entered, there must be ten values to enter.

Now that the INSERT statement is clear, time to move on and use it.

INSERT SQL Statement

The first method of inserting data is to use the INSERT statement as described previously. This example will insert one row into the ShareDetails.Shares table using Query Editor. When inserting the data, the row will be inserted immediately without any opportunity to roll back changes. This statement does not use any transaction processing to allow any changes to take place. You will also see with this example how Query Editor can aid you as a developer in building the SQL statement for inserting a row. Let's dive straight in and create the row.

Try It Out: Query Editor Scripting

1. Ensure that you have a Query Editor window open, connected to our ApressFinancial database, and that you are logged in with an account that has insert permissions on the ShareDetails.Shares table (this will be any member of the administrator's or database owner's role).

2. Right-click against the ShareDetails.Shares table, select Script Table As ➤ INSERT To ➤ New Query Editor Window.

3. This will bring up the following code. SSE covers itself concerning the use of reserved words, spaces in names, etc., by surrounding every object name with square brackets. It also fully qualifies the table name with the database name and schema owner, in this case, ShareDetails. Moving to the values, you can see the column name repeated so that when altering the values, if the table has a large number of columns, you know which column you are working with. The final part in the jigsaw is an indication of the data type and length to aid you as well.

```
INSERT INTO [ApressFinancial].[ShareDetails].[Shares]
        ([ShareDesc]
        ,[ShareTickerId]
        ,[CurrentPrice])
     VALUES
        (<ShareDesc, nvarchar(50),>
        ,<ShareTickerId, nvarchar(50),>
        ,<CurrentPrice, numeric,>)
```

4. We need to place a modification at the top of this code, just to ensure that Query Editor has a setting to allow double quotes to be used to surround strings. This was covered in Chapter 5 when discussing database options. To cover yourself though, you can always place the following code at the start of queries where quotation marks will be used. There is one hidden downfall that will be covered at the end.

```
SET QUOTED_IDENTIFIER OFF
```

5. By altering the code within the Query Editor pane, you will see that the next section of code actually inserts the data into the ShareDetails.Shares table. Notice that no GO statement is included at the end of this code. It is not necessary because there is only one INSERT and no other statements that need to form part of this same transaction.

```
SET QUOTED_IDENTIFIER OFF
GO
INSERT INTO [ApressFinancial].[ShareDetails].[Shares]
        ([ShareDesc]
        ,[ShareTickerId]
        ,[CurrentPrice])
     VALUES
        ("ACME'S HOMEBAKE COOKIES INC",
         'AHCI',
         2.34125)
```

6. Now that all the information has been entered into the Query Editor pane, it is time to execute the code. Press F5 or Ctrl+E, or click the Execute button on the toolbar. You should then see the following result, which indicates that there has been one row of data inserted into the table.

```
(1 row(s) affected)
```

This now sees the first row of information placed into the database in the ShareDetails.Shares table. It is simple and straightforward. All the columns have been listed and a value has been inserted. Because the name had a single quotation mark within it, it is simpler to surround the name with double quotation marks. However, to make sure that this string was not seen as an identifier, we have to switch that option off.

SSMSE has the ability to create template scripts for several T-SQL statements. Templates, which you saw earlier in the book, hold parameter placeholders that require modification to build up the whole statement. Template scripts differ from actual templates, as the information created within Query Editor for these templates is for one statement only. Therefore, what you are actually seeing is the template for a one-line script.

When using the scripting options within Query Editor, it is possible to build the script as you have just seen for inserting a row into the `ShareDetails.Shares` table, and save the T-SQL within a new Query Editor pane, to a file, or even to a clipboard. This would then allow the data to be reinserted instantaneously should the table be deleted. To an extent, scripting to files or a clipboard is not as useful as scripting to a Query Editor pane. By scripting to files or a clipboard, you would need to move back into these files to make the necessary changes for data insertion. As you saw, when the script is placed in the Query Editor pane, the table and the columns are listed, but obviously the values need to be altered. This would have to be completed in a file or a clipboard by reopening these contents and making modifications after the event.

The scripting template does build the whole `INSERT` statement and lists all the columns as well as—in the `VALUES` section of the statement—the name of the column and its data type definition. From there, it is easier to know what value is expected within the `INSERT` statement line.

The example mentions that using `SET QUOTED_IDENTIFIER OFF` does have one hidden downfall: in many cases, when using T-SQL statements, it is possible to surround reserved words with double quotation marks, rather than square brackets; however, with the `QUOTED_IDENTIFIER` set to `OFF`, you will only be able to surround reserved words with square brackets. If you had `QUOTED_IDENTIFIER` set to `ON`, you could not have put `ACME'S` in the name; the code would have to have been written with two single quotation marks. Therefore, the code would have had to look like the following:

```
INSERT INTO [ApressFinancial].[ShareDetails].[Shares]
          ([ShareDesc]
          ,[ShareTickerId]
          ,[CurrentPrice])
    VALUES
          ('ACME''S HOMEBAKE COOKIES INC',
           'AHCI',
           2.34125)
```

Now that you know how to construct an `INSERT` statement, it is time to look at how you need not define all the columns within a table.

It is not always necessary to define columns with a value when inserting rows of data. This next section looks at two of these methods: the use of default values and allowing a `NULL` value. As you have just seen in our first examples, we specified every column in the table within the `INSERT` statement. You are now probably wondering whether you have to specify every column every time a row is inserted into a table. The answer is no. However, there are a few areas to be aware of.

Default Values

The first method for avoiding having to enter a value is to set a column or a set of columns with a default value. We set up the `CustomerDetails.Customers` table to have a default value when creating the tables in Chapter 5. Default values are used when a large number of `INSERT`s for a column would have the same value entered each time. Why have the overhead of passing this information, which would be the column name plus the value, through to SSE, when SSE can perform the task quickly and simply for you? Network traffic would be reduced and accuracy ensured as the column information would be completed directly by SSE.

Although it has been indicated that default values are best for a large number of `INSERT`s, it can also be argued that this need not be the case. Some people feel that all that is required is a significant number of rows to be affected from a default value setting for the use of default values to be an advantage. It does come down to personal preference as to when you think setting a default value will be of benefit. However, if there are times when you wish a column to have an initial value when a row is inserted with a specific value, then it is best to use a default value.

In the example in the next section, where we build up our next set of `INSERT` statements, I will demonstrate how a default value will populate specific columns. When creating the `CustomerDetails.Customers` table, we created a column that is set up to be populated with a

default value: the DateAdded column. In this column, we call an SSE reserved function, GETDATE().
This function gets the date and time from the operating system and returns it to SSE. By having this
within a column default value, it is then inserted into a row when a row is added.

Using NULL Values

The next method for avoiding having to fill in data for every column is to allow NULL values in the
columns. We did this for some columns when defining the tables. Ensuring that each column's Allow
Nulls option is checked can ensure this is true for all our columns. If you take a look at Figure 8-1,
one of the columns in the ShareDetails.Shares table, ShareTickerId, does allow a NULL value to be
entered into it.

Table - ShareDetails.Shares		⬩
Column Name	Data Type	Allow Nulls
🔑 ShareId	bigint	☐
ShareDesc	nvarchar(50)	☐
ShareTickerId	nvarchar(50)	☑
CurrentPrice	numeric(18, 5)	☐
		☐

Figure 8-1. *NULLs selected on a column*

Therefore, the previous example could have placed data only in the ShareDesc and CurrentPrice
fields if we'd wanted, as ShareId is an IDENTITY column and is auto-filled. If the ShareDetails.Shares
row had only been inserted with those two columns, the statement would have looked like the fol-
lowing T-SQL:

```
INSERT INTO [ApressFinancial].[ShareDetails].[Shares]
         ([ShareDesc]
         ,[CurrentPrice])
     VALUES
         ("ACME'S HOMEBAKE COOKIES INC",
          2.34125)
```

Figure 8-2 shows what the data would have looked like had we used the preceding T-SQL
instead of the code in the previous section.

Table - ShareDetails.Shares	[ReadOnly]xp-h...ils.Shares.sql* 🔒		
ShareId	ShareDesc	ShareTickerId	CurrentPrice
1	ACME'S HOMEB...	*NULL*	2.34125

Figure 8-2. *Insert with NULL*

To see the same result as in Figure 8-2, you would view this table in SSMSE. This is covered
shortly, as unfortunately we are in the chicken-and-egg scenario of showing an area before it has
been discussed. As you can see, the columns that had no data entered have a setting of NULL. A NULL
setting is a special setting for a column. The value of NULL requires special handling within SSE or
applications that will be viewing this data. What this value actually means is that the information
within the column is unknown; it is not a numeric or an alphanumeric value. Therefore, because
you don't know if it is numeric or alphanumeric, you cannot compare the value of a column that has
a setting of NULL to the value of any other column, and this includes another NULL column.

▪**Note** One major rule involving NULL values: a primary key cannot contain any NULL values.

Try It Out: NULL Values and SSMSE Compared to T-SQL

1. Ensure that SSMSE is running and that you are logged in with an account that allows the insertion of rows. Any of our users can do this.

2. Expand the ApressFinancial node in Object Explorer so you can see the CustomerDetails. Customers table. Right-click this table and select Open Table.

3. In the main pane on the right, you should now see a grid similar to Figure 8-3. This grid would usually show all the rows of data that are within the table, but as this table contains no data, the grid is empty and ready for the first row to be entered. Notice that on the far left-hand side appears a star. It will change to an arrow shortly. This is the row marker and denotes which row the grid is actually pointing to and working with for insertion. The arrow denotes which row you are viewing, and when the marker changes to a pencil, it denotes that you are writing data in that row, ready for updating the table. Perhaps not so relevant in this instance, but very useful when several rows are displayed.

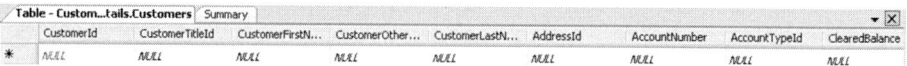

Figure 8-3. *No data held within the table*

4. It is a simple process to enter the information into the necessary columns as required. However, if you don't enter the data into the correct columns, or leave a column empty when in fact it should have data, you will receive an error message. The first column, CustomerId, is protected, as this is an IDENTITY column, but if you enter Mr into the CustomerTitleId column, you will see something similar to the message shown in Figure 8-4 when moving to another cell. This message is informing you that CustomerTitleId is expecting an integer data type and that what was entered was not of that type.

Figure 8-4. *Invalid data type*

5. Now press the down arrow, after altering CustomerTitleId to the correct data type, to indicate that you have finished creating this customer and wish to create the next. This of course means that some columns that have to be populated aren't, and SSE tells us so, as you see in Figure 8-5. We want to create a row that was full of NULL values, but we can't. The error message indicates that CustomerFirstName has not been set up to allow a NULL value, and we need to put some data in there.

Figure 8-5. *Trying to insert a row with NULL when NULLs are not allowed*

6. Clicking OK allows you back into the grid where the whole row can be populated with correct information. Notice that we can miss out placing any data in the `CustomerOtherInitials` column. After populating our grid, click the down arrow, and our grid should resemble Figure 8-6. The thing to notice is that although this is the first row entered, the `CustomerId` is set to 2. Whether insertion of a row is successful or not, an identity value is generated. Therefore, `CustomerId` 1 was generated when we received the second error as we were trying to move on to a new row. This can and will cause gaps within your numbering system. You can see how valuable using defaults as initial values for columns can be. Where the real benefit of using default values comes is in ensuring that specific columns are populated with the correct default values. As soon as we move off from the new row, the default values are inserted and ready to be modified. There is now a record of when the row was added, ideal for auditing. After we look at inserting a row with T-SQL, we will see what we might be able do about this.

Table - Custom...tails.Customers	Summary						
CustomerId	CustomerTitleId	CustomerFirstN...	CustomerOther...	CustomerLastN...	AddressId	AccountNumber	AccountTypeId
2	1	Vic	NULL	McGlynn	111	87612311	1
NULL	NULL	NULL	NULL	NULL	NULL	NULL	NULL

Figure 8-6. *The populated grid*

Note By having an `IDENTITY` column, every time a row is entered or an attempt is made to enter a row and all the *data entered* is of valid data types—whether this is through SSMSE or an `INSERT` statement—the column value within the table will be incremented by the `Identity Increment` amount.

7. Now open up a Query Editor window and enter the following code. This code will replicate the first part of this example in which we entered the wrong data type.

```
USE ApressFinancial
GO
INSERT INTO CustomerDetails.Customers (CustomerTitleId) VALUES ('Mr')
```

8. Now execute this by pressing Ctrl+E or F5, or clicking the Execute button on the toolbar. This code will generate an error because, once again, this is the wrong data type.

```
Msg 245, Level 16, State 1, Line 1
Conversion failed when converting the varchar value 'Mr' to data type int.
```

9. Change the code to replicate our second attempt at entering a row where the data type for the title is now correct but we are still missing other values:

```
USE ApressFinancial
GO
INSERT INTO CustomerDetails.Customers (CustomerTitleId) VALUES (1)
```

10. Now execute this by pressing Ctrl+E or F5, or clicking the Execute button on the toolbar. This code will generate a different error, informing us this time that we didn't allow a NULL into the CustomerFirstName column, and therefore we have to supply a value.

```
Msg 515, Level 16, State 2, Line 1
Cannot insert the value NULL into column 'CustomerFirstName', table
'ApressFinancial.CustomerDetails.Customers'; column does not allow nulls.
INSERT fails.
The statement has been terminated.
```

11. This final example will work successfully. However, note that the CustomerLastName is before that of the CustomerFirstName column. This demonstrates that it is not necessary to name the columns within the insertion in the same order as they are defined within the table. It is possible to place the columns in any order you desire.

```
INSERT INTO CustomerDetails.Customers
(CustomerTitleId,CustomerLastName,CustomerFirstName,
CustomerOtherInitials,AddressId,AccountNumber,AccountTypeId,
ClearedBalance,UnclearedBalance)
VALUES (3,'Mason','Jack',NULL,145,53431993,1,437.97,-10.56)
```

12. This time when you execute the code, you should see the following results, indicating the row has been inserted successfully:

```
(1 row(s) affected)
```

13. Now let's go back and view the data to see what has been entered. Find the CustomerDetails. Customers table in Object Explorer again. Right-click the table and select Open Table. The table now has two rows with two gaps in what we want as our ideal ascending sequence, as you see in Figure 8-7.

CustomerId	CustomerTitleId	CustomerFirstN...	CustomerOther...	CustomerLastN...	AddressId	AccountNumber
2	1	Vic	NULL	McGlynn	111	87612311
5	3	Jack	NULL	Mason	145	53431993

Figure 8-7. *Second customer inserted*

That is all there is to it. Just as simple as using SSMSE, but you did get more informative error messages. We now have a slight problem in that already there are two gaps in the table. This can be remedied easily within Query Editor, which we'll do in the next section.

DBCC CHECKIDENT

The DBCC statements can be used for many different operations, such as working with IDENTITY columns. If you find that when testing out IDENTITY columns you receive a number of errors, and the identity number has jumped up further than you wished, it is possible to reset the seed of the IDENTITY column so Query Editor starts again from a known point. The syntax for this statement is very simple:

```
DBCC CHECKIDENT ('table_name'[,{NORESEED |{RESEED[,new_reseed_value]}}])
```

The following elaborates on the three areas of the syntax that may need explanation:

- The name of the table that you wish to reset the identity value for is placed in single quotation marks.

- You can then use NORESEED to return back what SSE believes the current identity value should be; in other words, what the current maximum identity value is within the IDENTITY column.

- The final option is the one we are interested in. You can reseed a table automatically by simply specifying the RESEED option with no value. This will look at the table defined and will reset the value to the current maximum value within the table. Or optionally, you can set the column of the table to a specific value by separating the value and the option RESEED by a comma.

If you use RESEED and there are currently no rows in the table, but there had been in the past, the value will still be set to the last value entered, so take care.

Resetting the seed for an IDENTITY column though does have a danger, which you need to be aware of. If you reset the point to start inserting values for the IDENTITY column back past the greatest number on the given table, you will find that there is the potential of an error being produced. When a value that already exists is generated from an INSERT after resetting the IDENTITY column value, you will receive an error message informing you the value already exists. To give an example, you have a table with the values 1, 2, 5, 6, 7, and 8 and you reset the IDENTITY value back to 2. You insert the next row, which will correctly get the value 3, and the insertion will work. This will still work the same with the next insertion, which will receive the value 4. However, come to the next row, and there will be an attempt to insert the value 5, but that value already exists; therefore, an error will be produced. However, if you had reset the value to 8, the last value successfully entered, everything would be OK.

As we do not have the value 1 for the first row in the CustomerDetails.Customers table, it would be nice to correct this. It also gives a good excuse to demonstrate CHECKIDENT in action. The code that follows will remove the erroneous row entry and reset the seed of the IDENTITY column back to 0, to a value indicating that no rows have been entered. We will then via T-SQL reenter the customer information. Enter the following code, place the code into Query Editor, and execute it. The first line removes the row from CustomerDetails.Customers, and the second line resets the identity. Don't worry too much about the row deletion part, as deleting rows is covered in detail later in the chapter in the "Deleting Data" section.

```
DELETE FROM CustomerDetails.Customers
DBCC CHECKIDENT('CustomerDetails.Customers',RESEED,0)
INSERT INTO CustomerDetails.Customers
```

```
(CustomerTitleId,CustomerFirstName,CustomerOtherInitials,
CustomerLastName,AddressId,AccountNumber,AccountTypeId,
ClearedBalance,UnclearedBalance)
VALUES (1,'Vic',NULL,'McGlynn',111,87612311,1,4311.22,213.11)
INSERT INTO CustomerDetails.Customers
(CustomerTitleId,CustomerLastName,CustomerFirstName,
CustomerOtherInitials,AddressId,AccountNumber,AccountTypeId,
ClearedBalance,UnclearedBalance)
VALUES (3,'Mason','Jack',NULL,145,53431993,1,437.97,-10.56)
```

When the code is run, you should see the following information output to the query results pane:

```
(2 row(s) affected)
Checking identity information: current identity value '5', current column value '0'.
DBCC execution completed. If DBCC printed error messages, contact your system
administrator.

(1 row(s) affected)

(1 row(s) affected)
```

Column Constraints

A **constraint** is essentially a check that SSE places on a column to ensure that the data to be entered in the column meets specific conditions. This will keep out data that is not satisfying the condition and therefore avoid data inconsistencies. Constraints are used to keep database integrity by ensuring that a column only receives data within certain parameters.

```
[DateAdded] [datetime] NULL CONSTRAINT
[DF_Customers_DateAdded]  DEFAULT (getdate()),
```

Constraints are used to not only insert default values, but also validate data as well as primary keys. However, when using constraints within SSE, you do have to look at the whole picture, which is the user graphical system with the SSE database in the background. If you are using a constraint for data validation, some people will argue that perhaps it is better to check the values inserted within the user front-end application rather than in SSE. This has some merit, but what also has to be kept in mind is that you may have several points of entry to your database. This could be from the user application, a web-based solution, or other applications if you are building a central database. Many people will say that all validation, no matter what the overall picture is, should always be placed in one central place, which is the SSE database. Then there is only one set of code to alter if anything changes. It is a difficult choice and one that you need to look at carefully.

This part of the chapter will demonstrate how to add a constraint, specifically for data validation, through code in Query Editor. First of all, it is necessary to look at the code that is used to build the constraint.

ADD CONSTRAINT

You have two ways to add a constraint to a table. You saw the first when creating a default value as we built a table via SSMSE in Chapter 5.

To build a constraint via code, you need to use the ALTER TABLE statement, no matter what type of constraint this is. The ALTER TABLE statement can cover many different alterations to a table, but in this instance, the example just concentrates on adding a constraint. This makes the ALTER TABLE

statement easy, as the only real meat to the clause comes with the ADD CONSTRAINT syntax. The next example will work with the CustomerDetails.CustomerProducts table, and you will see three different types of constraints added, all of which will affect insertion of rows. It is worth reiterating the adding of a default value constraint again, as this will differ from the DateAdded column in the CustomerDetails.Customers table. Once the constraints have been added, you will see them all in action, and how errors are generated from erroneous data input.

Try It Out: Altering a Table for a Default Value in Query Editor

1. Ensure that Query Editor is running. Although all the examples deal with the CustomerDetails.Cus-tomerProducts table, each constraint being added to the table will be created one at a time, which will allow a discussion for each point to take place. In the Query Editor pane, enter the following code, which will add a primary key to the CustomerProducts table. This will place the CustomerFinancialProductId column within the key, which will be clustered.

```
USE ApressFinancial
GO
ALTER TABLE CustomerDetails.CustomerProducts
ADD CONSTRAINT PK_CustomerProducts
PRIMARY KEY CLUSTERED
(CustomerFinancialProductId) ON [PRIMARY]
GO
```

2. Next we add a CHECK constraint on the AmountToCollect column. The CustomerDetails.CustomerProducts table is once again altered, and a new constraint added called CK_CustProds_AmtCheck. This constraint will ensure that for all rows inserted into the CustomerDetails.CustomerProducts table from this point on, the score must be greater than 0. Notice as well that the NOCHECK option is mentioned, detailing that any rows already inserted will not be checked for this constraint. If they have invalid data, which they don't, then the constraint would ignore them and still be added.

```
ALTER TABLE CustomerDetails.CustomerProducts
WITH NOCHECK
ADD CONSTRAINT CK_CustProds_AmtCheck
CHECK ((AmountToCollect > 0))
GO
```

3. Moving on to the third constraint to add to the CustomerDetails.CustomerProducts table, we have a DEFAULT value constraint. In other words, this will insert a value of 0 to the Renewable column if no value is entered specifically into this column. This signifies that the premium collected is a one-off collection.

```
ALTER TABLE CustomerDetails.CustomerProducts WITH NOCHECK
    ADD CONSTRAINT DF_CustProd_Renewable
    DEFAULT (0)
    FOR Renewable
```

4. Execute the three batches of work by pressing F5 or Ctrl+E, or clicking the Execute button on the toolbar. You should then see the following result:

```
The statement(s) completed successfully.
```

5. There are two methods to check that the code has worked before adding in any data. Move to Object Explorer in Query Editor. This isn't refreshed automatically, so you do need to refresh it. You should then see the three new constraints added, two under the Constraints node and one under the Keys node, as well as a display change in the Columns node, as shown in Figure 8-8.

Figure 8-8. *CustomerDetails.CustomerProducts table details*

6. Another method is to move to SSMSE, find the `CustomerDetails.CustomerProducts` table, right-click it, and select Modify. This brings us into the Table Designer, where we can navigate to the necessary column to check out the default value, in this case `Renewable`. Also notice the yellow key against the `CustomerFinancialProductId` signifying that this is now a primary key, as shown in Figure 8-9.

Figure 8-9. *Default value constraint on column Renewable*

7. Move to the Table Designer toolbar and click the Manage Check Constraints button, shown here:

8. This will display the Check Constraints dialog box, shown in Figure 8-10, where we will see the `AmountToCollect` column constraint displayed. We can add a further constraint by clicking the Add button. Do so now.

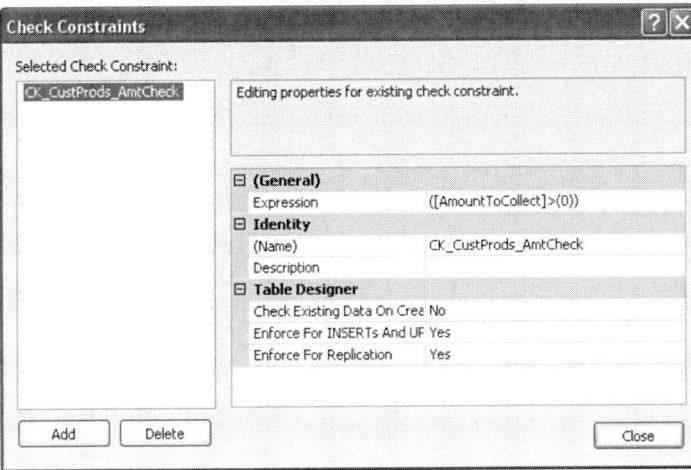

Figure 8-10. *Check Constraints dialog box*

9. This will alter the Check Constraints dialog box to allow a new check constraint to be added, as you see in Figure 8-11. This check will ensure that the `LastCollection` date is greater than the value entered in another column. Here we want to ensure that the `LastCollection` date is equal to or after the `LastCollected` date. Recall that `LastCollection` defines when we last took the payment, and `LastCollected` defines when the last payment should be taken.

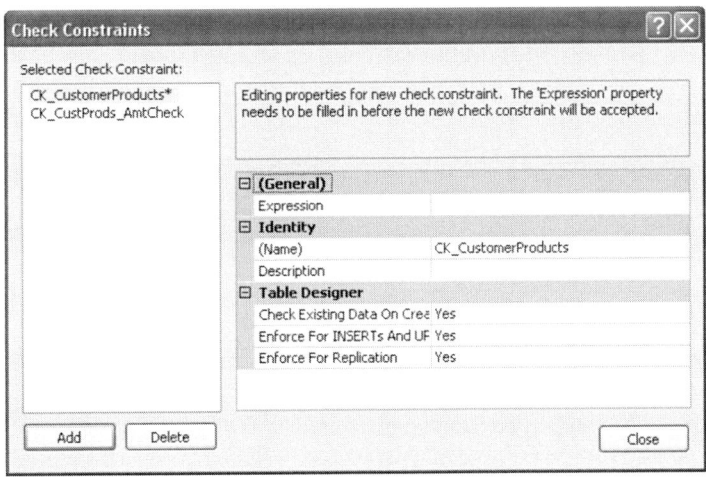

Figure 8-11. *Adding a new constraint in the Check Constraints dialog box*

10. The expression we want to add, which is the test the constraint is to perform, is not a value nor a system function like `GETDATE()`, but a test between two columns from a table, albeit the same table we are working with. This is as simple as naming the columns and the test you wish to perform. Also at the same time, change the name of the constraint to something meaningful. Your check constraint should look something like what appears in Figure 8-12. Afterwards, click Close, which will add the constraint to the list, although it has not yet been added to the table. It is not until the table is closed that this will happen, so do that now.

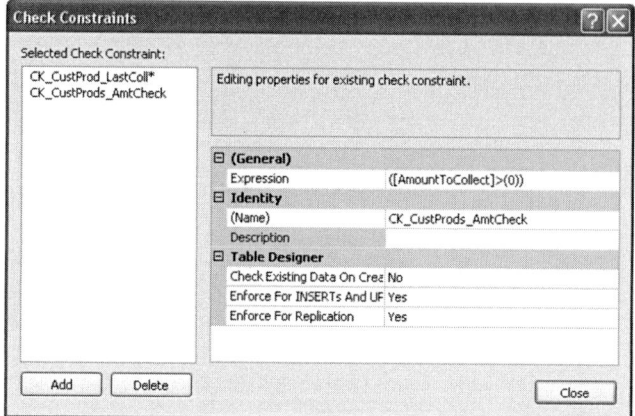

Figure 8-12. *LastColl constraint in the Check Constraints dialog box*

11. Now it's time to test the constraints to ensure that they work. First of all, we want to check the AmountToCollect constraint. Enter the following code, which will fail as the amount to collect is a negative amount.

```
INSERT INTO CustomerDetails.CustomerProducts
(CustomerId,FinancialProductId,AmountToCollect,Frequency,
LastCollected,LastCollection,Renewable)
VALUES (1,1,-100,0,'24 Aug 2005','24 Aug 2005',0)
```

12. When you execute the code in Query Editor, you will see the following result. Instantly you can see that the constraint check (CK_CustProds_AmtCheck) has cut in and the row has not been inserted.

```
Msg 547, Level 16, State 0, Line 1
The INSERT statement conflicted with the CHECK constraint
"CK_CustProds_AmtCheck". The conflict occurred in database
"ApressFinancial", table "CustomerDetails.CustomerProducts",
column 'AmountToCollect'.
The statement has been terminated.
```

13. We alter this now to have a positive amount, but change the LastCollection so that we break the CK_CustProd_LastColl constraint. Enter the following code:

```
INSERT INTO CustomerDetails.CustomerProducts
(CustomerId,FinancialProductId,AmountToCollect,Frequency,
LastCollected,LastCollection)
VALUES (1,1,100,0,'24 Aug 2005','23 Aug 2005')
```

14. When the preceding code is executed, you will see the following error message:

```
Msg 547, Level 16, State 0, Line 1
The INSERT statement conflicted with the CHECK constraint
"CK_CustProd_LastColl". The conflict occurred in database
"ApressFinancial", table "CustomerDetails.CustomerProducts".
The statement has been terminated.
```

Adding a constraint occurs through the ALTER TABLE statement as has just been demonstrated. However, the ADD CONSTRAINT statement is quite a flexible statement and can achieve a number of different goals.

The preceding example uses ADD CONSTRAINT to insert a primary key, which can be made up of one or more columns (none of which can contain a NULL value), and also to insert a validity check and a set of default values. The only option not covered in the example is the addition of a foreign key, but this is very similar to the addition of a primary key.

The first constraint added is the primary key, which we saw in Chapter 5. The second constraint definition builds a column check to ensure that the data entered is valid:

```
ADD CONSTRAINT constraint_name CHECK (constraint_check_syntax)
```

The syntax for a CHECK constraint is a simple true or false test. When adding in a constraint for checking the data, the information to be inserted is valid (true) or invalid (false) when the test is applied. As you will see, using mathematical operators to test a column against a single value or a range of values will determine whether the data can be inserted.

Notice in the example that the ADD CONSTRAINT statement is preceded with a WITH NOCHECK option on the ALTER TABLE statement. This informs SSE that any existing data in the table will not be validated when it adds the table alteration with the constraint, and that only data modified or inserted after the addition of the constraint will be checked. If you do wish the existing rows to be checked, you would use the WITH CHECK option. The advantage of this is that the existing data is validated against that constraint, and if the constraint was added to the table successfully, you know your data is valid. If any error was generated, you know that there was erroneous data, and that you need to fix that data before being able to add the constraint. This is just another method of ensuring that your data is valid.

Finally, for adding a default value, the ADD CONSTRAINT syntax is very simple.

```
ADD CONSTRAINT constraint_name
DEFAULT default_value
FOR column_to_receive_the_value
```

The only part of the preceding syntax that requires further explanation is the default_value area. default_value can be a string, a numeric, NULL, or a system function (for example, GETDATE(), which would insert the current date and time). So the default value does not have to be fixed; it can be dynamic.

Dealing with Several Rows at Once

It is now necessary to enter a few more customers so that a reasonable amount of data is contained within the CustomerDetails.Customers table to work with later in the book. We need to do the same with several other tables as well, such as TransactionDetails.TransactionTypes, CustomerDetails.CustomerTransactions, etc. This section will prove that no extra or specialized processing is required when inserting several rows. When working with data, there may be many times that several rows of data are inserted at the same time. This could be to initially populate a table, or when testing. In this sort of situation where you are repopulating a table, it is possible to save your query to a text file, which can then be reopened in Query Editor and executed without having to reenter the code. This is demonstrated at the end of the upcoming example.

Inserting Several Rows in a Query Batch

This next example will demonstrate inserting several rows. The work will be completed in batches. There is no transaction processing surrounding these INSERTs, and therefore each insertion will be treated as a single unit of work, which either completes or fails.

> ■**Note** A transaction allows a number of insertions or modifications to be treated as one unit, and if any insertion failed within the transaction, all the units would be returned back to their original value, and no insertions would take place. Transactions will be discussed in more detail in the "Transactions" section later in this chapter.

Try It Out: Insert Several Rows At Once

1. Ensure that Query Editor is up and running. In the Query Editor pane, enter the following code. Notice there are two GO statements within this set of INSERTs. Although each INSERT is its own self-contained unit of work, a GO statement also determines the end of a batch, or unit, of work. Therefore, the GO statements are superfluous if any error occurs with any of the INSERT statements.

```
INSERT INTO CustomerDetails.Customers
(CustomerTitleId,CustomerFirstName,CustomerOtherInitials,
CustomerLastName,AddressId,AccountNumber,AccountTypeId,
ClearedBalance,UnclearedBalance)
VALUES (3,'Bernie','I','McGee',314,65368765,1,6653.11,0.00)
GO
INSERT INTO CustomerDetails.Customers
(CustomerTitleId,CustomerFirstName,CustomerOtherInitials,
CustomerLastName,AddressId,AccountNumber,AccountTypeId,
ClearedBalance,UnclearedBalance)
VALUES (2,'Julie','A','Dewson',2134,81625422,1,53.32,-12.21)
GO
INSERT INTO CustomerDetails.Customers
(CustomerTitleId,CustomerFirstName,CustomerOtherInitials,
CustomerLastName,AddressId,AccountNumber,AccountTypeId,
ClearedBalance,UnclearedBalance)
VALUES (1,'Kirsty',NULL,'Hull',4312,96565334,1,1266.00,10.32)
```

2. Now just execute the code in the usual way. You will see the following output in the results pane. This indicates that three rows of information have been inserted into the database, one at a time.

```
(1 row(s) affected)

(1 row(s) affected)

(1 row(s) affected)
```

Retrieving Data

This section of the chapter will demonstrate how to view the data that has been placed in the tables so far. Many ways of achieving this are available, from using SSMSE through to T-SQL statements, and as you would expect, they will all be covered here.

The aim of retrieving data is to get the data back from SSE using the fastest retrieval manner possible. We can retrieve data from one or more tables through joining tables together within our query syntax; all of these methods will be demonstrated.

The simplest method of retrieving data is using SSMSE, and we look at this method first. With this method, you don't need to know any query syntax: it is all done for you. However, this leaves you with a limited scope for further work.

You can alter the query built up within SSMSE to cater to work that is more complex, but you would then need to know the SELECT T-SQL syntax; again, this will be explained and demonstrated. This can become very powerful very quickly, especially when it comes to selecting specific rows to return.

The results of the data can also be displayed and even stored in different media, like a file. It is possible to store results from a query and send these to a set of users, if so desired.

Initially, the data returned will be in the arbitrary order stored within SSE. This is not always suitable, so another aim of this chapter is to demonstrate how to return data in the order that you desire for the results. Ordering the data is quite an important part of retrieving meaningful results, and this alone can aid the understanding of query results from raw data.

Retrieving images is not as straightforward as retrieving normal rows of data, so I'll cover this in Chapter 12 along with other advanced T-SQL techniques.

Starting with the simplest of methods, let's look at SSMSE and how easy it is for us to retrieve rows. We have partially covered this earlier when inserting rows.

Using SSMSE to Retrieve Data

The first area that will be demonstrated is the simplest form of data retrieval, but it is also the least effective. Retrieving data using SSMSE is a very straightforward process, with no knowledge of SQL required in the initial stages. Whether it has to return all rows, or even when you want to return specific rows, using SSMSE makes this whole task very easy. This first example will demonstrate how flexible SSMSE is in retrieving all the data from the CustomerDetails.Customers table.

Try It Out: Retrieving Data Within SSMSE

1. Ensure that SSMSE is running. Navigate to the ApressFinancial database and click the Tables node; this should then list all the tables in the right-hand pane. Find the CustomerDetails.Customers table, right-click it to bring up the pop-up menu you have seen a number of times before, and select Open Table. This instantly opens up a new Query Editor pane like the one in Figure 8-13, which shows all the rows that are in the CustomerDetails.Customers table. But how did SSE get this data? Let's find out.

Table - Custom...tails.Customers		xp-home.Apres...ustomers.sql*		xp-home.Apres...leINSERTs.sql		xp-home.maste...erProducts.sql	
CustomerId	CustomerTitleId	CustomerFirstN...	CustomerOther...	CustomerLastN...	AddressId	AccountNumber	AccountTypeId
1	1	Vic	NULL	McGlynn	111	87612311	1
2	3	Jack	NULL	Mason	145	53431993	1
3	3	Bernie	I	McGee	314	65368765	1
4	2	Julie	A	Dewson	2134	81625422	1
5	1	Kirsty	NULL	Hull	4312	96565334	1

Figure 8-13. *CustomerDetails.Customers table retrieving data*

2. On the toolbar, you will see a button that, when pressed, will show the SQL code that was built to create this query:

3. Clicking the button alters the screen to that shown in Figure 8-14. This is the SQL syntax generated by SSMSE to provide the information requested.

	CustomerId	CustomerTitleId	CustomerFirstN...	CustomerOther...	CustomerLastN...	AddressId	AccountNumber	AccountTypeId
▶	1	1	Vic	NULL	McGlynn	111	87612311	1
	2	3	Jack	NULL	Mason	145	53431993	1
	3	3	Bernie	I	McGee	314	65368765	1
	4	2	Julie	A	Dewson	2134	81625422	1
	5	1	Kirsty	NULL	Hull	4312	96565334	1
✱	NULL	NULL	NULL	NULL	NULL	NULL	NULL	NULL

Table tabs: Table - Custom...ils.Customers* xp-home.Apres...ustomers.sql* xp-home.Apres...leINSERTs.sql xp-home.maste...erProducts.sql

```
SELECT    *
FROM      CustomerDetails.Customers
```

Figure 8-14. *Output displayed via SQL Window*

4. On the right you might see a window called the Properties window. If you don't see the screen as shown in Figure 8-15, you can open this up by pressing F4 or by clicking the View menu item. The point of interest at the moment is the section you see highlighted in Figure 8-15, Top Specification. Here we can enter the maximum number of rows that we wish to return.

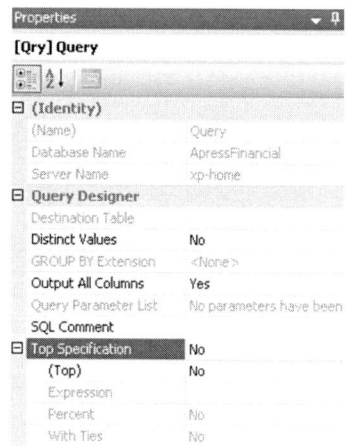

Figure 8-15. *Top Specification selected*

5. The order of the rows returned will be determined by the clustered index created on the table. However, if the table has no clustered index, the order the rows are returned in is arbitrary and there is no guarantee that the order will be the same when running the same query repeatedly. Changing Top Specification/(Top) to Yes will change the Properties window as shown in Figure 8-16.

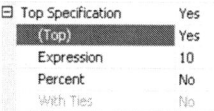

Figure 8-16. *Top Specification required*

6. We can enter any number we choose in the Expression property, but for this first time at least, enter **3** in the Expression box. This will return a maximum of three rows. If we entered a value of 100, we would only get five rows returned, as that is the maximum number of rows in the table at this moment in time. You would use this perhaps when you don't know the number of rows within a table, but you are only interested in a maximum number of 100 if there are more. This would be when you want to look at just a small selection of content in columns within a table. Now notice in the top-left corner of the results grid the following icon. This indicates that you have changed a property of the query and you need to do a refresh.

7. Right-click a blank area of the results grid and you will see the pop-up menu shown in Figure 8-17.

Figure 8-17. *Results grid pop-up menu*

8. Click Execute SQL to change the results grid to return just three rows of data, as shown in Figure 8-18.

CustomerId	CustomerTitleId	CustomerFirstN...	CustomerOther...	CustomerLastN...	AddressId
1	1	Vic	NULL	McGlynn	111
2	3	Jack	NULL	Mason	145
3	3	Bernie	I	McGee	314

Figure 8-18. *Three rows returned*

9. Again, by clicking the SQL button on the toolbar, the SQL code is exposed. Notice how this differs from the previous example in that TOP (3) has been placed after the SELECT statement.

```
SELECT     TOP (3) *
FROM       CustomerDetails.Customers
```

So now that you know how to return data from SSMSE, let's look at using T-SQL and probably the T-SQL statement you will use most often: the SELECT statement.

The SELECT Statement

If we wish to retrieve data for viewing from SSE using T-SQL statements, the SELECT statement is the one we need to use. This is quite a powerful statement, as it can retrieve data in any order, from any number of columns, from any table that we have the authority to retrieve data from, perform calculations on that data during data retrieval, and even include data from other tables! If the user does not have the authority to retrieve data from a table, you will receive an error message similar to that which you saw earlier in the chapter informing the user that permission is denied. SELECT has a lot more power than even the functions mentioned so far, but for the moment, let's concentrate on the fundamentals.

Let's take some time to inspect the simple syntax for a SELECT statement.

```
SELECT [ ALL | DISTINCT ]
[ TOP expression [ PERCENT ] [ WITH TIES ] ]
    {
                  *
        | { table_name | view_name | alias_name }.*
        | { column_name | [ ] expression | $IDENTITY | $ROWGUID }
        [ [ AS ] column_alias ]
        | column_alias = expression
    } [ ,...n ]
FROM table_name | view_name alias_name
WHERE filter_criteria
ORDER BY ordering_criteria
```

The following list breaks down the SELECT syntax, explaining each option. More explanation will be given throughout the chapter as well.

SELECT: Required—this informs SSE that a SELECT instruction is being performed; in other words, we just want to return a set of columns and rows to view.

ALL | DISTINCT: Optional—we want to return either all of the rows or only distinct, or unique, rows. Normally, you do not specify either of these options.

TOP expression/PERCENT/WITH TIES: Optional—you can return the top number of rows as defined by either the order of the data in the clustered index or, if the result is ordered by an ORDER BY clause, the top number from that order sequence. If there is no clustered index or no ordering, the rows will be returned in an arbitrary order. You can also add the word PERCENT to the end: this will mean that the top *n* percent of rows will be returned. If PERCENT is not specified, all the rows will be returned (unless specific column names are given). WITH TIES can only be used with an ORDER BY. If you specify you want to return TOP 10 rows, and the 11th row has the same value as the 10th row on those columns that have been defined in the ORDER BY, then the 11th row will also be returned. Same for subsequent rows, until you get to the point that the values differ.

*: Optional—by using the asterisk, you are instructing SSE to return all the columns from all the tables included in the query. This is not an option that should be used on large amounts of data or over a network, especially if it is busy. By using this, we are bringing back more information than is required. Wherever possible we should name the columns instead.

table_name.* | view_name.* | alias_name.*: Optional—similar to *, but you are defining which table, if the SELECT is working on more than one table. When working with more than one table, this is known as a JOIN, and this option will be demonstrated in Chapter 11 when we take a look at joins.

column_name: Optional but recommended; not required if * is used—this option is where we name the columns that we wish to return from a table. When naming the columns, it is always a good idea to prefix the column names with their corresponding table name. This becomes mandatory when we are using more than one table in our SELECT statement and instances where there may be columns within different tables that share the same name.

expression: Optional—we don't have to return columns of rows within a SELECT. We can return a value, a variable, or an expression.

$IDENTITY: Optional—will return the value from the IDENTITY column.

$ROWGUID: Optional—will return the value from the ROWGUID column.

AS: Optional—we can change the column header name when displaying the results by using the AS option.

FROM table_name | view_name: Required—we have to inform SSE where the information is coming from.

WHERE filter_clause: Optional—if we want to retrieve rows that meet specific criteria, we need to have a WHERE clause specifying the criteria to use to return the data. The WHERE clause tends to contain the name of a column on the left-hand side of a comparison operator (like =, <, or >) and either another column within the same table, or another table, a variable, or a static value. There are other options that the WHERE statement can contain, where more advanced searching is required, but on the whole these comparison operators will be the main constituents of the clause.

ORDER BY ordering_criteria: Optional—the data will be returned arbitrarily from the table if no ORDER BY clause is specified, which, if you have a clustered index built on the table, will be in that order; otherwise, it will be in the order in which they were inserted. However, you can alter the ordering by using the ORDER BY clause, which will determine the order of the rows returned, and you can specify whether each column is returned in ascending or descending order. Ascending, ASC, or descending, DESC, is defined for each column, not defined just once for all the columns within the ORDER BY. Sorting is completed once the data has been retrieved from SSE but before any statement like TOP.

Keep in mind that when building a SELECT statement, you do not have to name all the columns. In fact, you should only retrieve the columns that you do wish to see; this will reduce the amount of information sent over the network. There is no need to return information that will not be used.

Naming the Columns

When building a SELECT statement, it is not necessary to name every column if you don't want to see every column. You should only return the columns that you need. It is very easy to slip into using * to return every column, even when running one-time-only queries. Try to avoid this at all costs; typing out every column name takes time, but when you start dealing with more complex queries and a larger number of rows, the few extra seconds is worth it.

Now that you know not to name every column unless required, and to avoid using *, what other areas do you need to be aware of? First of all, it is not necessary to name columns in the same order that they appear in the table—it is quite acceptable to name columns in any order that you wish. There is no performance hit or gain from altering the order of the columns, but we may find that a different order of the columns might be better for any future processing of the data.

When building a SELECT statement and including the columns, if the final output is to be sent to a set of users, the column names within the database may not be acceptable. For example, if you are sending the output to the users via a file, they will see the raw result set. Or if you are using a tool such as Crystal Reports to display data from a SELECT statement within an SSE stored procedure, naming the columns would help there as well. The column names are less user friendly, and some column names will also be confusing for users; therefore, it would be ideal to be able to alter the names of the column headings. Replacing the SSE column headings with the new alias column headings desired, in either quotation marks or square bracket delimiters, is easily accomplished with the AS keyword. There is more on this in the next section.

Now that you know about naming the columns, let's take a look at how the SQL statement can return data.

The First Searches

This example will revolve around the `CustomerDetails.Customers` table, making it possible to demonstrate how all of the different areas mentioned previously can affect the results displayed.

Try It Out: The First Set of Searches

1. Ensure that Query Editor is running and that you are within the `ApressFinancial` database. In the Query Editor pane, enter the following SQL code:

```
SELECT * FROM CustomerDetails.Customers
```

2. Execute the code using Ctrl+E, F5, or the Execute button on the toolbar. You should then see something like the results shown in Figure 8-19.

	CustomerId	CustomerTitleId	CustomerFirstName	CustomerOtherInitials	CustomerLastName	AddressId	AccountNumber	AccountTyp...	ClearedB...
1	1	1	Vic	NULL	McGlynn	111	87612311	1	4311.22
2	2	3	Jack	NULL	Mason	145	53431993	1	437.97
3	3	3	Bernie	I	McGee	314	65368765	1	6653.11
4	4	2	Julie	A	Dewson	2134	81625422	1	53.32
5	5	1	Kirsty	NULL	Hull	4312	96565334	1	1266.00

Figure 8-19. *Customers listing*

3. This is a simple `SELECT` statement returning all the columns and all the rows from the `CustomerDetails.Customers` table. Let's now take it to the next stage where specific column names will be defined in the query, which is a much cleaner solution. In this instance from the `CustomerDetails.Customers` table, we would like to return a customer's first name, last name, and current account balances. This would mean naming `CustomerFirstName`, `CustomerLastName`, and `ClearedBalance` as the column names in the query. The code will read as follows:

```
SELECT CustomerFirstName,CustomerLastName,ClearedBalance
FROM CustomerDetails.Customers
```

4. Now execute this code, which will return the results shown in Figure 8-20. As you can see, not every column is returned.

	CustomerFirstName	CustomerLastName	ClearedBalance
1	Vic	McGlynn	4311.22
2	Jack	Mason	437.97
3	Bernie	McGee	6653.11
4	Julie	Dewson	53.32
5	Kirsty	Hull	1266.00

Figure 8-20. *Specific columns returned*

5. As you have seen from the examples so far, the column names, although well named from a design viewpoint, are not exactly suitable if we had to give this to a set of users. Using the same query as before, a couple of minor modifications are required to give the columns aliases. The first alias name is in quotes as it contains a space. Notice the last column also does not have `AS` specified because this keyword is optional.

```
SELECT CustomerFirstName As 'First Name',
CustomerLastName AS 'Surname',
ClearedBalance Balance
FROM CustomerDetails.Customers
```

6. Execute this and the displayed output changes—much more friendly column names, as you see in Figure 8-21.

	First Name	Surname	Balance
1	Vic	McGlynn	4311.22
2	Jack	Mason	437.97
3	Bernie	McGee	6653.11
4	Julie	Dewson	53.32
5	Kirsty	Hull	1266.00

Figure 8-21. *Friendly column names*

The first SELECT statement demonstrates the fact that in most SSE instances, whether we use upper- or lowercase doesn't matter to our queries; however, some language installations are case sensitive. When installing SSE, if we chose a SQL collation sequence that was case sensitive, as denoted by CS within the suffix of the collation name, SQL_Latin1_General_Cp437_CS_AS, for instance, the first SELECT query would generate an error. The collation sequence for SSE was chosen in Chapter 1 when we installed the application. Changing a collation sequence within SSE is a very difficult task that requires rebuilding parts of SSE, so this book won't move into that area.

Tip It is strongly recommended, and considered best practice, that you use the correct casing when using queries. Not only does this avoid confusion, but it also means that if you do switch to a case-sensitive installation, it will not be necessary to alter the query.

Moving back to the first query, this query will select all columns and all rows from the CustomerDetails.Customers table, ordered according to how the database sees it—as you can see in Figure 8-21, it has quite plainly done this.

Looking at the second and third query examples, the columns returned have been reduced to just three columns: the customer's first and last names and the cleared balance amounts. All the rows are still being returned. In the last example, notice that after two of the three columns, there is an AS keyword. This signifies that the following literal is to be used as the column heading; note that if we wish to use two words separated by spaces, we must surround these words by identifiers, whether they be quotation marks, as in our example, or square brackets.

Now that the basics of the SELECT statement have been covered, we will next look at the methods within Query Editor to display output in different manners.

Varying the Output Display

There are different ways of displaying the output: from a grid, as we have seen; from a straight text file; still within a Query Editor pane; or as pure text, just like a tabulated Word file. You may have found the results in the previous exercise laid out in a different format than shown previously, depending on how you initially set up Query Editor. In the results so far, you have seen the data as a grid. This next section will demonstrate tabular text output, otherwise known as **Results in Text**, as well as outputting the data to a file. Let's get right on with the first option, Results in Text.

Try It Out: Putting the Results in Text and a File

1. You should still be in Query Editor. From the Query menu option, select Results To ➤ Results in Text, or press Ctrl+T. Figure 8-22 shows the other options available from the Results To menu.

Figure 8-22. *Sending the results to different places*

2. If we run the same query as earlier (the code is detailed again here), we will be able to see the difference. Once the code is entered, execute it.

```
SELECT CustomerFirstName As 'First Name',
CustomerLastName AS 'Surname',
ClearedBalance Balance
FROM CustomerDetails.Customers
```

3. Examine the output, which should resemble Figure 8-23. As you can see, the output has changed a great deal. No longer is the output in a nice grid where the columns have been shrunk to a more manageable size, but each column's data takes up, and is displayed to, the maximum number of characters that each column could contain. This obviously stretches out the display, but from here we can see easily how large each column is supposed to be.

```
Results

First Name                                               Surname                                            Balance
-------------------------------------------------------- -------------------------------------------------- -----------
Vic                                                      McGlynn                                            4311.22
Jack                                                     Mason                                              437.97
Bernie                                                   McGee                                              6653.11
Julie                                                    Dewson                                             53.32
Kirsty                                                   Hull                                               1266.00

(5 row(s) affected)
```

Figure 8-23. *Results as text*

4. There will be times, though, when users require output to be sent to them. For example, they may wish to know specific details from a set of rows, and so you build a query and save the results to a file to send to them. Or perhaps they want output to perform some analysis of data within a Microsoft Excel spreadsheet. Again, this can be achieved from the Query menu by selecting Results To ➤ Results to File, or Ctrl+Shift+F. Specify sending results to a file and rerun the code. Once the code has been executed, a Save Results dialog box like the one in Figure 8-24 will appear; this could show any folder location initially—in this case, it shows the My Documents folder.

Figure 8-24. *Locating where to save the results*

5. So now that you know how to save to different locations, move back to displaying the output to a grid by pressing Ctrl+D.

You now know how to return data, but what happens if you don't want every row and you want to select which rows to display? We look at that next.

Limiting the Search: The Use of WHERE

You have a number of different ways to limit the search of rows within a query. Some of the most basic revolve around the three basic relational operators: <, >, and = (less than, greater than, and equal to). There is also the use of the keyword NOT, which could be included with these three operators; however, NOT does not work as in other programming languages that you may have come across: this will be demonstrated within the example in this section so you know how to use the NOT operator successfully.

All of these operators can be found in the WHERE clause of the SELECT statement used to reduce the number of rows returned within a query.

■**Note** You may come across some legacy code where you will find that the WHERE statement is used to join two tables together to make the results look as if they came from one table. For some databases, this is the "standard" way to join two tables; however, with SSE, the WHERE statement is purely used as a filter method.

Try It Out: The WHERE Statement

1. First of all to use a different table, let's enter some more rows in to the ShareDetails.Shares table. Enter and execute the following code:

```
INSERT INTO ShareDetails.Shares
(ShareDesc, ShareTickerId,CurrentPrice)
VALUES ('FAT-BELLY.COM','FBC',45.20)
INSERT INTO ShareDetails.Shares
(ShareDesc, ShareTickerId,CurrentPrice)
VALUES ('NetRadio Inc','NRI',29.79)
INSERT INTO ShareDetails.Shares
(ShareDesc, ShareTickerId,CurrentPrice)
VALUES ('Texas Oil Industries','TOI',0.455)
INSERT INTO ShareDetails.Shares
(ShareDesc, ShareTickerId,CurrentPrice)
VALUES ('London Bridge Club','LBC',1.46)
```

2. The requirement for this section is to find the current share price for FAT-BELLY.COM. We restrict the SELECT statement so that only the specific row comes back by using the WHERE statement, as can be seen in the following code:

```
SELECT ShareDesc,CurrentPrice
FROM ShareDetails.Shares
WHERE ShareDesc = 'FAT-BELLY.COM'
```

3. Execute this code, and you will see that the single row for FAT-BELLY.COM is returned, as shown in Figure 8-25.

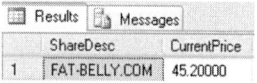

Figure 8-25. *The results of limiting the search*

4. To prove that we are working within an installation that is not case sensitive from a data perspective (unless you installed a different collation sequence to that described in Chapter 1), if you perform the following query, you will get the same results as displayed in Figure 8-25.

```
SELECT ShareDesc,CurrentPrice
FROM ShareDetails.Shares
WHERE ShareDesc = 'FAT-BELLY.COm'
```

Note As you can see, this may not always be what you want, because you may want your data to be case sensitive. If you do, query code will also become case sensitive.

5. You have seen the WHERE in action using the equals sign; it is also possible to use the other relational operations in the WHERE statement. The next query demonstrates how SSE takes the WHERE condition and starts returning rows after the given point. This query provides an interesting set of results. Enter the code as detailed here:

```
SELECT ShareDesc,CurrentPrice
FROM ShareDetails.Shares
WHERE ShareDesc > 'FAT-BELLY.COM'
AND ShareDesc < 'TEXAS OIL INDUSTRIES'
```

6. Once done, execute the code and check the results, which should resemble Figure 8-26.

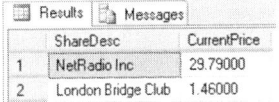

	ShareDesc	CurrentPrice
1	NetRadio Inc	29.79000
2	London Bridge Club	1.46000

Figure 8-26. *Shares output*

7. Let's now bring in another option in the WHERE statement that allows us to avoid returning specific rows. This can be achieved in one of two ways: the first is by using the less than and greater than signs; the second is by using the NOT operator. Enter the following code, which will return all rows except FAT-BELLY.COM. Run both sets of code at once. This will return two sets of output, known as **multiple result sets**.

```
SELECT ShareDesc,CurrentPrice
FROM ShareDetails.Shares
WHERE ShareDesc <> 'FAT-BELLY.COM'

SELECT ShareDesc,CurrentPrice
FROM ShareDetails.Shares
WHERE NOT ShareDesc = 'FAT-BELLY.COM'
```

8. Executing this code will produce the output shown in Figure 8-27. Notice how in neither set of output FAT-BELLY.COM has been listed.

	ShareDesc	CurrentPrice
1	ACME'S HOMEBAKE COOKIES INC	2.34125
2	NetRadio Inc	29.79000
3	Texas Oil Industries	0.45500
4	London Bridge Club	1.46000

	ShareDesc	CurrentPrice
1	ACME'S HOMEBAKE COOKIES INC	2.34125
2	NetRadio Inc	29.79000
3	Texas Oil Industries	0.45500
4	London Bridge Club	1.46000

Figure 8-27. *Multiple output*

As you have seen, it is possible to limit the number of rows to be returned via the WHERE clause; we can return rows up to a certain point, after a certain point, or even between two points with the use of an AND statement. It is also possible to exclude rows that are not equal to a specific value or range of values by using the NOT statement or the <> operator.

When the SSE data engine executes the T-SQL SELECT statement, it is the WHERE statement that is dealt with before any ordering of the data, or any limitation placed on it concerning the number of rows to return. The data is inspected where possible using an index, to determine whether a row stored in the relevant table matches the selection criteria within the WHERE statement, and if it does, to return it. If an index cannot be used, a full table scan will be performed to find the relevant information.

Table scans can present a large performance problem within your system, and you will find that if a query has to perform a table scan, data retrieval could be very slow, depending on the size of the table being scanned. If the table is small with only a small number of rows, a table scan is likely to retrieve data more quickly than the use of an index. However, table scanning and the speed of data retrieval will be the biggest challenge you will face as an SSE developer. With data retrieval, it is important to bear in mind that whenever possible, if you are using a WHERE clause to limit the rows returned, you should try to specify the columns from an index definition in this WHERE clause. By doing this, you will be giving the query the best chance for optimum performance.

As discussed in Chapter 7, getting the index right is crucial to fast data manipulation and retrieval. If you find you are forever placing the same columns in a WHERE clause, but those columns do not form part of an index, perhaps this is something that should be revisited to see whether any gain can come from having the columns be part of an index.

For any table, ensuring that the WHERE clause is correct is important. As has been indicated from a speed perspective, using an index will ensure a fast response of data. This gains greater importance with each table added, and even more importance as the size of each table grows.

Finally, by ensuring the WHERE statement filters out the correct rows, you will ensure that the required data is returned, the right results are displayed, and less data is sent across the network, as the processing is done on the server and not the client. Also, having the appropriate indexing strategy helps with this as well.

It is also possible to return a specific number of rows, or a specific percentage of the number of rows, as you saw when displaying rows in SSMSE. These statements are discussed next, with a short code example demonstrating each in action. First of all, we will look at a statement that does not actually form part of the SELECT statement itself.

SET ROWCOUNT n

SET ROWCOUNT n is a totally separate statement from the SELECT statement and can in fact be used with other statements within T-SQL. What this statement will do is limit or reset the number of rows that will be processed for the session that the statement is executed in.

■**Note** Caution should be exercised if you have any statements that also use a TOP statement, described in a moment.

The SET ROWCOUNT n function stops the processing of the SELECT statement, or even UPDATE and DELETE statements, described in the "Updating Data" and "Deleting Data" sections, respectively, once the number of rows defined has been reached. The difference between SET ROWCOUNT and SELECT TOP n is that the latter will perform one more internal instruction to that of the former. Processing halts immediately when the number of rows processed through SET ROWCOUNT is reached. However, by using the TOP statement, all the rows are returned internally, the TOP n rows are selected from that group internally, and these are then passed for display. Returning a limited number of

rows is useful when you want to look at a handful of data to see what values could be included, or perhaps you wish to return a few rows for sampling the data.

You can set the number of rows to be affected by altering the number, n, at the end of the SET ROWCOUNT function. This setting will remain in force only within the Query Editor pane in which the statement is executed, or within the stored procedure in which the statement is executed.

To reset the session so that all rows are taken into consideration, you would set the ROWCOUNT number to 0.

Try It Out: SET ROWCOUNT

1. In Query Editor, enter the following code into a new Query Editor pane; once entered, execute it.

```
SET ROWCOUNT 3
SELECT * FROM ShareDetails.Shares
SET ROWCOUNT 0
SELECT * FROM ShareDetails.Shares
```

2. You should see two result sets, as shown in Figure 8-28. The first will return three rows from the ShareDetails.Shares table. The second result set will return all rows from ShareDetails.Shares.

Figure 8-28. *Limiting the output via ROWCOUNT*

TOP n

This option, found within the SELECT statement itself, will return a specific number of rows from the SELECT statement, and is very much like the SET ROWCOUNT function for that reason. In fact, the TOP n option is the preferred option to use when returning a set number of rows, as opposed to the SET ROWCOUNT function. The reason behind this is that TOP n only applies to that query statement; however, by using SET ROWCOUNT n, you are altering all statements until you reset SSE to act on all rows through SET ROWCOUNT 0.

▓**Caution** Although it is possible to use TOP n without any ORDER BY statement, it is usual to combine TOP with ORDER BY. When no order is specified, the rows returned are arbitrary, and if you want consistent results, ordering will provide this. If you are not concerned about which rows are returned, you can avoid using ORDER BY.

Any WHERE statements and ORDER BY statements within the SELECT statement are dealt with first, and then, from the resultant rows, the TOP n function comes into effect. This will be demonstrated with the following example.

Try It Out: TOP n

1. In Query Editor, enter the following code into a new Query Editor pane; once entered, execute it.

```
SELECT TOP 3 * FROM ShareDetails.Shares
SET ROWCOUNT 3
SELECT TOP 2 * FROM ShareDetails.Shares
SET ROWCOUNT 2
SELECT TOP 3 * FROM ShareDetails.Shares
```

2. The code returns three result sets, as shown in Figure 8-29. Take a moment to peruse these result sets. The first set is just the top three rows that are taken from an arbitrary order SSE has chosen. The second will only return two rows, even though the ROWCOUNT is set to 3. The third result set takes into account the ROWCOUNT setting, as this is the lesser value this time. Therefore, again, only two rows are returned.

Figure 8-29. *A mixture of TOP and ROWCOUNT*

TOP n PERCENT

TOP n PERCENT is very similar to the TOP n clause with the exception that instead of working with a precise number of rows, it is a percentage of the number of rows that will be returned. Keep this in mind, as it is not a percentage of the number of rows within the table. Also, the number of rows is rounded up; therefore, as soon as the percentage moves over to include another row, SSE will include this extra row.

You see more of this option in Chapter 9, which discusses the building of views.

String Functions

A large number of system functions are available for manipulating data. This section looks purely at the string functions available for use within a T-SQL statement; later in the book, we will look at some more functions that are available to us. Following are the functions that are used in the next example:

LTRIM/RTRIM: These perform similar functionality. If you have a string with leading spaces, and you wish to remove those leading spaces, you would use LTRIM so that the returned varchar value would have a nonspace character as its first value. If you have trailing spaces, you would use RTRIM. You can only use this function with a data type of varchar, nvarchar, a data type that can be implicitly converted to these two data types, or a data type converted to varchar or nvarchar using the CAST SSE function.

CAST: A specialized function that will convert one data type to another data type. I don't cover this within the book. If you wish to convert data types, check on the statement in Books Online, which can be found by selecting Help in Query Editor.

LEFT/RIGHT: This function will return the leftmost or rightmost characters from a string. Passing in a second parameter to the function will determine the number of characters to return from whichever side of the string. The LEFT and RIGHT functions accept any data type except text or ntext expressions to perform the string manipulation, implicitly converting any noncharacter data type or varchar or nvarchar, and returning a varchar or nvarchar data type as a result.

Try It Out: String Functions

1. Enter the code that follows into an empty Query Editor window. Alter the output to text format, by pressing Ctrl+T. Notice the use of the + operator within the SELECT query. This will concatenate the strings defined within the query into one single string value.

Note Unlike with some programming languages, you cannot use the & character, as this has a totally different meaning in SSE.

```
SELECT CustomerFirstName + ' ' + CustomerLastName AS 'Name',
ClearedBalance Balance
FROM CustomerDetails.Customers
```

2. Execute this code, which produces the output in Figure 8-30.

	Name	Balance
1	Vic McGlynn	4311.22
2	Jack Mason	437.97
3	Bernie McGee	6653.11
4	Julie Dewson	53.32
5	Kirsty Hull	1266.00

Figure 8-30. *Concatenating results*

3. As you can see, it's a bit unwieldy. The Name column heading goes far wider than is required. There is a complex way of getting this right, but a much simpler method is to use the LEFT statement. The sum of the width of the two columns gives this column width displayed in the output; by using the LEFT statement, it is possible to achieve something better. Clear the Query Editor pane and enter the following code:

```
SELECT LEFT(CustomerFirstName + ' ' + CustomerLastName,50) AS 'Name',
ClearedBalance Balance
FROM CustomerDetails.Customers
```

4. Execute the preceding code. This produces the results shown in Figure 8-31. What the preceding query has done is to reduce the output to the first 50 characters starting from the left.

Figure 8-31. *Concatenating results and reducing the width*

5. The best way to remove all trailing spaces from the first name and surname concatenated columns is the `RTRIM` statement. The following code does this, although the output in the text layout doesn't. This is because SSE still doesn't know what the maximum size of the concatenation will be, and it has to believe that the maximum number of characters of the sum of the two columns could still be displayed. However, in truth, the amount of data returned will be minimal. Therefore, this is a great method of reducing the amount of data passed over a network.

```
SELECT RTRIM(CustomerFirstName + ' ' + CustomerLastName) AS 'Name',
ClearedBalance Balance
FROM CustomerDetail.Customers
```

In all of our examples thus far, as you know, rows are returned either in an arbitrary order or in the order of the clustered index if one has been defined. We look now at how this can be changed.

Order! Order!

Of course, retrieving the rows in the order of the clustered index may not always be what you desire. However, it is possible to change the order in which you return rows. This is achieved through the `ORDER BY` clause, which is part of the `SELECT` statement. The `ORDER BY` clause can have multiple columns, even with some being in ascending order and others in descending order.

If you should find that you are repeatedly using the same columns within an `ORDER BY` clause, or that the query is taking some time to run, you should consider having the columns within the query as an index. (Indexes were covered in Chapter 6.)

Ordering the data will of course increase processing time, but it is used as a necessity to display the data in the correct order. Ordering on varchar columns also takes longer than numeric columns.

Note Ordering takes place after the filtering of rows but before the TOP statement, so you could still be ordering a large set of rows before returning the top few you may need.

Let's now take a look at building a query that uses an ORDER BY clause.

Try It Out: Altering the Order

1. Clear the Query Editor pane and set the display option back to showing a grid by pressing Ctrl+D. Once complete, enter the following code into the Query Editor pane. This will return the data in the ascending (the default) order of the cleared balance of our customers.

```
SELECT LEFT(CustomerFirstName + ' ' + CustomerLastName,50) AS 'Name',
ClearedBalance Balance
FROM CustomerDetails.Customers
ORDER BY Balance
```

2. Execute the code; this will produce the results shown in Figure 8-32.

Figure 8-32. *Altering the order by balance*

3. We can also complete the same query, but have the cleared balance in descending order, rather than ascending order. This is simply done by placing DESC after the column name. Change your code as detailed here:

```
SELECT LEFT(CustomerFirstName + ' ' + CustomerLastName,50) AS 'Name',
ClearedBalance Balance
FROM CustomerDetails.Customers
ORDER BY Balance DESC
```

4. Execute the code; this will produce the results shown in Figure 8-33.

Figure 8-33. *Making the order in descending sequence*

The LIKE Operator

It is possible to use more advanced techniques for finding rows where a mathematical operation doesn't quite fit; for example, someone is trying to track down a customer, but doesn't know the customer's full name or does know the first part of his or her surname but doesn't know how to spell the full name.

Suppose you know that the surname ends in "Glynn" as in the first customer we added, but you don't know if it starts with Mc, Mac, or even M. So how would this be put into a query? There is a keyword that you can use as part of the WHERE statement, called LIKE. This will use pattern matching to find the relevant rows within an SSE table using the information provided.

The LIKE operator can come with one of four operators, which are used alongside string values that you want to find. Each of the four operators is detailed in the following list. They can be used together, and using one does not exclude using any others.

Note LIKE is NOT case sensitive.

%: This would be placed at the end and/or the beginning of a string. The best way to describe this is through an example; if you were searching the customers who had the letter "a" within their surname, you would search for "%a%", which would look for the letter "a," ignoring any letters before and after the letter "a," and just checking for that letter within the first name column.

_: This looks at a string, but only for a single character before or after the position of the underscore. Therefore, looking in the first name column for "_a" would return any customer who has two letters in his or her first name where the second letter is an "a." In our example, no rows would be returned. However, if you combined this with the % sign and search for "_a%", then you would get back Jason Atkins, Ian McMahon, and Ian Prentice. You would not get back Vic McGlynn, because "a" is not the second letter.

[]: This lets you specify a number of values or a range of values to look for. For example, if you were looking in the player's first name for the letters "c-f", you would use LIKE "%[c-f]%".

[^...]: Similar to the preceding option, this one lists those items that do not have values within the range specified.

The best way to learn how to use LIKE is to see an example.

Try It Out: The LIKE Operator

1. We are going to try and find Vic McGlynn via the CustomerLastName column. We know the name ends with Glynn. The code that follows will search all of the customer rows looking for anything prefixing Glynn.

   ```
   SELECT CustomerFirstName + ' ' + CustomerLastName
   FROM CustomerDetails.Customers
   WHERE CustomerLastName LIKE '%Glynn'
   ```

2. Execute the code; this will give the results shown in Figure 8-34.

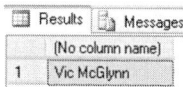

Figure 8-34. *Using the LIKE operator*

3. We can also go to extremes using the LIKE operator, for example, seeing which customers have the letter "n" anywhere in their name. The code for this is shown here:

```
SELECT CustomerFirstName + ' ' + CustomerLastName AS [Name]
FROM CustomerDetails.Customers
WHERE CustomerFirstName + ' ' + CustomerLastName LIKE '%n%'
```

4. When you execute this, you should get the results shown in Figure 8-35: four customers are returned, as they have an "n" somewhere in their name.

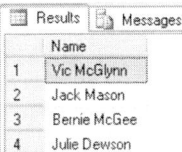

	Name
1	Vic McGlynn
2	Jack Mason
3	Bernie McGee
4	Julie Dewson

Figure 8-35. *Using LIKE to search for customers with "n" in their name*

5. Why would we want to go to such lengths? Would it not have been possible to use the Name alias, which is a combination of the first name and last name columns? Well, unfortunately not—the code we might expect to use would look something like the following:

```
SELECT CustomerFirstName + ' ' + CustomerLastName AS [Name]
FROM CustomerDetails.Customers
WHERE [Name] LIKE '%n%'
```

6. Execute this code. Instead of the success messages that we have become used to, an error message will be returned. We can only search on real column names, not aliases.

```
Msg 207, Level 16, State 1, Line 3
Invalid column name 'Name'.
```

Creating Data: SELECT INTO

The topic discussed in this section is quite an advanced area to be getting into, but it's not too advanced to be covered within this book. It is possible to create a new table within a database by using the INTO keyword, like that found in INSERT INTO, within a SELECT statement, providing, of course, you have the right database permissions to create tables in the first place. First of all, it is necessary to clarify the syntax of how the SELECT INTO statement is laid out; we simply add the INTO clause after the column names, but before the FROM keyword. Although the following section of code shows just one table name, it is possible to create a new table from data from one or more tables.

```
SELECT *|column1,column2,...
INTO new_tablename
FROM tablename
```

The INTO clause is crucial to the success of the creation of the new table. The SELECT statement will fail if there is a table already in existence with the same name for the same table owner. This will be demonstrated within the example.

The table generated will consist of the columns returned from the built SELECT statement, whether that is all the columns from the table mentioned within the FROM statement or a subset. The new table will also contain only the rows returned from the SELECT statement. To clarify, this statement is creating a new table using the structure within the SELECT statement. There will be no keys,

constraints, relationships, or in fact any other facet of SSE, except a new table. Hence creating tables using SELECT...INTO should only be done after some thought.

■**Note** If you really need to do this, it is faster to create the table using CREATE TABLE and then insert the data into it using the INSERT INTO...SELECT statement rather than SELECT ... INTO new_tablename FROM tablename.

Two tables can exist with the same name within a database, providing that they have different schemas. The tables in ApressFinancial all have the database owner as their owner, but it is possible for a CustomerDetails.Customers table to exist for an owner like VMcGlynn.

■**Note** Although possible, this is NOT recommended, as it causes confusion.

Let's look at the INTO statement in action.

Try It Out: SELECT INTO

1. In an empty Query Editor window, enter the following code:

```
SELECT CustomerFirstName + ' ' + CustomerLastName AS [Name],
ClearedBalance,UnclearedBalance
INTO CustTemp
FROM CustomerDetails.Customers
```

2. Execute the code. This will return the following message in the results pane:

```
(5 row(s) affected)
```

3. If you now move to Object Explorer on the left-hand side (if Object Explorer is no longer there, press F8) and complete a refresh, you should see a new table in the expanded Tables node, called CustTemp, as shown in Figure 8-36.

Figure 8-36. *New table created with SELECT INTO*

You should use the `INTO` clause with care. For instance, in this example, security has not been set up for the table, and we are also creating tables within our database that have not been through any normalization or development life cycle. It is also very easy to fill up a database with these tables if we are not careful. However, it is a useful and handy method for taking a backup of a table and then working on that backup while testing out any queries that might modify the data. Do ensure though that there is enough space within the database before building the table if you do use this technique.

Note It is best to avoid doing this in a production environment unless you really do need to keep the table permanently.

Who Can Add, Delete, and Select Data

In this chapter, we inserted a certain amount of data into tables. All went well, and the data was inserted and selected easily. This was due to using the same connection that created the database. This does not mean that anyone who has access to our database could also add data as easily as we can. In Chapter 1, we set up several different users for the system. To recap, here are the users and their authority:

- *RDewson*: Administrator/database owner

- *AJMason*: Database owner

- *VMcGlynn*: Administrator

- *sa*: Administrator, SSE's default system administrator login

We want to prove that not all users can or should be able to do anything against our data. If any new user is created and given authority to connect to `ApressFinancial`, then providing this user is not an administrator on the local machine, he or she will not be able to view, insert, or delete any data. We do have a limited user in AJMason, although we did give this user db_owner rights earlier in the book to demonstrate his connection. This was a short-term solution that is now no longer valid. We need to refine this user so that we can restrict exactly what the user can do. In the following example, we will remove the db_owner role and give AJMason `SELECT` permissions on the `ShareDetails.Shares` table only. You will then see this in action.

Try It Out: Refining Permissions

1. Within Object Explorer, expand the Security node, and then the Users node. Find `XP-Home\AJMason`, right-click it, and select Properties.

2. This will bring up the Database User dialog box. As shown in Figure 8-37, deselect the db_owner role that you will see as being checked.

Figure 8-37. *Removing database owner role*

3. Move to the Securables options for this user. In this area, shown in Figure 8-38, it is possible to define the exact privileges that this user can have, and can even pass on to other users. At present, the options are blank, but we can add objects by clicking the Add button.

Figure 8-38. *Securables tab, preparing to add objects*

4. This brings up the Add Objects dialog box. We want to refine who we are going to give the SELECT privilege to in the ShareDetails schema. Select this schema in the combo box as shown in Figure 8-39 and click OK.

Figure 8-39. *Defining only the ShareDetails schema*

5. When we return to the Securables dialog box, as shown in Figure 8-40, we will see two tables defined, `ShareDetails.SharePrices` and `ShareDetails.Shares`. Below that is a list of Explicit Permissions, which at the moment reflects the `ShareDetails.SharePrices` table. Nothing is selected; therefore, at present, if we applied the actions performed up to this point, AJMason would be unable to perform anything on this table (or any other table as we had removed db_owner).

Figure 8-40. *Detailing the ShareDetails schema objects*

6. Change the Securables list to the `ShareDetails.Shares` table. Then in the Explicit Permissions area, grant Select permissions by checking the check box, as shown in Figure 8-41. I will explain this section in more detail shortly. Once done, click OK, which will then apply these changes to AJMason.

Figure 8-41. *Allowing the user to only select from ShareDetails.Shares*

7. Now switch to AJMason on your computer, and connect to SQL Server Express. Create a new Query Editor window. If you do a SELECT * FROM ShareDetails.Shares, you will get an empty list. This is because you have SELECT permissions. However, if you try to INSERT some data, as shown in Figure 8-42, you will see an error.

Figure 8-42. *AJMason cannot insert data.*

Securables

There are three different options in the Securables dialog box for explicit permissions: GRANT, WITH GRANT, and DENY. Although we are looking at tables here, the same options are possible with other objects that we come across within SSE.

GRANT: This will grant the user or role access on that action on the defined table.

WITH GRANT: This will also grant the user or role access on that action on the defined table. However, anyone with this option can also pass on the permission to other users or roles.

DENY: This explicitly denies any user or role the action against the table.

By setting up different roles and placing users in those roles, you should now know which users can access the data, and how to set up groups of users to protect your data. Of course, it is normal practice to set up several roles within your database for each area of the business. There would be a role for supervisors, and perhaps another for line managers, and another for directors, and so on. It all depends on your database and the solution you are providing as to how many different roles are required. But from this it is simpler to control access to the data.

Updating Data

Now that data has been inserted into our database, and you have seen how to retrieve this information, it is time to look at how to modify the data, referred to as **updating the data**, and the different methods of deletion.

Ensuring that you update the right data at the right time is crucial to maintaining data integrity. You will find that when updating data, and also when removing or inserting data, it is best to group this work as a single, logical unit, called a **transaction**, thereby ensuring that if an error does occur, it is still possible to return the data back to its original state. This section describes how a transaction works and how to incorporate transactions within your code. When looking at transactions, we will only be taking an overview of them. We will look at the basics of a transaction and how it can affect the data.

Deleting data can take one of two forms. The first is where a deletion of the data is logged in the transaction log. This means that if there is a failure of some sort, the deletion can be backed out. The second is where the deletion of the data is minimally logged. Knowing when to use each of these actions can improve performance of deletions.

This discussion aims to ensure that you

- Know the syntax of the UPDATE statement.
- Are competent at updating data within a SQL Server table.
- Are aware of transactions and how to use them effectively within SSE.
- Understand the dangers when transactions are nested.
- Know the syntax for the DELETE statement.
- Know how to use this statement in T-SQL.
- Are aware of the pitfalls of the TRUNCATE statement.

First of all, let's take a look at the syntax for the UPDATE statement.

The UPDATE Statement

The UPDATE statement will update columns of information on rows within a single table returned from a query that can include selection and join criteria. The syntax of the UPDATE statement has similarities to the SELECT statement, which makes sense, as it has to look for specific rows to update, just as the SELECT statement looks for rows to retrieve. You will also find that before doing updates, especially more complex updates, you'll want to build up a SELECT statement first and then transfer the JOIN and WHERE details in to the UPDATE statement. The syntax that follows is in its simplest form. Once you become more experienced, the UPDATE statement can become just as complex and versatile as the SELECT statement.

```
UPDATE
    [ TOP ( expression ) [ PERCENT ] ]
    [[ server_name . database_name . schema_name .
    | database_name .[ schema_name ] .
    | schema_name .]
table_or_viewname
    SET
        { column_name = { expression | DEFAULT | NULL }
        | column_name { .WRITE ( expression , @Offset , @Length ) }
        | @variable = expression
        | @variable = column = expression [ ,...n ]
        } [ ,...n ]
    [FROM { <table_source> } [ ,...n ] ]
    [ WHERE { <search_condition>]
```

The first set of options we know from the SELECT statement. The tablename clause is simply the name of the table on which to perform the UPDATE. Moving on to the next line of the syntax, we reach the SET clause. It is in this clause that any updates to a column take place. One or more columns can be updated at any one time, but each column to be updated must be separated by a comma.

When updating a column, there are four choices that can be made for data updates. This can be through a direct value setting; a section of a value setting providing that the recipient column is varchar, nvarchar, or varbinary; the value from a variable; or a value from another column or even from another table. We can even include mathematical functions or variable manipulations in the right-hand clause, include concatenated columns, or manipulate the contents through STRING, DATE, or any other function. Providing that the end result sees the left-hand side having the same data type as the right-hand side, the update will then be successful. As a result, we cannot place a character value into a numeric data type field without converting the character to a numeric value.

If we are updating a column with a value from another column, the only value that it is possible to use is the value from the same row of information in another column, provided this column has an appropriate data type. When we say "same row," remember that when tables are joined together, this means that values from these joined tables can also be used as they are within the same row of information. Also, the expression could be the result of a subquery.

▨**Note** A **subquery** is a query that sits inside another query. We look at these in Chapter 12.

The FROM table source clause will define the table(s) used to find the data to perform the update on the table defined next to the UPDATE statement. Like SELECT statements, it is possible to create JOIN statements; however, you must define the table you are updating within the FROM clause.

Finally, the WHERE condition is exactly as in the SELECT statement, and can be used in exactly the same way. Note that omitting the WHERE clause will mean the UPDATE statement will affect every row in the table.

Updating Data Within Query Editor

To demonstrate the UPDATE statement, the first update to the data will be to change the name of a customer, replicating when someone changes their name due to marriage or deed, for example. This uses the UPDATE statement in its simplest form, by locating a single row and updating a single column.

Try It Out: Updating a Row of Data

1. Ensure that Query Editor is running and that you are logged in with an account that can perform updates. In the Query Editor pane, enter the following UPDATE statement:

```
UPDATE CustomerDetails.Customers
SET CustomerLastName = 'Brodie'
WHERE CustomerId = 1
```

2. It is as simple as that! Now that the code is entered, execute the code, and you should then see a message like this:

```
(1 row(s) affected)
```

3. Now enter a SELECT statement to check that Vic McGlynn is now Vic Brodie. For your convenience, here's the statement, and the results are shown in Figure 8-43:

```
SELECT * FROM CustomerDetails.Customers
WHERE CustomerId = 1
```

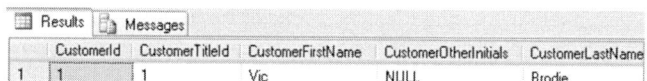

Figure 8-43. *Vic McGlynn is now Vic Brodie.*

4. Now here's a little trick that you should know, if you haven't stumbled across it already. If you check out Figure 8-44, you will see that the UPDATE code is still in the Query Editor pane, as is the SELECT statement. No, we aren't going to perform the UPDATE again! If you highlight the line with the SELECT statement by holding down the left mouse button and dragging the mouse, only the highlighted code will run when you execute the code again.

```
xp-home.Apress...ieveBrodie.sql    xp-home.Apres...eToBrodie.s
   UPDATE CustomerDetails.Customers
   SET CustomerLastName = 'Brodie'
   WHERE CustomerId = 1

   SELECT * FROM CustomerDetails.Customers
   WHERE CustomerId = 1
```

Figure 8-44. *How to execute only specific code*

■Note On executing the highlighted code, you should see only the values returned for the SELECT statement as
you saw previously, and no results saying that an update had been performed.

5. It is also possible to update data using information from another column within the table, or with the value
 from a variable. This next example will demonstrate how to update a row of information using the value
 within a variable, and a column from the same table. Notice how although the row will be found using the
 CustomerLastName column, the UPDATE statement is also updating that column with a new value. Enter
 the following code and then execute it:

```
DECLARE @ValueToUpdate VARCHAR(30)
SET @ValueToUpdate = 'McGlynn'
UPDATE CustomerDetails.Customers
    SET CustomerLastName = @ValueToUpdate,
        ClearedBalance = ClearedBalance + UnclearedBalance ,
        UnclearedBalance = 0
  WHERE CustomerLastName = 'Brodie'
```

6. You should then see the following output:

```
(1 row(s) affected)
```

7. Now to check what has happened. You may be thinking that the update has not happened because you are
 altering the column that is being used to find the row, but this is not so. The row is found, the update occurs,
 and then the row is written back to the table. Once the row is retrieved for update, there is no need for that
 value to be kept. Just check that the update occurred by entering and executing the following code:

```
SELECT CustomerFirstName, CustomerLastName,
ClearedBalance, UnclearedBalance
FROM CustomerDetails.Customers
WHERE CustomerId = 1
```

 You should now see the alteration in place, as shown in Figure 8-45.

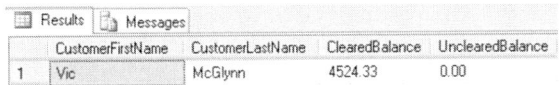

Figure 8-45. *Updating multiple columns*

8. Now let's move on to updating columns in which the data types don't match. SSE does a pretty good job when
 it can to ensure the update occurs, and these following examples will demonstrate how well SSE copes with
 updating an integer data type with a value in a varchar data type. The first example will demonstrate where
 a varchar value will successfully update a column defined as integer. Enter the following code:

```
DECLARE @WrongDataType VARCHAR(20)
SET @WrongDataType = '4311.22'
UPDATE CustomerDetails.Customers
    SET ClearedBalance = @WrongDataType
WHERE CustomerId = 1
```

9. Execute the code; you should see the following message when you do:

```
(1 row(s) affected)
```

10. The value `4311.22` has been placed into the `ClearedBalance` column for `CustomerId` `1`. SSE has performed an internal data conversion (known as an **implicit data type conversion**) and has come up with a `money` data type from the value within `varchar`, as this is what the column expects, and therefore can successfully update the column. Here is the output as proof:

```
SELECT CustomerFirstName, CustomerLastName,
ClearedBalance, UnclearedBalance
FROM CustomerDetails.Customers
WHERE CustomerId = 1
```

Figure 8-46 shows the results of updating the column.

	CustomerFirstName	CustomerLastName	ClearedBalance	UnclearedBalance
1	Vic	McGlynn	4311.22	0.00

Figure 8-46. *Updating a column with internal data conversion*

11. However, in this next example, the data type that SSE will come up with is a `numeric` data type. When we try to alter an integer-based data type, `bigint`, with this value, which we can find in the `AddressId` column, the `UPDATE` does not take place. Enter the following code:

```
DECLARE @WrongDataType VARCHAR(20)
SET @WrongDataType = '2.0'
UPDATE CustomerDetails.Customers
    SET AddressId = @WrongDataType
WHERE CustomerId = 1
```

12. Now execute the code. Notice when we do that SSE generates an error message informing us of the problem. Hence, never leave data conversions to SSE to perform. Try to get the same data type updating the same data type. We look at how to convert data within Chapter 12.

```
Msg 8114, Level 16, State 5, Line 3
Error converting data type varchar to bigint.
```

Updating data can be very straightforward, as the preceding examples have demonstrated. Where at all possible, either use a unique identifier, for example, the `CustomerId`, when trying to find a customer, rather than a name. There can be multiple rows for the same name or other type of criteria, but by using the unique identifier, you can be sure of using the right row every time. To place this in a production scenario, we would have a Windows-based graphical system that would allow you to find details of customers by their name, address, or account number. Once you found the right customer, instead of keeping those details to find other related rows, keep a record of the unique identifier value instead.

Getting back to the `UPDATE` statement and how it works, first of all SSE will filter out from the table the first row that meets the criteria of the `WHERE` statement. The data modifications are then made, and SSE moves on to try to find the second row matching the `WHERE` statement. This process is repeated until all the rows that meet the `WHERE` condition are modified. Therefore, if using a unique identifier, SSE will only update one row, but the `WHERE` statement looks for rows that have a `CustomerLastName` of McGlynn, in which case multiple rows could be updated. So choose your row selection criteria for updates carefully.

But what if you didn't want the update to occur immediately? There will be times when you will want to perform an update, and then check that the update is correct before finally committing the changes to the table. Or when doing the update, you want to check for errors or unexpected data updates. This is where transactions come in, and these are covered next.

Transactions

A **transaction** is a method through which developers can define a unit of work logically or physically that, when it completes, leaves the database in a consistent state. A transaction forms a single unit of work, which must pass the ACID test before it can be classified as a transaction. The ACID test is an acronym for Atomicity, Consistency, Isolation, and Durability:

Atomicity: In its simplest form, all data modifications within the transaction must be both accepted and inserted successfully into the database, or none of the modifications will be performed.

Consistency: Once the data has been successfully applied, or rolled back to the original state, all the data must remain in a consistent state, and the data must still maintain its integrity.

Isolation: Any modification in one transaction must be isolated from any modifications in any other transaction. Any transaction should see data from any other transaction either in its original state or once the second transaction has completed. It is impossible to see the data in an intermediate state.

Durability: Once a transaction has finished, all data modifications are in place and can only be modified by another transaction or unit of work. Any system failure (hardware or software) will not remove any changes applied.

Transactions within a database are a very important topic, but also one that requires a great deal of understanding. This chapter covers the basics of transactions only. To really do justice to this area, we would have to deal with some very complex and in-depth scenarios, covering all manner of areas such as triggers, nesting transactions, and transaction logging, which is beyond the scope of this book.

A transaction can be placed around any data manipulation, whether it is an update, insertion, or deletion, and can cater to one row or many rows, and also many different statements. There is no need to place a transaction around a SELECT statement unless you are doing a SELECT...INTO, which is of course modifying data. This is because a transaction is only required when data manipulation occurs such that changes will either be committed to the table or discarded. A transaction could cover several UPDATE, DELETE, or INSERT statements, or indeed a mixture of all three. However, there is one very large warning that goes with using transactions.

▨Caution Be aware that when creating a transaction, you will be keeping a hold on the whole table, pages of data, or specific rows of information in question, and depending upon how your SSE database is set up to lock data during updates, you could be stopping others from updating any information, and you could even cause a **deadlock**, also known as a **deadly embrace**. If a deadlock occurs, SSE chooses one of the deadlocks and kills the process; there is no way of knowing which process SSE will select.

A deadlock is where two separate data manipulations, in different transactions, are being performed at the same time. However, each transaction is waiting for the other to finish the update before it can complete its update. Neither manipulation can be completed because each is waiting for the other to finish. A deadlock occurs, and it can (and will) lock the tables and database in

question. So, for example, transaction 1 is updating the customers table followed by the customer transactions table. Transaction 2 is updating the customer transactions table followed by the customers table. A lock would be placed on the customers table while those updates were being done by transaction 1. A lock would be placed on the customer transactions table by transaction 2. Transaction 1 could not proceed because of the lock by transaction 2, and transaction 2 could not proceed due to the lock created by transaction 1. Both transactions are "stuck." So it is crucial to keep the order of table updates the same, especially where both could be running at the same time.

It is also advisable to keep transactions as small, and as short, as possible, and under no circumstances hold onto a lock for more than a few seconds. We can do this by keeping the processing within a transaction to as few lines of code as possible, and then either roll back (that is, cancel) or commit the transaction to the database as quickly as possible within code. With every second that you hold a lock through a transaction, you are increasing the potential of trouble happening. In a production environment, with every passing millisecond that you hold onto a piece of information through a lock, you are increasing the chances of someone else trying to modify the same piece of information at the same time and the possibility of the problems that would then arise.

There are two parts that make up a transaction, the start of the transaction and the end of the transaction, where you decide whether you want to commit the changes or revert back to the original state. We will now look at the definition of the start of the transaction, and then the T-SQL statements required to commit or roll back the transaction. The basis of this section is that only one transaction is in place, and that you have no nested transactions. Nested transactions are much more complex and should only really be dealt with once you are proficient with SSE. The statements we are going through in the upcoming text assume a single transaction; the COMMIT TRAN section changes slightly when the transaction is nested.

BEGIN TRAN

The T-SQL statement BEGIN TRAN denotes the start of the transaction processing. From this point on, until the transaction is ended with either COMMIT TRAN or ROLLBACK TRAN, any data modification statements will form part of the transaction.

It is also possible to suffix the BEGIN TRAN statement with a name of up to 32 characters in length. If you name your transaction, it is not necessary to use the name when issuing a ROLLBACK TRAN or a COMMIT TRAN statement. The name is there for clarity of the code only.

COMMIT TRAN

The COMMIT TRAN statement will commit the data modifications to the database permanently, and there will be no going back once this statement is executed. This function should only be executed when all changes to the database are ready to be committed.

ROLLBACK TRAN

If you wish to remove all the database changes that have been completed since the beginning of the transaction, say, for example, because an error had occurred, then you could issue a ROLLBACK TRAN statement.

So, if you were to start a transaction with BEGIN TRAN and then issue an INSERT that succeeds, and then perhaps an UPDATE that fails, you could issue a ROLLBACK TRAN to roll back the transaction as a whole. As a result, you roll back not only the UPDATE changes, but also, because they form part of the same transaction, the changes made by the INSERT, even though that particular operation was successful.

To reiterate, keep transactions small and short. Never leave a session with an open transaction by having a `BEGIN TRAN` with no `COMMIT TRAN` or `ROLLBACK TRAN`. Ensure that you do not cause a deadly embrace.

If you issue a `BEGIN TRAN`, then you MUST issue a `COMMIT TRAN` or `ROLLBACK TRAN` transaction as quickly as possible; otherwise, the transaction will stay around until the connection is terminated.

Locking Data

The whole area of locking data, how locks are held, and how to avoid problems with them, is a very large complex area and not for the fainthearted. However, it is necessary to be aware of locks, and at least have a small amount of background knowledge on them so that when you design your queries, you stand a chance of avoiding problems.

The basis of locking is to allow one transaction to update data, knowing that if it has to roll back any changes, no other transaction has modified the data since the first transaction did.

To explain this with an example, if you have a transaction that updates the `CustomerDetails.Customers` table, and then moves on to update the `TransactionDetails.Transactions` table, but hits a problem when updating the `TransactionDetails.Transactions` table, the transaction must be safe in the knowledge that it is only rolling back the changes it made, and not changes by another transaction. Therefore, until all the table updates within the transaction are either successfully completed or have been rolled back, the transaction will keep hold of any data inserted, modified, or deleted.

However, one problem with this approach is that SSE may not just hold the data that the transaction has modified. Keeping a lock on the data that has just been modified is called **row-level locking**. On the other hand, SSE may be set up to lock the database, which is known as **database-level locking**, found in areas such as backups and restores. The other levels in between are row, page, and table locking, and so you could lock a large resource, depending on the task being performed.

This is about as deep as I will take this discussion on locks, so as not to add any confusion or create a problematic situation. Dealing with locks is handled automatically by SSE, but it is possible to make locking more efficient by developing an effective understanding of the subject, and then customizing the locks within your transactions.

Updating Data: Using Transactions

Now, what if, in the first update query of this chapter, we had made a mistake or an error occurred? For example, say we chose the wrong customer, or even worse, omitted the `WHERE` statement, and therefore all the rows were updated. These are unusual errors, but quite possible. More common errors could result from where more than one data modification has to take place and succeed, and the first one succeeds but a subsequent modification fails. By using a transaction, we would have had the chance to correct any mistakes easily, and could then revert to a consistent state. Of course, this next example is nice and simple, but by working through it, the subject of transactions will hopefully become a little easier to understand and appreciate.

Try It Out: Using a Transaction

1. Make sure Query Editor is running for this first example, which will demonstrate `COMMIT TRAN` in action. There should be no difference from an `UPDATE` without any transaction processing, as it will execute and update the data successfully. However, this should prove to be a valuable exercise, as it will also demonstrate the naming of a transaction. Enter the following code:

```
SELECT 'Before',ShareId,ShareDesc,CurrentPrice
  FROM ShareDetails.Shares
 WHERE ShareId = 3
BEGIN TRAN ShareUpd
UPDATE ShareDetails.Shares
   SET CurrentPrice = CurrentPrice * 1.1
 WHERE ShareId = 3
COMMIT TRAN
SELECT 'After',ShareId,ShareDesc,CurrentPrice
  FROM ShareDetails.Shares
 WHERE ShareId = 3
```

Notice in the preceding code that the COMMIT TRAN does not use the name associated with the BEGIN TRAN. The label after the BEGIN TRAN is simply that, a label, and it performs no functionality. It is therefore not necessary to then link up with a similarly labeled COMMIT TRAN.

2. Execute the code. Figure 8-47 shows the results, which list out the ShareDetails.Shares table before and after the transaction.

Figure 8-47. *Updating with transaction label and a COMMIT TRAN*

3. We are now going to work through a ROLLBACK TRAN. We will take this one stage at a time so that you fully understand and follow the processes involved. Note in the following code that the WHERE statement has been commented out with --. By having the WHERE statement commented out, hopefully you'll have already guessed that every row in the ShareDetails.Shares table is going to be updated. The example needs you to execute all the code at once, so enter the following code into your Query Editor pane, and then execute it. Note we have three SELECT statements this time—before, during, and after the transaction processing.

```
SELECT 'Before',ShareId,ShareDesc,CurrentPrice
  FROM ShareDetails.Shares
-- WHERE ShareId = 3
BEGIN TRAN ShareUpd
UPDATE ShareDetails.Shares
   SET CurrentPrice = CurrentPrice * 1.1
-- WHERE ShareId = 3
SELECT 'Within the transaction',ShareId,ShareDesc,CurrentPrice
  FROM ShareDetails.Shares
ROLLBACK TRAN
SELECT 'After',ShareId,ShareDesc,CurrentPrice
  FROM ShareDetails.Shares
-- WHERE ShareId = 3
```

4. The results, as you see in Figure 8-48, show us exactly what has happened. Take a moment to look over these results. The first list shows the full set of rows in the ShareDetails.Shares table prior to our UPDATE. The middle rowset shows us the BEGIN transaction where we have updated every share, and the final listing shows the data restored back to its original state via a ROLLBACK TRAN.

(No column name)	ShareId	ShareDesc	CurrentPrice	
1	Before	1	ACME'S HOMEBAKE COOKIES INC	2.34125
2	Before	2	FAT-BELLY.COM	45.20000
3	Before	3	NetRadio Inc	32.76900
4	Before	4	Texas Oil Industries	0.45500
5	Before	5	London Bridge Club	1.46000

(No column name)	ShareId	ShareDesc	CurrentPrice	
1	Within the transaction	1	ACME'S HOMEBAKE COOKIES INC	2.57538
2	Within the transaction	2	FAT-BELLY.COM	49.72000
3	Within the transaction	3	NetRadio Inc	36.04590
4	Within the transaction	4	Texas Oil Industries	0.50050
5	Within the transaction	5	London Bridge Club	1.60600

(No column name)	ShareId	ShareDesc	CurrentPrice	
1	After	1	ACME'S HOMEBAKE COOKIES INC	2.34125
2	After	2	FAT-BELLY.COM	45.20000
3	After	3	NetRadio Inc	32.76900
4	After	4	Texas Oil Industries	0.45500
5	After	5	London Bridge Club	1.46000

Figure 8-48. *Updating with transaction label and a ROLLBACK TRAN*

Nested Transactions

Let's look at one last example before moving on. It is possible to nest transactions inside one another. I touch on this enough for you to have a good understanding on nested transactions, but this is not a complete coverage, as it can get very complex and messy if you involve save points, stored procedures, triggers, and so on. The aim of this section is to give you an understanding of the basic but crucial points of how nested transactions work.

Nested transactions can occur in a number of different scenarios. For example, you could have a transaction in one set of code in a stored procedure, which calls a second stored procedure that also has a transaction. We will look at a simpler scenario where we just keep the transactions in one set of code.

What you need to be clear about is how the ROLLBACK and COMMIT TRAN statements work in a nested transaction. First of all, let's see what we mean by nesting a simple transaction. The syntax is shown here, and you can see that two BEGIN TRAN statements occur before you get to a COMMIT or a ROLLBACK:

```
BEGIN TRAN
    Statements
    BEGIN TRAN
        Statements
    COMMIT|ROLLBACK TRAN
COMMIT|ROLLBACK TRAN
```

As each transaction commences, SSE adds 1 to a running count of transactions it holds in a system variable called @@TRANCOUNT. Therefore, as each BEGIN TRAN is executed, @@TRANCOUNT increases by 1. As each COMMIT TRAN is executed, @@TRANCOUNT decreases by 1. It is not until @@TRANCOUNT is at a value of 1 that you can actually commit the data to the database. The code that follows might help you to understand this a bit more.

Enter and execute this code and take a look at the output, which should resemble Figure 8-49. The first BEGIN TRAN increases @@TRANCOUNT by 1, as does the second BEGIN TRAN. The first COMMIT TRAN marks the changes to be committed, but does not actually perform the changes because

@@TRANCOUNT is 2. It simply creates the correct BEGIN/COMMIT TRAN nesting and reduces @@TRANCOUNT by 1. The second COMMIT TRAN will succeed and will commit the data, as @@TRANCOUNT is 1.

```
BEGIN TRAN ShareUpd
    SELECT '1st TranCount',@@TRANCOUNT
    BEGIN TRAN ShareUpd2
        SELECT '2nd TranCount',@@TRANCOUNT
        COMMIT TRAN ShareUpd2
    SELECT '3rd TranCount',@@TRANCOUNT
COMMIT TRAN -- It is at this point that data modifications will be committed
SELECT 'Last TranCount',@@TRANCOUNT
```

Figure 8-49. *Showing the @@TRANCOUNT*

■**Note** After the last COMMIT TRAN, @@TRANCOUNT is at 0. Any further instances of COMMIT TRAN or ROLLBACK TRAN will generate an error.

If in the code there is a ROLLBACK TRAN, the data is immediately rolled back no matter where you are within the nesting, and @@TRANCOUNT is set to 0. Therefore, any further ROLLBACK TRAN or COMMIT TRAN instances will fail, so you do need to have error handling, which we look at in Chapter 11.

Try to avoid nesting transactions where possible, especially when one stored procedure calls another stored procedure within a transaction. It is not "wrong," but it does require a great deal of care.

Now that updating data has been completed, the only area of data manipulation left is row deletion, which we look at now.

Deleting Data

Deleting data can be considered very straightforward, especially compared to all of the other data manipulation functions covered previously, particularly transactions and basic SQL. However, mistakes made when deleting data are very hard to recover from. Therefore, you must treat deleting data with the greatest of care and attention to detail, and especially test any joins and filtering via a SELECT statement before running the delete operation.

Deleting data without the use of a transaction is almost a final act: the only way to get the data back is to reenter it, restore it from a backup, or retrieve the data from any audit tables that had the data stored in them when the data was created. Deleting data is not like using the recycle bin on a Windows machine: unless the data is within a transaction, it is lost. Keep in mind that even if you

use a transaction, the data will be lost once the transaction is committed. That's why it's very important to back up your database before running any major data modifications.

This section of the chapter will demonstrate the DELETE T-SQL syntax and then show how to use this within Query Editor. It is also possible to delete rows from the results pane within SSMSE, which will also be demonstrated.

However, what about when you want to remove all the rows within a table, especially when there could be thousands of rows to remove? You will find that the DELETE statement takes a very long time to run, as each row to delete is logged in the transaction log, thus allowing transactions to be rolled back. Luckily, there is a statement for this scenario, called TRUNCATE, which is covered in the section "Truncating a Table" later in the chapter. However, caution should be exercised when using this statement, and you'll see why later.

First of all, it is necessary to learn the simple syntax for the DELETE statement for deleting rows from a table. Really, things don't come much simpler than this.

DELETE Syntax

The DELETE statement is very short and sweet. To run the statement, simply state the table you wish to delete rows from, as shown here:

```
DELETE
[FROM] tablename
WHERE where_condition
```

The FROM is optional, so your syntax could easily read

```
DELETE tablename
WHERE where_condition
```

There is nothing within this statement that has not been covered in other chapters. The only area that really needs to be mentioned is that rows can only be deleted from one table at a time, although when looking for rows to delete, you can join to several tables, as you can with SELECT and UPDATE.

Now that you've seen the DELETE syntax, let's dive right in with an example.

Using the DELETE Statement

Recall we created a table with the SELECT INTO statement called CustTemp. Rather than delete data from the main tables created so far, we'll use this temporary table in this section of the book.

We'll use transactions a great deal here to avoid having to keep inserting data back into the table. It's a good idea to use transactions for any type of table modification in your application. Imagine that you're at the ATM and you are transferring money from your savings account to your checking account. During that process, a transaction built up of many actions is used to make sure that your money doesn't credit one system and not the other. If an error occurs, the entire transaction rolls back, and no money will move between the accounts.

Let's take a look at what happens if you were to run this statement:

```
BEGIN TRAN
    DELETE CustTemp
```

When this code runs, SSE opens a transaction and then tentatively deletes all the rows from the CustTemp table. The rows are not actually deleted until a COMMIT TRAN statement is issued. In the interim, though, SSE will place a lock on the rows of the table, or if this was a much larger table, SSE may decide that a table lock (locking the whole table to prevent other modifications) is better. Because of this lock, all users trying to modify data from this table will have to wait until a COMMIT TRAN or ROLLBACK TRAN statement has been issued and completed. If one is never issued, users will

be blocked. This problem is one of a number of issues frequently encountered in applications when analyzing performance issues. Therefore, never have a BEGIN TRAN without a COMMIT TRAN or ROLLBACK TRAN.

So, time to start deleting rows.

Try It Out: Deleting Rows

1. Enter the following statements in an empty Query Editor pane. This will remove all the rows from our table within a transaction, prove the point by trying to list the rows, and then roll back the changes so that the rows are put back into the table.

```
BEGIN TRAN
    SELECT * FROM CustTemp
    DELETE CustTemp
    SELECT * FROM CustTemp
ROLLBACK TRAN
SELECT * FROM CustTemp
```

2. Execute the code. You should see the results displayed in Figure 8-50. Notice that the number of rows in the CustTemp table before the delete is 5, then after the delete the row count is tentatively set to 0. Finally, after the rollback, it's set back to 5. If we do not issue a ROLLBACK TRAN statement in here, we would see 0 rows, but other connections would be blocked until we did.

	Name	ClearedBalance	UnclearedBalance
1	Vic McGlynn	4311.22	213.11
2	Jack Mason	437.97	-10.56
3	Bernie McGee	6653.11	0.00
4	Julie Dewson	53.32	-12.21
5	Kirsty Hull	1266.00	10.32

	Name	ClearedBalance	UnclearedBalance

	Name	ClearedBalance	UnclearedBalance
1	Vic McGlynn	4311.22	213.11
2	Jack Mason	437.97	-10.56
3	Bernie Mc...	6653.11	0.00
4	Julie Dews...	53.32	-12.21
5	Kirsty Hull	1266.00	10.32

Figure 8-50. *Deletion of all rows that were rolled back*

3. Let's take this a stage further and only remove the last three rows of the table. Again, this will be within a transaction. Alter the preceding code as indicated in the following snippet. Here we are using the TOP statement to delete three random rows. Why random? SSE only stores rows in a definite order if they are covered by a clustered index. No other index, or no index, can guarantee the order in which SSE stores other rows. This is not the best way to delete rows, as in virtually all cases you will want to control the deletion.

```
BEGIN TRAN
    SELECT * FROM CustTemp
    DELETE TOP (3) CustTemp
    SELECT * FROM CustTemp
ROLLBACK TRAN
SELECT * FROM CustTemp
```

4. Execute the code, which should produce the results shown in Figure 8-51.

Figure 8-51. *Deletion of three rows that were rolled back*

Truncating a Table

All delete actions caused by DELETE statements are recorded in the transaction log. Each time a row is deleted, a record is made of that fact. If you are deleting millions of rows before committing your transaction, your transaction log can grow quickly. Recall from earlier in the chapter the discussions about transactions; now think about this a bit more. What if the table you are deleting from has thousands of rows? That is a great deal of logging going on within the transaction log. But what if the deletion of these thousands of rows was, in fact, cleaning out all the data from the table to start afresh? Or perhaps this is some sort of transient table? Performing a DELETE would seem a lot of overhead when you don't really need to keep a log of the data deletions anyway. If the action failed for whatever reason, you would simply retry removing the rows a second time. This is where the TRUNCATE TABLE statement comes into its own.

By issuing a TRUNCATE TABLE statement, you are instructing SSE to delete every row within a table, without any logging or transaction processing taking place. In reality, minimal data is logged about what data pages have been deallocated and therefore removed from the database. This is in contrast to a DELETE statement, which will only deallocate and remove the pages from the table if it can get sufficient locks on the table to do this. The deletion of the rows can be almost instantaneous, and a great deal faster than using the DELETE statement. This occurs not only because of the differences with what happens with the transaction log, but also because of how the data is locked at the time of deletion. Let's clarify this point before progressing.

When a DELETE statement is issued, each row that is to be deleted will be locked by SSE so that no modifications or other DELETE statements can attempt to work with that row. Deleting hundreds or thousands of rows is a large number of actions for SSE to perform, and it will take time to locate each row and place a lock against it. However, a TRUNCATE TABLE statement locks the whole table. This is one action that will prevent any data insertion, modification, or deletion from taking place.

■**Note** A TRUNCATE TABLE statement will delete data from a table with millions of rows in only a few seconds, whereas using DELETE to remove all the rows on the same table would take several minutes.

The syntax for truncating a table is simple:

```
TRUNCATE TABLE [{database.schema_name.}]table
```

■**Caution** Use the TRUNCATE TABLE statement with extreme caution: there is no going back after the transaction is committed; you cannot change your mind. Also, every row is removed; you cannot use this statement to selectively remove some of the rows.

One "side effect" to the TRUNCATE TABLE clause is that it reseeds any identity columns. For example, say that you have a table with an identity column that is currently at 2,000,000. After truncating the table, the first inserted piece of data will produce the value 1 (if the seed is set to 1). If you issue a DELETE statement to delete the rows from the table, the first piece of data inserted after the table contents have been deleted will produce a value of 2,000,001, even though this newly inserted piece of data may be the only row in the table!

One of the limitations with the TRUNCATE TABLE statement is that you cannot issue it against tables that have foreign keys referencing them. For example, the CustomerDetails.Customers table has a foreign key referencing the TransactionDetails.Transactions table. If you try to issue the following statement:

```
TRUNCATE TABLE CustomerDetails.Customers
```

you will receive the following error message:

```
Msg 4712, Level 16, State 1, Line 1
Cannot truncate table 'CustomerDetails.Customers' because it is being referenced
by a FOREIGN KEY constraint.
```

Dropping a Table

Another way to quickly delete the data in a table is to just delete the table and re-create it. Don't forget that if you do this, you will need to also re-create any constraints, indexes, and foreign keys. When you do this, SSE will deallocate the table, which is minimally logged. To drop a table in SSE, issue the following statement:

```
DROP TABLE table_name
```

As with TRUNCATE TABLE, DROP TABLE cannot be issued against a table that has a foreign key referencing it. In this situation, either the foreign key constraint referencing the table or the referencing table itself must first be dropped before it is possible to drop the original table.

Summary

This chapter has taken a look at how to insert and then retrieve data on a set of tables in simplest form. Later in the book, we will return to retrieving data with more complex language as well as when working with data from more than one table within a single query.

We have taken a look at NULL values and default values, as well as how to insert data within columns defined with these settings and constraints. You should also be comfortable with getting information from tables using different searching, filtering, and ordering criteria.

Updating data can go wrong, and does, especially when you are working in a live environment and you wish to update data that is in flux. In such a scenario, getting the correct rows of information to update and then actually updating them is akin to a fine art.

Therefore, surrounding any of your work with a transaction will prevent any costly and potentially irretrievable mistakes from taking place, so always surround data modifications or deletions with a transaction. With data inserts, it is not quite so critical that you surround your work with a transaction, although it is recommended. For example, if you are inserting data within a test environment and the data insertion is easily removed if you have part of the insertion wrong, perhaps it's not overly critical to use a transaction; although to be safe, really and truly, I still recommend that you use a transaction.

Updating columns within a table is very straightforward. As long as the data type defined for the column to update is the same as, or is compatible with, the data type of the column, variable, or static value that is being used to update this column, then you will have no problem. For example, you can update a varchar column with char data type values. However, it is not possible to update an integer column with a varchar value that has a noninteger value without converting the varchar value to an integer. But don't be fooled, you can update a varchar with an integer or a numeric data type.

The DELETE statement in this chapter completes the statements for data retrieval and manipulation. From this point, SSE is your oyster, and there should be no stopping you now. Deleting data is a very straightforward process, perhaps too straightforward, and with no recycle bin, you really do have to take care. Having to reinstate data is a lot harder than having to remove rows inserted incorrectly or changing back modifications completed in error.

Whenever deleting data (no matter how few the rows are), it is strongly recommended that a transaction be used, as this chapter has demonstrated, and also that a SELECT statement be included to check your work.

Finally, the removal of every row within a table was also shown, along with the dire warnings if you got it wrong. Really, only use the TRUNCATE TABLE statement in development or with the utmost extreme care within production.

So where can you go from here? The next chapter will look at views of data.

CHAPTER 9

■■■

Building a View

A **view** is a virtual table that, in itself, doesn't contain any data or information. All it contains is the query that the user defines when creating the view. A view is actually a named query against one or more tables. Views are used as a security measure (to restrict users to specific columns or rows), as a method of joining data from multiple tables (and presenting it as if it resides in one table), and by returning summary data instead of detailed data. Another use for a view is to provide a method of accessing the underlying data in a manner that provides the end user with a business layout. For example, you will see within this chapter a view that shows customer details along with enriched transaction details, thus making it easier for anyone interrogating your data, even someone who has no knowledge of the underlying data model, to access useful information.

Building a simple view is a straightforward process and can be completed in SQL Server Management Studio Express (SSMSE) or a Query Editor pane using T-SQL. Each of these tools has two options to build a view, and this chapter covers all four options so that you become conversant with building a view no matter which tool is currently at hand.

To give things a bit more bite in this chapter, a query within a query, known as a **subquery**, will also be demonstrated, along with how to create a subquery to create a column in a result set.

Finally, placing an index on a view can speed up data retrieval, but it also can create performance problems. Building an index on a view is not quite as straightforward as building an index on a table.

The aim of this chapter is to

- Make you aware of what a view is

- Inform you as to how views can improve a database's security

- Show how to encrypt views so that underlying tables cannot be seen

- Demonstrate building a view using

 - SSMSE View Designer

 - SSMSE Create a View Wizard

 - A Query Editor pane and T-SQL

- Show how to join two tables within a view

- Demonstrate subqueries within a view

- Build an index on a view and explain why you would or would not do this

What Is a View?

There will be times when you want to group together data from more than one table, or perhaps only allow users to see specific information from a particular table, where some of the columns may contain sensitive or even irrelevant data. A view can take one or more columns from one or more

tables and present this information to a user, without the user accessing the actual underlying tables. A view protects the data layer while allowing access to the data. All of these scenarios can be seen as the basis and reason for building a view rather than another method of data extraction. If you are familiar with MS Access, views are similar to Access queries. Because a view represents data as if it were another table, a virtual table in fact, it is also possible to create a view of a view.

Let's take a look at how a view works. As you know, we have a customer table that holds information about our customers such as their first name, last name, account number, and balances. There will be times when you want your users to have access to only the first and last names, but not to the other sensitive data. This is where a view comes into play. You would create a view that returns only a customer's first and last name but no other information.

Creating a view can give a user enough information to satisfy a query he or she may have about data within a database without that user having to know any T-SQL commands. A view actually stores the query that creates it, and when you execute the view, the underlying query is the code that is being executed. The underlying code can be as complex as required, therefore leaving the end user with a simple SELECT * command to run with perhaps a small amount of filtering via a simple WHERE statement.

From a view, in addition to retrieving data, you can also modify the data that is being displayed, delete data, and in some situations insert new data. There are several rules and limitations for deleting, modifying, and inserting data from multitable views, some of which will be covered in the "Indexing a View" section later in the chapter.

However, a view is not a tool for processing data like a stored procedure is. A view is only able to hold one query at a time. Therefore, a view is more like a query than a stored procedure. Just as with a stored procedure or a query within a Query Editor pane, you can include tables from databases that are running on different servers. Providing the user ID has the necessary security credentials, it is possible to include tables from several databases.

To summarize, a view is a virtual table created by a stored SQL statement that can span multiple tables. Views can be used as a method of security within your database, and they provide a simpler front end to a user querying the data.

Later in the chapter, you will see how to build a view and how all of these ideas are put into practice. Before we get to that, let's look in more depth at how a view can be used as a security vehicle.

■**Note** Creating a view when there is more than one table to retrieve data from on a regular basis is quite often a sensible solution, even more so when you wish to use views as a method of simplifying the database schema and abstracting the database data into a presentation layer for users.

Using Views for Security

Security is always an issue when building your database. So far, the book has covered the different database-provided roles, when to use them, how to set up different types of roles, and how useful they are. By restricting all users from accessing or modifying the data in the tables, you will then force everyone to use views and stored procedures to complete any data task. (There will be more on stored procedures in the next chapter.)

However, by taking a view on the data and assigning which role can have select access, update access, and so on, you are protecting not only the underlying tables, but also particular columns of data. This is all covered in the discussions involving security in this chapter.

Security encompasses not only the protection of data, but also the protection of your system. At some point as a developer, you will build a view, and then someone else will come along and remove or alter a column from an underlying table that was used in the view. This causes problems;

however, this chapter will show you how to get around this situation and secure the build of a view so that this sort of thing doesn't happen.

Imagine that you have a table holding specific security-sensitive information alongside general information—an example would be where you perhaps work for the licensing agency for driver licenses and alongside the name and address, there is a column to define the number of fines that have had to be paid. As you can see, this is information that should not be viewed by all employees within the organization. So, what do you do?

The simplest answer is to create a view on the data where you exclude the columns holding the sensitive data. In this way, you can restrict access on the table to the bare minimum of roles or logins, and leave either a view or a stored procedure as the only method of data retrieval allowed. This way, the information returned is restricted to only those columns that a general user is allowed to see.

It is also possible to place a `WHERE` statement within a view to restrict the rows returned. This could be useful when you don't wish all employee salaries to be listed; perhaps excluding the salaries of the top executives would be advised!

All these methods give you, as a developer, a method for protecting the physical data lying in the base tables behind the views. Combine this with what you learned with roles, and restricting table access, and you can really tighten the security surrounding your data. With more and more companies embracing initiatives like Sarbanes-Oxley, where security should be so tight a company can be defined as having secure data, views are a great method of moving toward this goal.

Another method of securing views is to encrypt the view definition, which we explore next.

Encrypting View Definitions

As well as restricting access to certain tables or columns within a database, views also give the option of encrypting the SQL query that is used to retrieve the data. Once a view is built and you are happy that it is functioning correctly, you would release that view to production; it is at this point that you would add the final area of security—you would encrypt the view.

The most common situation where you will find views encrypted is when the information returned by the view is of a privileged nature. To expand further, not only are you using a view to return specific information, you also don't wish anyone to see how that information was returned, for whatever reason. You would therefore encrypt the SQL code that makes up the view, which would mean that how the information was being returned would not be visible.

There is a downside to encrypting a view: once the process of encryption is completed, it is difficult to get back the details of the view. There are tools on the Internet that can decrypt an encrypted view. When you encrypt a view, the view definition is not processed via encryption algorithms, but is merely obfuscated, in other words, changed so that prying eyes cannot see the code. These tools can return the obfuscation back to the original code. Therefore, if you need to modify the view, you will find that it is awkward. Not only would you have to use a tool, but you would have to delete the view and re-create it, as it would not be editable. So, if you build a view and encrypt it, you should make sure that you keep a copy of the source somewhere. This is why it is recommended that encrypted views should be used with care, and really should only be placed in production, or at worst, in user testing.

Always keep a copy of the original view, before encryption, in the company's source control system, for example, SourceSafe, and make sure that regular backups are available.

Now that we have touched upon the security issues behind views, it is time to start creating views for the database solution that we are building together.

Creating a View: SSMSE

The first task for us is to create a view using SSMSE. This is perhaps the simplest solution, as it allows us to use "drag and drop" to build the view. This may be the slowest method for creating a new view, but it does give us the greatest visual flexibility for building the view, and this may also be the best method for dealing with views that already exist and require only minor modifications.

The View Designer can aid you in the design of a view, or modification of any view already built. For example, it can assist if you are trying to build a complex view from a simple view, or it can even be used as a trial-and-error tool while you are gaining your T-SQL knowledge.

However, enough of the background—let's take a look at how the View Designer works. In this example, we will be making a view of ShareDetail.Shares.

Try It Out: Creating a View in SSMSE

1. Ensure that SSMSE is running, and that the ApressFinancial database is expanded.

2. Find the Views node, and right-click it—this brings up the pop-up menu shown in Figure 9-1; from there select New View.

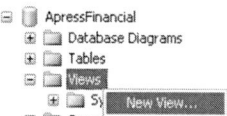

Figure 9-1. *Creating a new view*

3. The next screen you will see is the View Designer, with a modal dialog box on top presenting a list of tables that you can add to make the view. The background is pretty empty at the moment (move the dialog box around if you need to). It is within the View Designer that you will see all of the information required to build a view. There are no tables in the view at this time, so there is nothing for the View Designer to show. For those of you who are familiar with Access, you will see that the View Designer is similar to the Access Query Designer, only a bit more sophisticated! We want to add our table, so moving back to the modal dialog box, shown in Figure 9-2, select ShareDetails.Shares which appears onscreen as Shares (SharesDetails), click Add, and then click Close to remove the dialog box.

Figure 9-2. *Selecting the tables for your view*

4. Take a moment to see how the View Designer has changed, as illustrated in Figure 9-3. Notice that the background Query Designer area has been altered, the ShareDetails.Shares table has been added, and the beginnings of a SELECT statement now appear about two thirds of the way down the screen. By adding a table, the Query Designer is making a start to the view you wish to build.

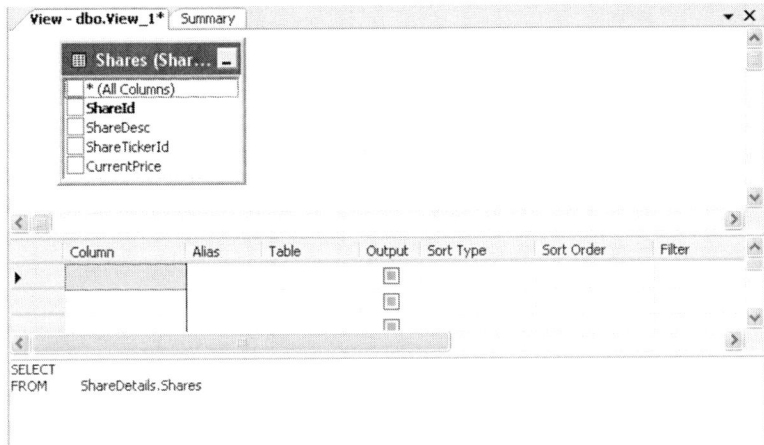

Figure 9-3. *The basic view*

5. There are four separate parts to the View Designer, each of which can be switched on or off for viewing via the toolbar buttons on top. Take a look at these toolbar buttons, as shown close up in Figure 9-4. The first button brings up the top pane, the diagram pane, where you can see the tables involved in the view and can access them via the leftmost toolbar button. The next button accesses the criteria pane, where you can filter the information you want to display. The third button accesses the SQL pane, and the fourth button accesses the results pane. As with Query Editor, here you also have the ability to execute a query through the Execute button (the one with the red exclamation point). The final button relates to verifying the T-SQL. When building the view, although the T-SQL is created as you build up the view, you can alter the T-SQL code, and this button will verify any changes.

Figure 9-4. *View toolbar buttons*

6. We will see the ShareDetails.Shares table listed in the top part of the Query Designer (the diagram pane) with no check marks against any of the column names, indicating that there are not yet any columns within the view. What we want is a view that will display the share description, the stock market ticker ID, and the current price. If we wanted all the columns displayed, we could click the check box next to * (All Columns), but for our example, just place checks against the last three columns, as shown in Figure 9-5. Notice as you check the boxes how the two areas below the table pane alter. The middle grid pane lists all the columns selected and gives you options for sorting and giving the column an alias name. The bottom part is the underlying query of the columns selected. The finished designer will look as shown in Figure 9-5.

Figure 9-5. *Our view with the columns selected*

7. We are going to change the details in the column grid now to enforce sorting criteria and to give the column aliases. This means that if a user just does SELECT * from the view, he or she will receive the data in the order defined by the view's query and also that some of the column names will have been altered from those of the underlying table. We want to ensure that the shares come out from the view in name order ascending. Move to the Sort Type column and click in the row that corresponds to ShareDesc. Select Ascending as shown in Figure 9-6.

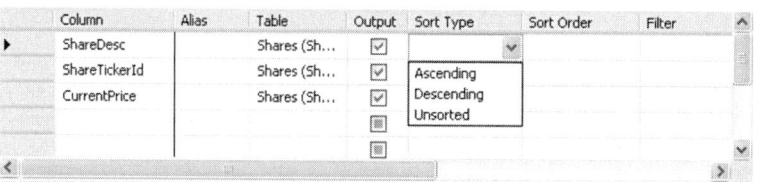

Figure 9-6. *Placing an order on the data*

8. In the next column, if we were defining more than one column to sort, we would define the order to sort the columns in. However, we still need to add the aliases, which are found in the second column of the grid. Notice the third option, CurrentPrice. To make this column more user friendly, we make the name Latest Price, with a space. When we type this and tab out of the column, it becomes [Latest Price], as you see in Figure 9-7; SSE places the square brackets around the name for us because of the space.

Column	Alias
ShareDesc	Description
ShareTickerId	Ticker
CurrentPrice	[Latest Price]

Figure 9-7. *Alias with identifier*

9. Scrolling to the right of the screen would allow us to define a filter for the view as well. This is ideal if we want to restrict what a user can see. Although sort orders can be changed by the T-SQL that calls the view, filters placed within the view cannot return more data than the view allows. So going back to our salary example mentioned earlier, this would be where we would restrict users to not seeing the MD's salary. In our example, we will only list those shares that have a current price, in other words where `CurrentPrice` is greater than 0, as shown in Figure 9-8.

Figure 9-8. *Filtering the data*

10. Notice the Query Editor pane, which now has the filter within it as well as the sorting order. Also take a look at the diagram pane and how the table display has been altered, as you see in Figure 9-9.

Figure 9-9. *The table with the view options applied*

11. Moving back to the T-SQL in the SQL pane, what about the `TOP (100) PERCENT` clause? Where did that come from? First of all, if you specify an order in a view, by default SSE will place the `TOP (100) PERCENT` clause within the SQL. It can be used if the table is very large and you don't want to allow users to return all the data on a production system, as it would tie up resources. You can also remove that clause from the Query Editor pane if you want; this will unlink your query from the designer and the Properties window, but you would also need to remove the `ORDER BY`. A final point to notice is how the column aliases are defined. The physical column is named followed by `AS` and then the alias.

Note The `AS` when defining aliases is optional.

```
SELECT TOP (100) PERCENT
  ShareDesc AS Description,
  ShareTickerId AS Ticker,
  CurrentPrice AS [Latest Price]
FROM ShareDetails.Shares
WHERE     (CurrentPrice > 0)
ORDER BY Description
```

12. If you wish to remove the `TOP` clause, it would be better to do this within the Properties window, shown in Figure 9-10, usually found on the bottom right of SSMSE; however, you would also need to remove the sorting. If it's not there, it can be found by selecting View ➤ Toolbox from the menu or by pressing F4. Within the properties, we can give the view a description—very useful—but we can also remove the `TOP` clause by setting `Top Specification` to No. We can also define whether this view is read-only by setting `Update Specification` to No.

Figure 9-10. *The properties of a view*

13. We do need to change some of the properties in the view definition, as shown in Figure 9-11. First of all, it is better to give the view a description. Also, like a table, a view should belong to a schema. This can be from an existing schema, or if you have a view traversing more than one table, you may have a schema to cater to that scenario. In our case, it fits into the `ShareDetails` schema.

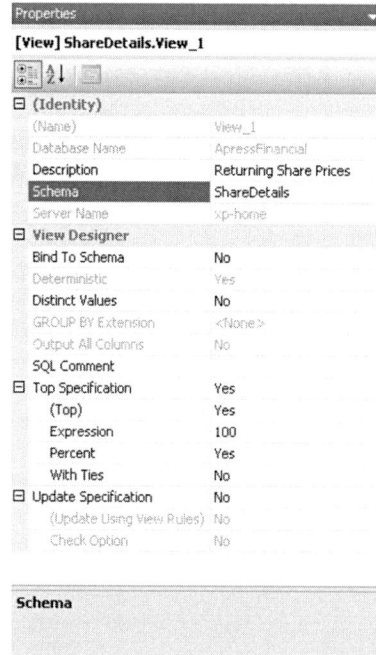

Figure 9-11. *Populated properties of a view*

14. We think the view is complete, but we need to test it out. By executing the query with the Execute button (the one sporting the red exclamation point), we will see the results in the results pane.

15. Now that the view is complete, it is time to save it to the database. Clicking the close button will bring up a dialog box asking whether you want to save the view. Click Yes to bring up a dialog box in which you give the view a name. You may find while starting out that there is a benefit to prefixing the name of the view with something like vw_ so that you know when looking at the object that it's a view. Many organizations do use this naming standard; however, it is not compulsory, and SSMSE makes it clear what each object is. The naming standard comes from a time when tools did not make it clear what object belonged to which group of object types. Once you have the name you wish, as shown in Figure 9-12, click OK.

Figure 9-12. *Naming the view*

16. This will bring us back to SSMSE, where we will see the view saved (see Figure 9-13).

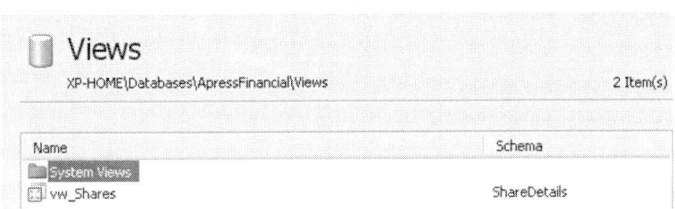

Figure 9-13. *Finding a view in Object Explorer*

We have now created our first view on the database. However, this method of building a view could be seen as a bit slow and cumbersome for something so simple. What if we wanted to combine two tables, or a view and another table?

Creating a View Using a View

Creating a view that uses another view is as straightforward as building a view based on a table. The downside of building a view with a view is that it cannot be indexed for faster execution. Therefore, depending on what the T-SQL of the final view is, data retrieval may not be as fast as it could be with an index. Also, by having a view within a view, you are adding increased complexity when debugging or profiling performance. Therefore, consider including the T-SQL from the selected view in this new view.

In this example, we will build a view of share prices using the vw_Shares view created previously. In reality, we would use the ShareDetails.Shares table along with ShareDetails.SharesPrices for the reasons just discussed.

Try It Out: Creating a View with a View

1. From Object Explorer, find Views, right-click, and select New View. The Add Table dialog box comes up as before (see Figure 9-14). From the Table tab, select `SharePrices(ShareDetails)`.

Figure 9-14. *Adding a table*

2. Move to the Views tab; there should only be one view, shown in Figure 9-15, as that is all we have created. Select the view, click Add, and then click Close.

Figure 9-15. *Adding a view*

3. The View Designer will now look similar to Figure 9-16, with two tables and the SQL showing a new type of join, a `CROSS JOIN`.

Figure 9-16. *With more than one object, how the basic view looks*

■**Note** A CROSS JOIN will take every row in one table and join it with every row in the second table. We look at these in Chapter 12.

4. We want to place an INNER JOIN between the table and the view where for each share we get all the share prices only. At this moment in time, we cannot do this, as vw_Shares does not have a share ID column. We therefore have to modify the vw_Shares view. Keep what you have built in the View Designer, and move back to Object Explorer. Find vw_Shares, right-click, and this time select Modify as shown in Figure 9-17.

Figure 9-17. *Modifying a view for a join*

5. From the View Designer, click the ShareId column, as shown in Figure 9-18. This will then include the ShareId column in the view as the last column. You can use the criteria pane to move this column if you wish.

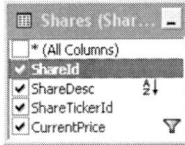

Figure 9-18. *Selecting the column*

6. Close this dialog box, which will bring up the Save Changes dialog box, as shown in Figure 9-19. Click Yes to save the changes.

Figure 9-19. *Saving the modifications*

7. We can now move back to our original View Designer, and you can now see the new column in the view, as shown in Figure 9-20; there should be no need to refresh the screen.

Figure 9-20. *The view with the "new" column*

8. It is very easy to link the two tables together by dragging a column from one table to a column in another table. This is very similar to how the relationships are built in the Database Designer, as you saw earlier in the book. First of all, click the ShareId column in the vw_Shares view. Keeping the mouse button down, drag the mouse pointer from the vw_Shares view over to the ShareId column in the ShareDetails.SharePrices table and then release it. The View Designer should now look like Figure 9-21. We have not really created a relationship in the truest sense of the word—this is simply the relation between the columns for the purpose of this query. You can see one gray line, which shows which fields are used for the join.

Figure 9-21. *The view with the JOIN completed*

9. Select `Price` and `PriceData` from the `ShareDetails.SharePrices` table and `Description` from the `vw_Shares` view, as shown in Figure 9-22.

Figure 9-22. *How the view with a JOIN looks*

10. The final part to this view creation is to build the sort orders. We want the result to be in the order of ascending description, but we want the most recent price first and the first price last. Figure 9-23 shows the criteria pane with these options.

11. Moving to the SQL pane, note the code shows the columns, as well as the `INNER JOIN` of the two `ShareId` columns, and finally the ordering of the data.

```
SELECT TOP (100) PERCENT
ShareDetails.SharePrices.Price, ShareDetails.SharePrices.PriceDate,
ShareDetails.vw_Shares.Description
FROM ShareDetails.SharePrices INNER JOIN ShareDetails.vw_Shares ON
   ShareDetails.SharePrices.ShareId = ShareDetails.vw_Shares.ShareId
ORDER BY ShareDetails.vw_Shares.Description,
ShareDetails.SharePrices.PriceDate DESC
```

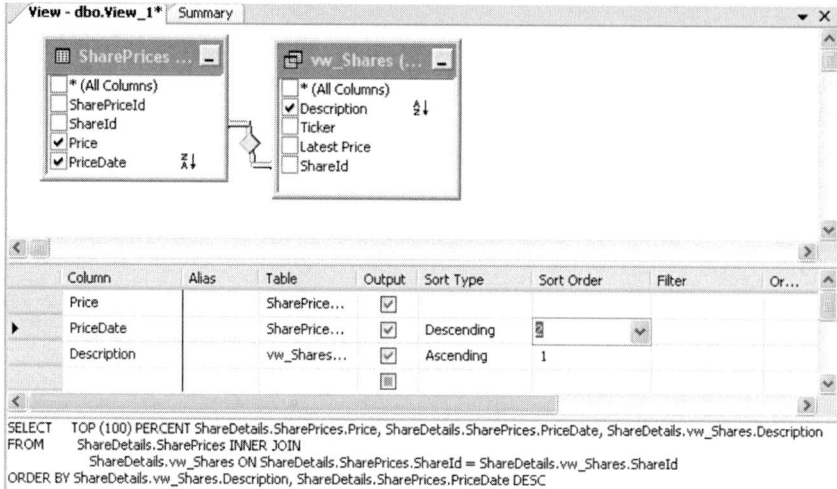

Figure 9-23. *Sorting with ascending and descending items*

12. Before we can execute to test the view, we need to add some data to the `ShareDetails.SharePrices` details. Just because we are completing one action doesn't preclude us from performing another action within another Query Editor window. Click the New Query button on the toolbar if it is not visible, and then from the menu select File ➤ New ➤ Query with Current Connection. In the code window that comes up, enter the following code and then execute to insert the data:

```
INSERT INTO ShareDetails.SharePrices (ShareId, Price, PriceDate)
VALUES (1,2.155,'1 Aug 2005 10:10AM')
INSERT INTO ShareDetails.SharePrices (ShareId, Price, PriceDate)
VALUES (1,2.2125,'1 Aug 2005 10:12AM')
INSERT INTO ShareDetails.SharePrices (ShareId, Price, PriceDate)
VALUES (1,2.4175,'1 Aug 2005 10:16AM')
INSERT INTO ShareDetails.SharePrices (ShareId, Price, PriceDate)
VALUES (1,2.21,'1 Aug 2005 11:22AM')
INSERT INTO ShareDetails.SharePrices (ShareId, Price, PriceDate)
VALUES (1,2.17,'1 Aug 2005 14:54')
INSERT INTO ShareDetails.SharePrices (ShareId, Price, PriceDate)
VALUES (1,2.34125,'1 Aug 2005 16:10')
INSERT INTO ShareDetails.SharePrices (ShareId, Price, PriceDate)
VALUES (2,41.10,'1 Aug 2005 10:10AM')
INSERT INTO ShareDetails.SharePrices (ShareId, Price, PriceDate)
VALUES (2,43.22,'2 Aug 2005 10:10AM')
INSERT INTO ShareDetails.SharePrices (ShareId, Price, PriceDate)
VALUES (2,45.20,'3 Aug 2005 10:10AM')
```

13. We can now navigate back to the View Designer window. Execute the code within the view's code window by pressing the Execute button, and you should see the results displayed in Figure 9-24.

	Price	PriceDate	Description
▶	2.34125	01/08/2005 16:...	ACME'S HOMEBAKE COOKIES INC
	2.17000	01/08/2005 14:...	ACME'S HOMEBAKE COOKIES INC
	2.21000	01/08/2005 11:...	ACME'S HOMEBAKE COOKIES INC
	2.41750	01/08/2005 10:...	ACME'S HOMEBAKE COOKIES INC
	2.21250	01/08/2005 10:...	ACME'S HOMEBAKE COOKIES INC
	2.15500	01/08/2005 10:...	ACME'S HOMEBAKE COOKIES INC
	45.20000	03/08/2005 10:...	FAT-BELLY.COM
	43.22000	02/08/2005 10:...	FAT-BELLY.COM

Figure 9-24. *Viewing test results*

14. Assign the view to the `ShareDetails` schema in the view's Properties window, as shown in Figure 9-25.

Figure 9-25. *Setting the view schema*

15. The final action is to save the view. As before, click the close button and save the view as `vw_SharePrices`.

Now we have built views using the designer; time to build one with T-SQL.

CREATE VIEW Syntax

Very quickly you will find that creating a view using T-SQL is the better way forward. It is just as fast as building a view using the designer.

```
CREATE VIEW [ schema_name . ] view_name [ (column [ ,...n ] ) ]
[ WITH <view_attribute> [ ,...n ] ]
AS select_statement [ ; ]
[ WITH CHECK OPTION ]
<view_attribute> ::= {[ ENCRYPTION ][ SCHEMABINDING ][ VIEW_METADATA ]}
```

The basic CREATE VIEW syntax is very simple and straightforward. The following syntax is the most basic syntax of the CREATE VIEW statement and is the one used most often.

```
CREATE VIEW [database_name.][schema_name.]view_name
WITH {ENCRYPTION | SCHEMABINDING}
AS
SELECT_statement
```

Taking a look at the first section of the syntax, notice that the name of the view can be prefixed with the name of the schema and the name of the database to which it belongs; however, the database name and the schema are optional. Providing that we are in the correct database and are logged in with the ID we wish to create the view for, the database_name and schema_name options are not required, as they will be assumed from the connection details. For production views, rather than views used purely by a single SSE user, it is recommended that they be built by the database owner. If the view is built by a nondatabase owner, then when someone tries to execute the view, that user will need to prefix the name of the view with the login of the person who created it.

Following on from these options, we build the query, typically formed with a SELECT statement that makes up the view itself. As you saw in the previous example, the SELECT statement can cover one or many tables or views, many columns, and as many filtering options using the WHERE statement as you wish. You can also order the data in a view; however, if you recall from our earlier example, to place an ORDER BY clause on a SELECT statement within a view, it is necessary to use the TOP statement. We specified the TOP 100 PERCENT in our first example to get around this problem. Failure to do so will result in an error, and the view will not be created. We also cannot reference any temporary variable or temporary table within a view, or create a new table from a view by using the INTO clause. To clarify, it is not possible to have a SELECT column INTO newtable.

The ENCRYPTION option will take the view created and encrypt the schema contained so that the view is secure and no one can see the underlying code or modify the contents of the SELECT statement within. However (I know I keep repeating this, but it is so important), do keep a backup of the contents of the view in a safe place in development in case any modifications are required.

■**Tip** Encrypting views may seem like a good idea to hide even further the schema of your database from potential users; however, use encrypted views with caution, and always keep a backup of the source in a safe and secure environment. People have been known to keep a printout of the view just in case the source became corrupt. Use encrypted views sparsely, and only when really required.

The SCHEMABINDING option ensures that any column referenced within the view cannot be dropped from the underlying table without dropping the view built with SCHEMABINDING first. This, therefore, keeps the view secure with the knowledge that there will be no run-time errors when columns have been altered or dropped from the underlying table, and the view is not altered in line with those changes. If you try to remove a column from the table that is contained within a schema bound view, for example, you will receive an error. There is one knock-on effect when using SCHEMABINDING: all tables or other views named within the SELECT statement must be prefixed with the name of the schema of the table or view, even if the owner of these objects is the same as the schema of the view.

Go back to the two options that will be used less often, the first being WITH CHECK OPTION. If the view is being used as the basis of completing updates to the underlying table, any modification call, such as UPDATE/DELETE/INSERT, will still make the data visible through the view.

> **Note** Even with `WITH CHECK OPTION` defined, if the data is modified directly in the table, it won't be verified against any views defined with the underlying tables. Also, if the view uses `TOP`, then `WITH CHECK OPTION` cannot be used.

The final possible option, `VIEW_METADATA`, exposes the view's metadata if you are calling the view via ODBC, OLE DB, etc.—in other words, from a program that is external to SSE.

Now that you are aware of the basic syntax for creating a view, the next example will take this knowledge and build a new view for the database.

Creating a View: A Query Editor Pane

Another method for creating views is by using T-SQL code in a Query Editor pane, in my experience the fastest and best option. This can be a faster method for building views than using SSMSE, especially as you become more experienced with T-SQL commands. This section will demonstrate the T-SQL syntax required to create a view, which you will soon see is very straightforward.

The `SELECT` statement forms the basis for most views, so this is where most of the emphasis is placed when developing a view. By getting the `SELECT` statement correct and retrieving the required data, it can then be easily transformed into a view. This is how the view in the following example is created, so let's look at building a view using T-SQL and a Query Editor pane. In the following example, we will create a view that returns a list of transactions for each customer with some customer information.

Try It Out: Creating a View in a Query Editor Pane

1. Ensure that SSMSE is running and that there is an empty Query Editor pane. First of all, let's get the T-SQL correct. We need to link in three tables, the `CustomerDetails.Customers` table to get the name and address, the `TransactionDetails.Transactions` table so we can get a list of transactions for the customer, and finally the `TransactionDetails.TransactionTypes` table so that each transaction type has its full description. The code is as follows:

```
SELECT c.AccountNumber,c.CustomerFirstName,c.CustomerOtherInitials,
tt.TransactionDescription,t.DateEntered,t.Amount,t.ReferenceDetails
FROM CustomerDetails.Customers c
JOIN TransactionDetails.Transactions t ON t.CustomerId = c.CustomerId
JOIN TransactionDetails.TransactionTypes tt ON
  tt.TransactionTypeId = t.TransactionType
ORDER BY c.AccountNumber ASC, t.DateEntered DESC
```

2. Once done, execute the code by pressing F5 or Ctrl+E, or clicking the Execute button.

3. We can now wrap the `CREATE VIEW` statement around our code. Execute this code to store the view in the `ApressFinancial` database. As there is an `ORDER BY` clause, we need to add to the query a `TOP` statement, so we have `TOP 100 Percent`.

```
CREATE VIEW CustomerDetails.vw_CustTrans
AS
SELECT TOP 100 PERCENT
c.AccountNumber,c.CustomerFirstName,c.CustomerOtherInitials,
tt.TransactionDescription,t.DateEntered,t.Amount,t.ReferenceDetails
FROM CustomerDetails.Customers c
```

```
JOIN TransactionDetails.Transactions t ON t.CustomerId = c.CustomerId
JOIN TransactionDetails.TransactionTypes tt ON
  tt.TransactionTypeId = t.TransactionType
ORDER BY c.AccountNumber ASC, t.DateEntered DESC
```

This view is a straightforward view with no ENCRYPTION or SCHEMABINDING options. The one complication within the view concerns the ORDER BY clause: one of the stipulations for this view is that it returns the data of financial transactions with the most recent transaction first. Therefore, an ORDER BY statement is required on the DateEntered column to return the rows in descending order. To avoid receiving an error message when building the view, it has been necessary to place a TOP option within the SELECT statement; in the case of the example, a TOP 100 PERCENT statement has been chosen so that all the rows are returned.

The remainder of the SELECT statement syntax is very straightforward.

Creating a View: SCHEMABINDING

The following example will bind the columns used in the view to the actual tables that lie behind the view, so that if any column contained within the view is modified, an error message will be displayed and the changes will be canceled. The error received will be shown so that we can see for ourselves what happens.

First of all, let's build the view before going on to discuss the background. This view is going to list products for customers, therefore linking the Customers.CustomerProducts and CustomerDetails.FinancialProducts tables.

Try It Out: Creating a View with SCHEMABINDING

1. Create a new Query Editor pane and connect it to the ApressFinancial database. We can then create the T-SQL that will form the basis of our view.

```
SELECT c.CustomerFirstName + ' ' + c.CustomerLastName AS CustomerName,
c.AccountNumber, fp.ProductName, cp.AmountToCollect, cp.Frequency,
cp.LastCollected
FROM CustomerDetails.Customers c
JOIN CustomerDetails.CustomerProducts cp ON cp.CustomerId = c.CustomerId
JOIN CustomerDetails.FinancialProducts fp ON
  fp.ProductId = cp.FinancialProductId
```

2. We need some test data within the system to test this out. This is detailed in the following code. Enter this code and execute it.

```
INSERT INTO CustomerDetails.FinancialProducts (ProductId,ProductName)
VALUES (1,'Regular Savings')
INSERT INTO CustomerDetails.FinancialProducts (ProductId,ProductName)
VALUES (2,'Bonds Account')
INSERT INTO CustomerDetails.FinancialProducts (ProductId,ProductName)
VALUES (3,'Share Account')
INSERT INTO CustomerDetails.FinancialProducts (ProductId,ProductName)
VALUES (4,'Life Insurance')
INSERT INTO CustomerDetails.CustomerProducts
(CustomerId,FinancialProductId,
AmountToCollect,Frequency,LastCollected,LastCollection,Renewable)
```

```
VALUES (1,1,200,1,'31 October 2005','31 October 2025',0)
INSERT INTO CustomerDetails.CustomerProducts
(CustomerId,FinancialProductId,
AmountToCollect,Frequency,LastCollected,LastCollection,Renewable)
VALUES (1,2,50,1,'24 October 2005','24 March 2008',0)
INSERT INTO CustomerDetails.CustomerProducts
(CustomerId,FinancialProductId,
AmountToCollect,Frequency,LastCollected,LastCollection,Renewable)
VALUES (2,4,150,3,'20 October 2005','20 October 2005',1)
INSERT INTO CustomerDetails.CustomerProducts
(CustomerId,FinancialProductId,
AmountToCollect,Frequency,LastCollected,LastCollection,Renewable)
VALUES (3,3,500,0,'24 October 2005','24 October 2005',0)
```

3. Test out that the T-SQL works as required by executing it. The results you get returned should look similar to Figure 9-26.

	CustomerName	AccountNumber	ProductName	AmountToCollect	Frequency	LastCollected
1	Vic McGlynn	87612311	Regular Savings	200.00	1	2005-10-31 00:00:00.000
2	Vic McGlynn	87612311	Bonds Account	50.00	1	2005-10-24 00:00:00.000
3	Bernie McGee	65368765	Share Account	500.00	0	2005-10-24 00:00:00.000
4	Jack Mason	53431993	Life Insurance	150.00	3	2005-10-20 00:00:00.000

Figure 9-26. *Testing SCHEMABINDING T-SQL*

4. We now need to create the CREATE VIEW. First of all, we are completing a test to see whether the view already exists within the system catalogs. If it does, we DROP it. Then we define the view using the WITH SCHEMABINDING clause. The other change to the T-SQL is to prefix the tables we are using with the schema that the tables come from. This is to ensure that the SCHEMABINDING is successful and can regulate when a column is dropped.

```
IF EXISTS(SELECT TABLE_NAME FROM   INFORMATION_SCHEMA.VIEWS
     WHERE  TABLE_NAME = N'vw_CustFinProducts'
        AND  TABLE_SCHEMA = N'CustomerDetails')
   DROP VIEW CustomerDetails.vw_CustFinProducts
GO
CREATE VIEW CustomerDetails.vw_CustFinProducts WITH SCHEMABINDING
AS
SELECT c.CustomerFirstName + ' ' + c.CustomerLastName AS CustomerName,
c.AccountNumber, fp.ProductName, cp.AmountToCollect, cp.Frequency,
cp.LastCollected
FROM CustomerDetails.Customers c
JOIN CustomerDetails.CustomerProducts cp ON cp.CustomerId = c.CustomerId
JOIN CustomerDetails.FinancialProducts fp ON
  fp.ProductId = cp.FinancialProductId
```

5. Once done, execute the code by pressing F5 or Ctrl+E, or clicking the Execute button. You should then see the following message:

```
The command(s) completed successfully.
```

6. Now that our vw_CustFinProducts view is created, which we can check by looking in Object Explorer, it is possible to demonstrate what happens if we try to alter a column used in the view and so affect one of the underlying tables. Enter the following code, and then execute it:

```
ALTER TABLE CustomerDetails.Customers
ALTER COLUMN CustomerFirstName nvarchar(100)
```

7. You will then see in the results pane two error messages: the first shows that an alteration has been attempted on the `CustomerDetails.Customers` table and has been disallowed and names the view stopping this, and the second shows that the alteration failed.

```
Msg 5074, Level 16, State 1, Line 1
The object 'vw_CustFinProducts' is dependent on column 'CustomerFirstName'.
Msg 4922, Level 16, State 9, Line 1
ALTER TABLE ALTER COLUMN CustomerFirstName failed because one or more
objects access this column.
```

■**Caution** You will find that in most cases when building views, the SCHEMABINDING option will be a good option to have on a view, ensuring that a view that works today will always work. It would only be when someone deliberately removed your view from the system to complete table changes, and then didn't correctly put it back, that you would find that a view has stopped working. Herein lies yet another scenario for keeping the code of encrypted views at hand: if you have encrypted views, along with SCHEMABINDING, and someone wishes to alter an underlying table, you had better have the code available!

Indexing a View

Views can be indexed just as tables can be indexed. Rules in choosing columns to make indexes on a view are similar to those for a table. There are also some major requirements you need to meet before you can index a view. I will show you these first so that you are aware which views can be indexed and what you have to do with your view.

When building indexes on views, the first index to be created must be a unique clustered index. Once such an index has been built, additional nonclustered indexes on this view can then be created. This can also be taken further, in that if we have a view with subsequent indexes on it, and we drop the unique clustered index, all of the other indexes will automatically be dropped. Also, if we drop the view, as we would expect, the indexes are also dropped.

The view that the index is to build on must only contain tables and cannot contain views. The tables must all come from one database, and the view must also reside in that database and have been built with the SCHEMABINDING option.

As you saw when creating our database, certain options can be switched on or off. The following options must be set to ON while creating an index. These options need only be set to ON for that session and therefore would precede the CREATE INDEX statement.

```
SET ANSI_NULLS ON
SET ANSI_PADDING ON
SET ANSI_WARNINGS ON
SET CONCAT_NULL_YIELDS_NULL ON
SET QUOTED_IDENTIFIER ON
```

On top of this, the NUMERIC_ROUNDABORT option must be set to OFF.

```
SET NUMERIC_ROUNDABORT OFF
```

Finally, the view itself cannot have text, ntext, or image columns defined in it. In Chapter 11, we look at how to group data through a clause called GROUP BY. If you have grouping within your view, the columns used to group data are the only columns that can be in the first index.

Although these seem like they could be quite restrictive requirements, the upside is that index-ing views also come with major speed implications. If a view remains without an index, every time that the view is executed, the data behind the view, including any joins, is rebuilt and executed. However, as the first index is a clustered index, this is similar to a clustered table index, and the data will be retrieved at index-creation time and stored in that order. Also, like table indexes, when the data is modified, the index will receive the updates as well. Therefore, if SSE can use the clustered index, there will be no need to run the query again.

SSE will use any indexes that you have on the tables when building the views. Indexing a view is most beneficial when the data in the underlying tables is not changing frequently and when the view is executed often. Keep in mind that a view is taking information from other tables and is not a table itself, and therefore any updates to the underlying tables will not be reflected in the view until the view is rerun.

By placing an index on a view, the columns named within the index are stored within the data-base, as are all of the columns defined for the view, along with the data rows. Therefore, any changes to the raw data within the native tables will also be reflected in the data stored for the view. Keep in mind the performance issues with this. Every data change in the tables used in the views requires SSE to evaluate the effect the change has on the view. This requires more processing by SSE, causing a slowdown in performance. Temper this perceived gain of using an index with the downside of the extra processing required to keep the data up to date in two places for the table and two places for the index for those columns involved in the view.

Now that you are aware of the pros and cons of building indexes on views, and how they differ from indexes for tables, it is time to build an index on our view.

The aim of this index is to locate a row in the view quickly. We want to be able to find all the prod-ucts for a customer based on his or her account number. Notice that we are not using CustomerId here. First of all, that column is not within the view, so it is unavailable for selection anyway, but we have to cater to when a customer phones up and supplies the account number. This customer will be unaware of his or her ApressFinancial internal CustomerId. Building the index is very quick and very simple, especially since you already know the basics from building indexes earlier in the book.

Try It Out: Indexing a View

1. The view we want to index is vw_CustFinProducts, as we know that was created with SCHEMABINDING. The unique clustered index will be on the AccountNumber, as we know that this will be unique. In a Query Editor query pane, enter the following code:

```
CREATE UNIQUE CLUSTERED INDEX ix_CustFinProds
ON CustomerDetails.vw_CustFinProducts (AccountNumber,ProductName)
```

2. Execute this code. When you do, you might get an error. The error I received was as follows:

```
Msg 1935, Level 16, State 1, Line 1
Cannot create index. Object 'vw_CustFinProducts' was created with the
following SET options off: 'ANSI_NULLS, QUOTED_IDENTIFIER'.
```

3. As was mentioned when discussing the options required to index a view, we didn't have these two options set to ON. We therefore have to re-create the view. From Object Explorer, right-click and select Script View As ➤ CREATE To ➤ New Query Editor Window, as you see in Figure 9-27.

Figure 9-27. *Scripting the view*

4. This brings up the code in a new Query Editor pane. Modify the two SET options and add in a DROP VIEW statement so that we can re-create the view. Executing the code should be successful.

```
USE [ApressFinancial]
GO
/****** Object:  View [CustomerDetails].[vw_CustFinProducts]
Script Date: 08/07/2005 12:31:54 ******/
SET ANSI_NULLS ON
GO
SET QUOTED_IDENTIFIER ON
GO
DROP VIEW  CustomerDetails.vw_CustFinProducts
GO
CREATE VIEW [CustomerDetails].[vw_CustFinProducts] WITH SCHEMABINDING
AS
SELECT c.CustomerFirstName + ' ' + c.CustomerLastName AS CustomerName,
c.AccountNumber, fp.ProductName, cp.AmountToCollect,
cp.Frequency, cp.LastCollected
FROM CustomerDetails.Customers c
JOIN CustomerDetails.CustomerProducts cp ON cp.CustomerId = c.CustomerId
JOIN CustomerDetails.FinancialProducts fp ON
  fp.ProductId = cp.FinancialProductId

GO
SET ANSI_NULLS OFF
GO
SET QUOTED_IDENTIFIER OFF
```

5. We can then move back to our pane with the CREATE INDEX statement. Executing that code should be successful now as well.

The index on a view has now been successfully created. As you can see, there are a number of restrictions, but not to the point that no index can exist. You just have to think about what you are doing, and if you have a query in your view that contains an item from the preceding list and you wish to create an index, you just have to find a way around it.

Summary

This chapter covered what a view is, how views can improve a database's security, how to encrypt a view, building a view using SSMSE and a Query Editor pane, how to join two tables within a view, and how to index a view.

Having seen three different methods to build a view, you should have found a method that suits you and your style of working. You may find that as time moves on, the tool used alters, as do the methods within that tool. Never discount any tool or option within SSE and banish it to the

annals of history: always keep each option and tool in mind, for one day that area may be your savior. When starting out, switch between each method for building a view so that you are fully conversant with each method.

Finally, being aware of the differences between indexes on tables and indexes in views is crucial to a successful and well-performing view. If you are unsure, try out the view with and then without an index within your development environment.

In the next chapter, we look at another security object within SSE called a stored procedure.

CHAPTER 10

■ ■ ■

Stored Procedures

It is time to look at how to place T-SQL into a **stored procedure** that can be run as often as required.

While you may save queries on a disk drive somewhere, you have not stored them within SSE itself yet, nor have you combined them into multiple units of work. Often, however, you need to execute multiple queries in series from SSE and may need to execute these many times. To do this, you should use stored procedures rather than the code in its "raw" state. SSE assumes that a stored procedure will be run more than once. Therefore, when it is executed for the first time, a query plan is created for it, detailing how best to execute the query. It is also possible, just like any other database object, to assign security to a stored procedure, so that only specific users can run it, lending added security compared to a one-time-only query saved to a hard drive.

The aim of this chapter is to build a simple stored procedure that will insert a single row into a table. We'll then look at error handling and controlling the flow of execution within our procedure.

This chapter will

- Describe what a stored procedure is
- Explain the advantages of stored procedures over views
- Cover the basic syntax for creating a stored procedure
- Show how to use variables in stored procedures
- Show how to control flow within a stored procedure

What Is a Stored Procedure?

In simplest terms, a stored procedure is a collection of compiled T-SQL statements and commands that are directly accessible by SSE. The commands placed within a stored procedure are executed as one single unit, or **batch**, of work—the benefit of this is that network traffic is greatly reduced, as multiple T-SQL statements are not forced to travel over the network individually. Only the name of the stored procedure to execute and arguments to it need to be transmitted. In addition to executing SELECT, UPDATE, INSERT, and DELETE statements, stored procedures can call other stored procedures, control the flow of execution, and call aggregate functions or perform other calculations.

Any developer with access rights to create objects within SSE can build a stored procedure. There are also hundreds of system-stored procedures, all of which start with a prefix of sp_, within SSE.

■**Caution** Under no circumstances should you attempt to modify any system-stored procedure that belongs to SSE, as this could corrupt not only your database but also other databases, requiring you to perform a full restore.

There is little point in building a stored procedure just to run a set of T-SQL statements only once; conversely, a stored procedure is ideal when you wish to run a set of T-SQL statements many times. The reasons for choosing a stored procedure are similar to those that would persuade you to choose a view rather than letting users access table data directly. Stored procedures also supply benefits; for example, SSE will always cache a stored procedure plan in memory, and it is likely to remain in cache and be reused, whereas ad hoc SQL plans created when running ad hoc T-SQL may or may not be stored in the procedure cache. The latter may lead to bloating of the procedure cache with lots of very similar plans for similar batches, as SSE won't match plans that use the same basic code but have different parameter values, for example.

Stored procedures give your application a single proven interface for accessing or manipulating your data. This means that you keep data integrity, make the correct modifications or selections to the data, and ensure that users of the database do not need to know structures, layouts, relationships, or connected processes required to perform a specific function. We can also validate any data input and ensure that the data brought into the stored procedure is correct.

Just like with views and tables, we can grant very specific execute permission for users of stored procedures (the only permission available on a stored procedure is EXECUTE).

To prevent access to the source code, you can encrypt stored procedures, although this really ought to be used in only the most required cases. The code itself isn't actually encrypted, it is only obfuscated, which means it is possible to decrypt the code if required. Therefore, it isn't a total prevention of viewing the code, but it does stop stray eyes. It also limits what can be seen in a tool called **SQL Server Profiler**, found in fuller versions of SQL Server, which is used to profile performance of stored procedures, code, etc., thus causing difficulty in checking what is happening if there is a problem. Therefore, to reiterate, you need to carefully justify any "encryption" you wish to do.

CREATE PROCEDURE Syntax

Begin a stored procedure with a CREATE PROCEDURE statement. The CREATE PROCEDURE syntax offers a great many flexible options and extends T-SQL with some additional commands. The syntax generally appears as follows:

```
CREATE PROCEDURE procedure_name
[ { @parameter_name} datatype [= default_value] [OUTPUT]]
[ { WITH [RECOMPILE | ENCRYPTION | RECOMPILE, ENCRYPTION } ]
AS
[BEGIN]
    statements
[END]
```

First of all, it is necessary to inform SSE which action you wish to perform. Obviously, we wish to create a stored procedure, and so we need to supply a CREATE PROCEDURE statement.

The next part of the syntax is to give the procedure a name. It would be advisable, just as it is with any SSE object, to adhere to a naming standard. Everyone has their own standard within their installation, but if you prefix the name with sp_, a very common naming convention, you will know what that object is. However, this is not something I recommend for two reasons. The first is that stored procedures prefixed by sp_ are seen as system-stored procedures. The second is that you can hit some unnecessary compile locks due to system-stored procedure lookups. Therefore, do avoid this naming convention.

Many people adopt a different naming convention whereby the prefix defines what the stored procedure will do; for example, an update would have a prefix of up, a deletion dt, and a selection sl. There are many different prefixes you could use, but once you have decided on your standard, you should stick with it.

Some procedures may require information to be provided in order for them to do their work; this is achieved by passing in a parameter. For example, passing in a customer number to a stored procedure would provide the necessary information to allow creation of a list of transactions for a statement. More than one parameter can be passed in: all you do is separate them with a comma.

Any parameter defined must be prefixed with an @ sign. Not all procedures will require parameters, and so this is optional; however, if you do wish to pass in parameters to a stored procedure, name the parameters and follow them with the data type and, where required, the length of the data to pass in. For example, the following specifies a parameter of name L_Name, with a varchar data type of length 50.

```
@L_Name varchar(50)
```

You can also specify a default parameter value in the event that a user does not provide one at execution time. The value specified must be a constant value, like 'DEFAULT' or 24031964, or it can be NULL. It is not possible to define a variable as a default value, since the procedure cannot resolve this when the procedure is built. For example, if your application is commonly, but not exclusively, used by the marketing department, you could make the department parameter optional by setting a default of marketing:

```
@department varchar(50) = 'marketing'
```

Thus, in this example, if you were from marketing, you would not need to provide the department input. If you were from information services, however, you could simply provide an input for department that would override the default.

It is also possible to return a value or a number of values from a stored procedure using a parameter to pass the information out. The parameter would still be defined as if it were for input, with one exception and one extra option. First of all, the exception: it is not possible to define a default value for this parameter. If you try to do so, no errors will be generated, but the definition will be ignored. The extra syntax option that is required is to suffix the parameter with the keyword OUTPUT. This must follow the data type definition:

```
@calc_result varchar(50) OUTPUT
```

You are not required to place OUTPUT parameters after the input parameters; they can be intermixed. Conventionally, however, try to keep the OUTPUT parameters until last, as it will make the stored procedure easier to understand.

Tip Output parameters can also be input parameters, and therefore can be used to pass a value in as well as retrieve a value out.

Before continuing, one last thing about parameters needs to be discussed, and it has to do with executing the procedure and working with the defined parameters. When it comes to executing a stored procedure that has input parameters, you have two ways to run it.

The first method is to name the stored procedure and then pass the input values for the parameters in the same order that they are defined. SSE will then take each comma-delimited value set and assign it to the defined variable. However, this does make an assumption that the order of the parameters does not change, and that any default value-defined parameters are also set with a value.

The second, and preferred, method of executing a stored procedure is to name the parameter, and follow this with the value to pass in. We are then ensuring that, at execution time, it doesn't matter what order the stored procedure has named the parameters, because SSE will be able to match the parameter defined with the parameter defined within the stored procedure. We then don't need to define a value for parameters that already have default values. Also, if the stored

procedure needs to be expanded, for backward compatibility, any new parameters can be defined with default values, therefore removing the need to change every calling code. There will be examples of each of the two different methods of passing in values to parameters within this chapter.

Next come two options that define how the stored procedure is built. First of all, just as a reminder, a stored procedure, when first run without an existing plan in the procedure cache, is compiled into an execution plan, which is an internal data structure in SSE that describes how it should go about performing the operations requested within the stored procedures. SSE stores the compiled code for subsequent executions, which saves time and resources.

However, the RECOMPILE option on a stored procedure dictates to SSE that every time the stored procedure is run, the whole procedure is recompiled. Typically, when a parameter can greatly affect the number of rows returned, you may want to add the RECOMPILE option to a stored procedure to force the optimizer to produce the best plan every time, i.e., you want to avoid reuse of a plan that may not be very good for certain parameter values.

The second of the two options is the ENCRYPTION keyword. It is possible to encrypt (well, obfuscate at least) a stored procedure so that the contents of the stored procedure cannot be viewed easily. Keep in mind that ENCRYPTION does not secure the data, but rather protects the source code from inspection and modification. Both ENCRYPTION and RECOMPILE are preceded by the WITH keyword and can be employed together when separated by a comma:

```
CREATE PROCEDURE sp_do_nothing
    @nothing int
    WITH ENCRYPTION, RECOMPILE
AS
    SELECT something FROM nothing
```

The keyword AS defines the start of the T-SQL code, which will be the basis of the stored procedure. AS has no other function, but is mandatory within the CREATE PROCEDURE command defining the end of all variable definitions and procedure creation options. Once the keyword AS is defined, you can then start creating your T-SQL code.

It is then possible to surround your code with a BEGIN...END block. I tend to do this as a matter of course so that there is no doubt where the start and end of the procedure lie.

Returning a Set of Rows

One method of achieving output from a stored procedure is to return a set of rows, also known as a **result set**. A result set may contain zero, one, or many rows as a single batch of output. This is achieved through the use of the SELECT statement within a stored procedure—what is selected is returned as the output of the stored procedure. Don't be fooled into thinking, though, that we can only return one result set within a stored procedure, as this is not true: we can return as many result sets as we wish.

In this chapter, you will see single result sets of data returned and how these look within Query Editor. Returning single, or even multiple, result sets should not really concern you at this stage, but it is of more concern to developers in languages such as C#, VB .NET, and so on. Multiple result sets will only concern you when you move on to more advanced stored procedures with multiple queries.

Creating a Stored Procedure: SSMSE

Now that you have seen some of the merits of a stored procedure over other methods of working with data, it is time to create the first stored procedure in this chapter. This stored procedure will be built in SSMSE to insert a customer into the CustomerDetails.Customers table from the information passed to it. This is also the first part in our overall security solution. By using a stored procedure to

enter the data into the underlying table, we will be in control of what data is entered, as the data can be validated and verified. You can also remove all access from the table and leave the stored procedure to serve as the only method of inserting data (you would also have stored procedures that update, delete, and retrieve data). We will look at this toward the end of the chapter.

Try It Out: Creating a Stored Procedure Using SSMSE

1. Navigate to the `ApressFinancial` database and right-click Stored Procedures. From the pop-up menu, select New Stored Procedure.

2. This opens a Query Editor pane with code from a basic stored procedure template, the template called `Create Stored Procedure (New Menu)` to be exact. You can either alter the procedure by changing the template options (by clicking Ctrl+Shift+M) or just write the code from scratch. As we have chosen to create a stored procedure via Object Explorer, we will use the template this time. Figure 10-1 shows the template options that can be changed.

Figure 10-1. *A stored procedure's blank template*

3. The first three options, shown in Figure 10-2, are not part of the stored procedure syntax; they are extra options used within the comments of the stored procedure. The first option is very useful because it will probably be a dbo account that adds the stored procedure to the database, and therefore it will be hard to track who the actual creator of the stored procedure was. It may be that only one account "releases" all the code to production for deployment. The second option, Create Date, is not quite as relevant, as this can be found by interrogating system views. The Description option is excellent and should form part of every stored procedure, as it will allow a short description of what the stored procedure is trying to achieve. Never go into too much detail in a description, because not everyone has good discipline in updating the comments when the stored procedure changes. However, a short "what we are trying to achieve" description is perfect.

Parameter	Type	Value
Author		Robbin Dewson
Create Date		15 Sep 2006
Description		This is to insert a cust...

Figure 10-2. *First set of template options filled*

4. We can now move to the template options that form part of the CREATE PROCEDURE syntax. The first option is the name. I have called this apf_insCustomer to define that it's a stored procedure in the ApressFinancial database and that we are inserting a row in the CustomerDetails.Customers table. Then we can insert two parameters, as this is what the template is set up for. The first two parameters will be used to populate CustomerFirstName and CustomerLastName. We will look at the rest in a moment. The parameter values do not have to be the same name as the columns they will be working with, but it is best to have similar names. The data type and data length should be defined as the same type and length as the columns they will be used for. Failure to do this could lead to problems with data truncation if you make the parameter columns too long, for example. We also remove the values in the default options. Your template options should now look similar to what you see in Figure 10-3.

Figure 10-3. *The remaining parameters*

5. Click OK. The code will now look like the following:

```
-- =================================================
-- Template generated from Template Explorer using:
-- Create Procedure (New Menu).SQL
--
-- Use the Specify Values for Template Parameters
-- command (Ctrl-Shift-M) to fill in the parameter
-- values below.
--
-- This block of comments will not be included in
-- the definition of the procedure.
-- =================================================
SET ANSI_NULLS ON
GO
SET QUOTED_IDENTIFIER ON
GO
-- =================================================
-- Author:  Robin Dewson
-- Create date: 15 Sep 2006
-- Description:  This is to insert a customer
-- =================================================
CREATE PROCEDURE apf_InsertCustomer
        -- Add the parameters for the stored procedure here
        @FirstName varchar(50) = ,
        @LastName varchar(50) =
```

```
AS
BEGIN
    -- SET NOCOUNT ON added to prevent extra result sets from
    -- interfering with SELECT statements.
    SET NOCOUNT ON;

    -- Insert statements for procedure here
    SELECT @FirstName, @LastName
END
GO
```

6. We can now define the remaining parameters. There are one or two points to make before we progress. First of all, the parameters can be in any order, although it is best to try and group parameters together. The second point is that parameters like @CustTitle, @AddressId, @AccountNumber, and @AccountTypeId in this example are showing the numeric reference values that would come from values defined in a graphical front end. You may be wondering why the stored procedure is not generating these values from other information passed. For example, why is the stored procedure not producing the title ID from Mr., Miss, etc.? It is likely that the operator using the front end had a combo box with a list of possible values to choose from, with IDs corresponding to titles. In the case of the address, the ID would link back to an external address database, so rather than holding the whole address, we could receive just the ID selected when the operator used the address lookup. The code with the remaining parameters is shown here:

```
CREATE PROCEDURE CustomerDetails.apf_InsertCustomer
    -- Add the parameters for the function here
    @FirstName varchar(50) ,
    @LastName varchar(50),
    @CustTitle int,
    @CustInitials nvarchar(10),
    @AddressId int,
    @AccountNumber nvarchar(15),
    @AccountTypeId int
```

7. Moving on to the remaining section of the stored procedure, we will take the values of our parameters and use these as input to the relevant columns. The remaining code for the stored procedure is as follows:

```
AS
BEGIN
    -- SET NOCOUNT ON added to prevent extra result sets from
    -- interfering with SELECT statements.
    SET NOCOUNT ON;

    INSERT INTO CustomerDetails.Customers
    (CustomerTitleId,CustomerFirstName,CustomerOtherInitials,CustomerLastName,
    AddressId,AccountNumber,AccountTypeId,ClearedBalance,UnclearedBalance)
    VALUES (@CustTitle,@FirstName,@CustInitials,@LastName,
    @AddressId,@AccountNumber,@AccountTypeId,0,0)

END
GO
```

8. When you execute the preceding code, providing you have made no typing mistakes, you should see the following output:

```
Command(s) completed successfully.
```

9. This will have added the stored procedure to the database. We can check this. Move back to Object Explorer, right-click Stored Procedures, and select Refresh. After the refresh, you should see the stored procedure in Object Explorer, as shown in Figure 10-4.

```
□ 📁 Programmability
   □ 📁 Stored Procedures
      ⊞ 📁 System Stored Procedures
      ⊞ 📄 CustomerDetails.apf_InsertCustomer
```

Figure 10-4. *Object Explorer with stored procedure listed*

10. We have completed our first developer-built stored procedure within the system. Inserting data using the stored procedure will now be demonstrated so you can see the procedure in action. To execute this stored procedure, we need to specify its name and pass the data in with parameters. There are two ways we can progress. The first method is to pass the data across in the same order as the parameters defined within the stored procedure as follows:

```
CustomerDetails.apf_InsertCustomer 'Henry','Williams',
1,NULL,431,'22067531',1
```

11. If you execute this, you should see the following output:

```
(1 row(s) affected)
```

12. However, there is a downside to this method: if someone alters the stored procedure and places a new parameter in the middle of the existing list or changes the order of the parameters, or perhaps you don't know the order of the parameters, then you are at risk for errors. The preferred method is to name the parameters and the values as shown in the next example. Notice as well that the order has changed.

```
CustomerDetails.apf_InsertCustomer @CustTitle=1,@FirstName='Julie',
@CustInitials='A',@LastName='Dewson',@AddressId=6643,
@AccountNumber='SS865',@AccountTypeId=6
```

13. Again, if you execute this, you should see the same results:

```
(1 row(s) affected)
```

You can check that the two customers have been entered if you wish.

Let's take a look at two different methods for executing procedures next.

Different Methods of Execution

There are two different methods of executing a stored procedure. The first is to just call the stored procedure, as you saw in the preceding example. The second is to use the EXEC(UTE) command. Both have the end result of invoking the stored procedure, but which is better for you to use depends on the particular situation.

No EXEC

It is possible to call a stored procedure without prefixing the stored procedure name with the EXEC(UTE) statement. However, the stored procedure call must be the first statement within a batch of statements if you wish to exclude this statement.

With EXEC

As I have just indicated, if the stored procedure call is the second or subsequent statement within a batch, you must prefix the stored procedure with the EXEC(UTE) statement. On top of this, if you are calling a stored procedure within another stored procedure, you would need to prefix the call with the EXEC(UTE) statement.

Using RETURN

One method of returning a value from a stored procedure to signify an error is to use the RETURN statement. This statement immediately stops a stored procedure and passes control back out of it. Therefore, any statements after the RETURN statement will not be executed.

It is not compulsory to have a RETURN statement within your code; it is only really necessary when you wish to either return an error code or exit from a stored procedure without running any further code from that point. A logical RETURN is performed at the end of a stored procedure, returning a value of 0.

By default, 0 is returned if no value is specified after the RETURN statement, which means that the stored procedure was successful. Any other integer value could mean that an unexpected result occurred and that you should check the return code, although it is possible to return the number of rows affected by the stored procedure, for example. Notice that the word "error" wasn't mentioned, as it may be valid for a nonzero return code to come out of a stored procedure.

In this example, we will create a stored procedure that will return two output parameters back to the calling procedure or code, indicating the cleared and uncleared balances of a specific customer.

We will also use the RETURN option to indicate whether the customer ID passed to the stored procedure finds no rows. Note that this is not an error, as the stored procedure code will be working as expected.

So you are probably wondering when to use output parameters and when to use RETURN. Output parameters are used to return information back to a calling set of code and can handle any data type. On the other hand, a RETURN can only return an integer numeric value and is used more often for indicating success or failure.

Try It Out: Using RETURN and Output Parameters

1. The Template Explorer contains a template set up for output parameters. Navigate to this template, shown in Figure 10-5, and double-click it.

Figure 10-5. *Template Explorer with OUTPUT stored procedure*

2. This will open up a new Query Editor pane with the basics of the relevant stored procedure, which is shown, reformatted, in the following code block. Take a moment to peruse this code. First of all, the first batch within the template sets up checks to see whether the stored procedure already exists, and if it does, deletes the procedure through the DROP PROCEDURE command. After running DROP PROCEDURE, just like after dropping any object, all of the permissions associated with that object are lost when we re-create it as we discussed earlier.

```
-- =============================================
-- Create stored procedure with OUTPUT parameters
-- =============================================
-- Drop stored procedure if it already exists
IF EXISTS (
  SELECT *
    FROM INFORMATION_SCHEMA.ROUTINES
   WHERE SPECIFIC_SCHEMA = N'<Schema_Name, sysname, Schema_Name>'
     AND SPECIFIC_NAME = N'<Procedure_Name, sysname, Procedure_Name>'
)
   DROP PROCEDURE <Schema_Name, sysname, Schema_Name>.<Procedure_Name,
     sysname, Procedure_Name>
GO

CREATE PROCEDURE <Schema_Name, sysname, Schema_Name>.<Procedure_Name,
    sysname, Procedure_Name>
    <@param1, sysname, @p1> <datatype_for_param1, , int> =
        <default_value_for_param1, , 0>,
    <@param2, sysname, @p2> <datatype_for_param2, , int>  OUTPUT
AS
    SELECT @p2 = @p2 + @p1
GO

-- =============================================
-- Example to execute the stored procedure
-- =============================================
DECLARE <@variable_for_output_parameter, sysname, @p2_output>
<datatype_for_output_parameter, , int>

EXECUTE <Schema_Name, sysname, Schema_Name>.<Procedure_Name, sysname,
Procedure_Name> <value_for_param1, , 1>, <@variable_for_output_parameter,
sysname, @p2_output> OUTPUT

SELECT <@variable_for_output_parameter, sysname, @p2_output>
GO
```

3. Now that you have seen the code, it is time to update the template parameters. Again, we find that the template is not ideal for our final solution, as we only have one input parameter and two output parameters. However, we have populated the template parameters we need. This stored procedure will belong to the CustomerDetails schema. We have one integer input parameter for the customer ID, followed by the first of our output parameters for cleared balances. Once you have entered these settings, as shown in Figure 10-6, click OK.

Figure 10-6. *Template values for OUTPUT stored procedure*

4. Let's look at the code that was generated. The first section of code checks whether the stored procedure exists. If it does, we delete it using the DROP PROCEDURE statement.

```
-- ===============================================
-- Create stored procedure with OUTPUT parameters
-- ===============================================
-- Drop stored procedure if it already exists
IF EXISTS (
  SELECT *
    FROM INFORMATION_SCHEMA.ROUTINES
    WHERE SPECIFIC_SCHEMA = N'CustomerDetails'
      AND SPECIFIC_NAME = N'apf_CustBalances'
)
    DROP PROCEDURE CustomerDetails.apf_CustBalances
GO
```

5. Move on to the second section, which creates the contents of the stored procedure; we'll go through each part of it in turn. This stored procedure takes three parameters: an input parameter of @CustId, and two output parameters that will be passed back to either another stored procedure or a program, perhaps written in C#, etc. Don't worry, it is possible to use Query Editor to see the value of the output parameter. When defining parameters in a stored procedure, there is no need to specify that a parameter is set for input, as this is the default; however, if we do need to define a parameter as an output parameter, we have to insert OUTPUT as a suffix to each parameter.

■**Tip** If we define an OUTPUT parameter but do not define a value within the stored procedure, it will have a value of NULL.

```
CREATE PROCEDURE CustomerDetails.apf_CustBalances
    @CustId int,
    @ClearedBalance money OUTPUT,
    @UnclearedBalance money OUTPUT
AS
```

6. Take a look at the next section of code, which is very similar to what we have covered several times earlier in the book where we are assigning values to variables.

```
SELECT @ClearedBalance = ClearedBalance,
@UnclearedBalance = UnclearedBalance
  FROM Customers
 WHERE CustomerId = @CustId
```

7. The final section of the stored procedure returns a value from a system global variable, @@ERROR. We'll look at this variable in the next chapter, but in essence this variable returns a number if an error occurred. From this the calling code can tell whether there have been problems and can then decide whether to ignore any values in the OUTPUT parameter.

```
    RETURN @@Error
GO
```

8. This completes the stored procedure definition. The template continues defining how to execute the stored procedure. The first part of this section defines the variables that hold the output values and the return value. We do not need to define a variable for the input value, although you could if it was required. Then we move to the EXECUTE section of code. When a value is returned from a stored procedure, it is set on the left-hand side of the stored procedure call and is not a parameter value. Then the stored procedure is defined with the three parameters. Note that each output parameter has to have the OUTPUT keyword after it. The final section of the code is a SELECT statement displaying the values returned and the output parameter.

```
-- ============================================
-- Example to execute the stored procedure
-- ============================================
DECLARE @ClearedBalance Money, @UnclearedBalance Money
DECLARE @RetVal int

EXECUTE @RetVal=CustomerDetails.apf_CustBalances 1, @ClearedBalance OUTPUT,
@UnclearedBalance OUTPUT

SELECT @RetVal AS ReturnValue, @ClearedBalance AS ClearedBalance,
@UnclearedBalance AS UnclearedBalance
GO
```

9. Now that the template has been altered with the changes we need, execute the template by pressing Ctrl+E or F5, or clicking the Execute button on the toolbar. This will create the stored procedure and run the examples at the end to demonstrate the procedure. Of course, we can run this section of code as many times as we want because the whole scenario, from dropping and losing the stored procedure through to re-creating the stored procedure, is all there, ready for us. The stored procedure will pass back its output parameter value to the @ClearedBalance and @UnclearedBalance variables defined within the execution batch and the return value to the @RetVal variable. From there, once the variables are set, the values can be printed out using a SELECT statement. This will produce the output shown in Figure 10-7 in the results pane.

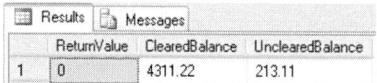

	ReturnValue	ClearedBalance	UnclearedBalance
1	0	4311.22	213.11

Figure 10-7. *Results after running the OUTPUT stored procedure*

We have now built two very basic stored procedures in which we are performing an INSERT and a SELECT. Next we look at control of flow.

Controlling the Flow

When working on a stored procedure, there will be times when it is necessary to control the flow of information through it. The main control of flow is handled with an IF...ELSE statement. You can also control the flow with a WHILE...BREAK statement.

Note The GOTO statement can also control the flow of a stored procedure. You can use this statement to jump to a label within a stored procedure, but this can be a dangerous practice and really is something that should be avoided. For example, it might be better to nest the stored procedure calls.

Controlling the flow through a stored procedure will probably be required when a procedure does anything more than work with one T-SQL statement. The flow will depend on your procedure taking an expression and making a true or false decision, and then taking two separate actions depending on the answer from the decision.

IF...ELSE

At times a logical expression will need to be evaluated that results in either a true or false answer. This is where an IF...ELSE statement is needed. There are many ways of making a true or false condition, and most of the possibilities involve relational operators such as <, >, =, and NOT; however, these can be combined with string functions, other mathematical equations, or comparisons between values in local variables, or even systemwide variables. It is also possible to place a SELECT statement within an IF...ELSE block, as long as a single value is returned.

A basic IF...ELSE statement would perhaps look like the following:

```
IF A=B
    Statement when True
ELSE
    Statement when False
```

IF...ELSE statements can also be nested and would look like the following; this example also shows you how to include a SELECT statement within an IF decision.

```
IF A=B
    IF (SELECT ClearedBalance FROM Customers WHERE CustomerId = 1) > $20000
        Statement2 when True
    ELSE
        Statement2 when False
ELSE
    Statement when False
```

As you can see, there is only one statement within each of the IF...ELSE blocks. If you wish to have more than one line of executable code after the IF or the ELSE, you must include another control-of-flow statement, the BEGIN...END block. Before we can try this out, let's take a look at how to code for multiple statements within an IF...ELSE block.

BEGIN...END

If you wish to execute more than one statement in the IF or ELSE code block, you need to batch the statements up. To batch statements together within an IF...ELSE, you must surround them with a BEGIN...END block. If you try to have more than one statement after the IF, the second and subsequent statements will run no matter what the setting of the IF statement is.

So if you have

```
DECLARE @VarTest
SET @VarTest = 2
IF @VarTest=1
SELECT 1
SELECT 2
```

the SELECT 2 statement would run no matter what value you have for @VarTest. If you only want SELECT 2 to run when @VarTest is 1, you would code the example, thus placing the code you want to run within the BEGIN...END block.

```
DECLARE @VarTest
SET @VarTest = 2
IF @VarTest=1
BEGIN
    SELECT 1
    SELECT 2
END
```

If you use an ELSE statement after a second or subsequent statement after an IF that has no BEGIN...END block, you would get an error message. Therefore, the only way around this is to use BEGIN...END.

WHILE...BREAK Statement

The WHILE...BREAK statement is a method of looping around the same section of code from zero to multiple times based on the answer from a Boolean test condition, or until explicitly informed to exit via the keyword BREAK.

The syntax for this command is as follows:

```
WHILE Boolean_expression
    { sql_statement | statement_block }
    [ BREAK ]
    { sql_statement | statement_block }
    [ CONTINUE ]
    { sql_statement | statement_block }
```

The code defined for the WHILE statement will execute while the Boolean expression returns a value of True. You can have other control-of-flow statements such as an IF...ELSE block within your WHILE block. This is where BREAK and CONTINUE could be used if required. You may wish to test a condition and, if it returns a particular result, BREAK the loop and exit the WHILE block. The other option that can be used is the CONTINUE statement. This moves processing straight to the WHILE statement again and will stop any execution of code that is defined after it. The best way to illustrate these concepts is to show a simple example of these three options in action.

Try It Out: WHILE...BREAK

1. The first option demonstrates how to build a WHILE loop and then test the value of a variable. If the test returns True, we will break out of the loop; if it returns False, we will continue processing. Within the example, there are two SELECT statements before and after an IF...ELSE statement. In this example, the first SELECT will show the values of the variables, but the IF test will either stop the loop via BREAK or will move the code back to the WHILE statement via the CONTINUE statement. Either of these actions will mean that the second SELECT will not execute.

```
DECLARE @LoopCount int, @TestCount int
SET @LoopCount = 0
SET @TestCount = 0
WHILE @LoopCount < 20
BEGIN
    SET @LoopCount = @LoopCount + 1
    SET @TestCount = @TestCount + 1
    SELECT @LoopCount, @TestCount
    IF @TestCount > 10
        BREAK
    ELSE
        CONTINUE
    SELECT @LoopCount, @TestCount
END
```

2. When the code is executed, we don't actually make it around the 20 loops due to the value of @TestCount causing the break. The output is shown in Figure 10-8.

Figure 10-8. *WHILE with BREAK and CONTINUE*

3. If we change the code to remove the ELSE CONTINUE statement, the second SELECT statement will be executed. The two rows changed have been highlighted. We are not going to execute the two lines because they have been commented out by prefixing the code with two hyphens, --.

```
DECLARE @LoopCount int, @TestCount int
SET @LoopCount = 0
SET @TestCount = 0
WHILE @LoopCount < 20
```

```
BEGIN
    SET @LoopCount = @LoopCount + 1
    SET @TestCount = @TestCount + 1
    SELECT @LoopCount, @TestCount
    IF @TestCount > 10
        BREAK
    ELSE
        CONTINUE
    SELECT @LoopCount, @TestCount
END
```

A snapshot of some of the output from this is shown in Figure 10-9.

Figure 10-9. *WHILE with BREAK only*

The third statement we'll look at in this section is the CASE statement. While not a control-of-flow statement for your stored procedure, it can control the output displayed based on decisions.

CASE Statement

When a query has more than a plain true or false answer—in other words, when there are several potential answers—you should use the CASE statement.

A CASE statement forms a decision-making process within a SELECT or an UPDATE statement. It is possible to set a value for a column within a result set based on a CASE statement and the resultant value. Obviously, with this knowledge, a CASE statement cannot form part of a DELETE statement.

Several parts of a CASE statement can be placed within a stored procedure to control the statement executed depending on each scenario. Two different syntax structures exist for the CASE statement depending on how you want to test a condition or what you want to test. Let's take a look at all the parts to the first CASE statement syntax:

```
CASE expression
WHEN value_matched THEN
    statement
```

```
[[WHEN value_matched2 THEN]
    [Statement2]]
...
...
...
[[ELSE]
    [catch_all_code]
END
```

First of all, you need to define the expression that is to be tested. This could be the value of a variable, or a column value from within the T-SQL statement, or any valid expression within SSE. This expression is then used to determine the values to be matched in each WHEN statement.

You can have as many WHEN statements as you wish within the CASE condition, and you do not need to cover every condition or possible value that could be placed within the condition. Once a condition is matched, only the statements within the appropriate WHEN block will be executed. Of course, only the WHEN conditions that are defined will be tested. However, you can cover yourself for any value within the expression that has not been defined within a WHEN statement by using an ELSE condition. This is used as a catchall statement. Any value not matched would drop into the ELSE condition, and from there you could deal with any scenario that you desire.

The second syntax is where you don't define the expression prior to testing it and each WHEN statement performs any test expression you desire.

```
CASE
    WHEN Boolean_expression THEN result_expression
    [ ...n ]
    [
    ELSE else_result_expression
    ]
END
```

As just indicated, CASE statements form part of a SELECT, an UPDATE, or an INSERT statement, therefore possibly working on multiple rows of data. As each row is retrieved from the table, the CASE statement kicks in, and instead of the column value being returned, it is the value from the decision-making process that is inserted instead. This happens after the data has been retrieved and just before the rows returned are displayed in the results pane. The actual value is returned initially from the table and is then validated through the CASE statement; once this is done, the value is discarded if no longer required.

Now that you are familiar with CASE statements, we can look at them in action.

Try It Out: Using the CASE Statement

1. Our first example will demonstrate the first CASE syntax, where we will take a column and test for a specific value. The results of this test will determine which action will be performed. We will prepopulate the TransactionDetails.TransactionTypes table first so that you can see how populating this table and the CASE statement work.

```
INSERT INTO TransactionDetails.TransactionTypes
(TransactionDescription,CreditType,AffectCashBalance)
VALUES ('Deposit',1,1)
INSERT INTO TransactionDetails.TransactionTypes
(TransactionDescription,CreditType,AffectCashBalance)
```

```
VALUES ('Withdrawal',0,1)
INSERT INTO TransactionDetails.TransactionTypes
(TransactionDescription,CreditType,AffectCashBalance)
VALUES ('BoughtShares',1,0)
SELECT TransactionDescription,
CASE CreditType
WHEN 0 THEN 'Debiting the account'
WHEN 1 THEN 'Crediting the account'
END
FROM TransactionDetails.TransactionTypes
```

2. Execute this code, and you should see the output shown in Figure 10-10.

Figure 10-10. *Simple CASE statement output*

3. A customer can have a positive or negative ClearedBalance. The CASE statement that follows will demonstrate this by showing either In Credit or Overdrawn. In this case, we want to use the second CASE syntax. We cannot use the first syntax, as we have an operator included within the test and we are not looking for a specific value. The code is defined as follows:

```
SELECT CustomerId,
CASE
WHEN ClearedBalance < 0 THEN 'OverDrawn'
WHEN ClearedBalance > 0 THEN ' In Credit'
ELSE 'Flat'
END, ClearedBalance
FROM CustomerDetails.Customers
```

4. Execute the code. This produces output similar to what you see in Figure 10-11.

	CustomerId	(No column name)	ClearedBalance
1	1	In Credt	4311.22
2	2	In Credt	437.97
3	3	In Credt	6653.11
4	4	In Credt	53.32
5	5	In Credt	1266.00
6	12	Flat	0.00
7	13	Flat	0.00

Figure 10-11. *Searched CASE statement output*

Bringing It All Together

Now that you have seen the control-of-flow statements, we can bring all of this together in our most complex set of code so far. The aim of this stored procedure is to take a "from" and "to" date, which can be over any period, and return the movement of a particular customer's transactions that have affected the cash balance. This mimics your bank statement when it says whether you have spent more than you have deposited.

This example includes one topic that is not covered until the next chapter: joining data from more than one table together. For the moment, just accept that when you see the statement JOIN, all it is doing is taking data from another table and allowing you to work with it.

So let's build that example.

Try It Out: Bringing It All Together

■Note In this example, we are performing a loop around rows of data within a table. This example demonstrates some of the functionality just covered with decisions and control of flow. SSE works best with sets of data, rather than a row at a time. However, there will be times that row-by-row processing like this happens. In SSE, you have the option to write .NET-based stored procedures, and this example would certainly be considered a candidate for this treatment. Our example works with one row at a time, where you would have a running total of a customer's balance so that you can calculate interest to charge or to pay.

1. First of all, let's create our stored procedure. We have our CREATE PROCEDURE statement that we enter in an empty Query Editor pane, and then we name the procedure with our three input parameters.

```
CREATE PROCEDURE CustomerDetails.apf_CustMovement @CustId bigint,
@FromDate datetime, @ToDate datetime
AS
BEGIN
```

2. We then need three internal variables. This stored procedure will return one row of transactions at a time while we are still in the date range. As we move through each row, we need to keep a running balance of the amounts for each transaction. We know that the data in the TransactionDetails.Transactions table has an ascending TransactionId as each transaction is entered, so the next transaction from the one returned must have a higher value. Therefore, we can store the transaction ID in a variable called @LastTran and use that in our filtering. Once the variables are declared, we then set them to an initial value. We use @StillCalc as a test for the WHILE loop. This could be any variable as we are using the CONTINUE and BREAK statements to determine when we should exit the loop.

```
DECLARE @RunningBal money, @StillCalc Bit, @LastTran bigint
SELECT @StillCalc = 1, @LastTran = 0, @RunningBal = 0
```

3. We tell the loop to continue until we get no rows back from our SELECT statement. Once we get no rows, we know that there are no more transactions in the date range.

```
WHILE @StillCalc = 1
BEGIN
```

4. Our more complex SELECT statement will return one row where the TransactionId is greater than the previous TransactionId returned; the transaction would affect the customer's cash balance; and the transaction is between the two dates passed in. If there is a transaction, we add or subtract the value from the @RunningBal variable. We use a CASE statement to decide whether we need to make the value a negative value for adding to the variable.

```
SELECT TOP 1 @RunningBal = @RunningBal + CASE
                WHEN tt.CreditType = 1 THEN t.Amount
                ELSE t.Amount * -1 END,
                @LastTran = t.TransactionId
  FROM CustomerDetails.Customers c
  JOIN TransactionDetails.Transactions t ON t.CustomerId = c.CustomerId
  JOIN TransactionDetails.TransactionTypes tt ON
    tt.TransactionTypeId = t.TransactionType
  WHERE t.TransactionId > @LastTran
    AND tt.AffectCashBalance = 1
    AND DateEntered BETWEEN @FromDate AND @ToDate
  ORDER BY DateEntered
```

5. If we get a row returned, we continue the loop. Once we get no rows returned, we know that there are no further transactions in the date range.

```
IF @@ROWCOUNT > 0
   -- Perform some interest calculation here...
   CONTINUE
ELSE
   BREAK
END

SELECT @RunningBal AS 'End Balance'
END
GO
```

6. We can now create the stored procedure and test our results. The example is going to check whether Vic McGlynn, customer ID 1, has had a positive or negative movement on her cash balance in the month of August 2005. The code to find this out follows. First of all, we insert some TransactionDetails.Transactions rows to test it out. We also prefix the stored procedure with an EXEC(UTE) statement, as this is part of a batch of statements.

```
INSERT INTO TransactionDetails.Transactions
(CustomerId,TransactionType,DateEntered,Amount,RelatedProductId)
VALUES (1,1,'1 Aug 2005',100.00,1)
INSERT INTO TransactionDetails.Transactions
(CustomerId,TransactionType,DateEntered,Amount,RelatedProductId)
VALUES (1,1,'3 Aug 2005',75.67,1)
INSERT INTO TransactionDetails.Transactions
(CustomerId,TransactionType,DateEntered,Amount,RelatedProductId)
VALUES (1,2,'5 Aug 2005',35.20,1)
INSERT INTO TransactionDetails.Transactions
(CustomerId,TransactionType,DateEntered,Amount,RelatedProductId)
VALUES (1,2,'6 Aug 2005',20.00,1)
EXEC  CustomerDetails.apf_CustMovement 1,'1 Aug 2005','31 Aug 2005'
```

7. Execute the preceding code, which should return a value that we expect, as shown in Figure 10-12.

Figure 10-12. *Complex stored procedure output*

Summary

In this chapter, you have met stored procedures, which are collections of T-SQL statements and commands compiled and ready to be executed by SSE. You have learned the advantages of a stored procedure over an ad hoc query, encountered the basic CREATE PROCEDURE syntax, and created some simple stored procedures.

The basics of building a stored procedure are very simple and straightforward. Therefore, building a stored procedure within Query Editor may be as attractive as using a template. As stored procedures are sets of T-SQL statements combined together, you will tend to find that you build up your query, and then at the end surround it with a CREATE PROCEDURE statement.

Probably the largest area of code creation outside of data manipulation and searching will be through control-of-flow statements. We look at other areas, particularly error handling, in Chapter 11, to advance your T-SQL knowledge.

■■■

T-SQL Essentials

Now that you know how to build and work with SSE objects, and insert, update, and delete data as well as retrieve it, we can now move on to more of the T-SQL essentials required to complete your programming knowledge.

Potentially the most important area covered by this chapter is error handling. After all, no matter how good your code is, if it cannot cope when an error occurs, it will be hard to keep the code stable and reliable. There will always be times that the unexpected happens, either from strange input data to something happening in the server. However, this is not the only area of interest. We will be looking at joining tables together, performing aggregations of data, and grouping data together. Finally, there will be times that you wish to hold data either in a variable or within a table that you only want to exist for a short period. Quite a great deal to cover, but this chapter and the next will be the stepping stones that move you from a novice to a professional developer.

This chapter will therefore look at the following:

- Joining two or more tables to see more informational results

- Having a method of storing information on a temporary basis via variables

- How to hold rows of information in a nonpermanent table

- How to aggregate values

- Organizing output data into groups of relevant information

- Returning on unique and distinct values

- Looking at and using system functions

- Error handling: how to create your own errors, trap errors, and make code secure

Using More Than One Table

Throughout this book, the SELECT and UPDATE statements have only dealt with and covered the use of one table. However, it is possible to have more than one table to retrieve data from within our SELECT or UPDATE statement, but we must keep in mind that the more tables included in the query, the more detrimental the effect on the query's performance. When we include subsequent tables, there must be a link of some sort between the two tables, known as a **join**. A join will take place between at least one column in one table and a column from the joining table. The columns involved in the join do not have to be in any key within the tables involved in the join. However, this is quite uncommon, and if you do find you are joining tables, there is a high chance that a relationship exists between them, which would mean you do require a primary key and a foreign key. This was covered in Chapter 3.

It is possible that one of the columns on one side of the join is actually a concatenation of two or more columns. As long as the end result is one column, this is acceptable. Also, the two columns

that are being joined do not have to have the same name, as long as they both have similar data types. For example, you can join a char with a varchar. What is not acceptable is that one side of the JOIN names a column, and on the other side is a variable or literal that is really a filter that would be found in a WHERE statement.

Joining two tables together can become quite complicated. The most basic join condition is a straight join between two tables, which is called an INNER JOIN. An INNER JOIN will join the two tables, and where there is a join of data using the columns from each of the two tables, then the data will be returned. For example, if there is a share in the ShareDetails.Shares table that has no price and you are joining the two tables on the share ID, you would only see output where there is a share with a share price. You will see this in action in this chapter.

It is possible to return all the rows from one table where there is no join. This is known as an OUTER JOIN. Depending on which table you want the rows always to be returned from, this will either be a LEFT OUTER JOIN or a RIGHT OUTER JOIN. Taking our shares example, we could use an OUTER JOIN so that even when there is no share price, we can still list the share. This example will also be demonstrated later in this chapter.

The final type of join is the scariest and most dangerous join. If you wish for every row in one table to be joined with every row in the joining table, you would use a CROSS JOIN. So if you had 10 rows in one table and 12 rows in the other table, you would see returned 120 rows of data (10 × 12). As you can imagine, this type of join just needs two small tables to produce even a large amount of output.

Although not the most helpful of syntax demonstrated within the book, the syntax for joining two tables is as follows:

```
FROM tablea
[FULL[INNER|OUTER|CROSS]] JOIN tableb
{ON tableb.column1 = tablea.column2 {AND|OR tableb.column...}}
```

The best way to look at the syntax is within a described example. We will use two tables to demonstrate the INNER JOIN in this example, ShareDetails.Shares and ShareDetails.SharePrices.

Joining two tables could not be simpler. All the columns in both tables are available to be returned through the query, and so we can list the columns desired as normal. However, if there are two columns of the same name, they must be prefixed with the name, or the alias name, of the table from which the information is derived.

■**Note** It is recommended that whenever a join does take place, whether the column name is unique or not, that all columns be prefixed with the table or alias name. This saves time if the query is expanded to include other tables, but it also clarifies exactly where the information is coming from.

Try It Out: Joining Two Tables

1. The first join we will look at is the INNER JOIN. This is where we have two tables and we want to list all the values where there is a join. In this case, we want to list all the shares where there is a share price, and we want to see every share price for that share. Notice that we don't need to define the word INNER. This is presumed if nothing else is specified. Also take note that, like columns, we have defined aliases for the table names. This makes prefixing columns easier. We are joining the two tables on ShareId, as this is the linking column between the two tables. Enter the following code:

   ```
   SELECT s.ShareDesc,sp.Price,sp.PriceDate
     FROM ShareDetails.Shares s
     JOIN ShareDetails.SharePrices sp ON sp.ShareId = s.ShareId
   ```

2. Once you have executed the code, you should see the output that appears in Figure 11-1. There is no output for ShareIds 3, 4, and 5, as they have no share price.

	ShareDesc	Price	PriceDate
1	ACME'S HOMEBAKE COOKIES INC	2.34125	2005-08-01 16:10:00.000
2	ACME'S HOMEBAKE COOKIES INC	2.17000	2005-08-01 14:54:00.000
3	ACME'S HOMEBAKE COOKIES INC	2.21000	2005-08-01 11:22:00.000
4	ACME'S HOMEBAKE COOKIES INC	2.41750	2005-08-01 10:16:00.000
5	ACME'S HOMEBAKE COOKIES INC	2.21250	2005-08-01 10:12:00.000
6	ACME'S HOMEBAKE COOKIES INC	2.15500	2005-08-01 10:10:00.000
7	FAT-BELLY.COM	45.20000	2005-08-03 10:10:00.000
8	FAT-BELLY.COM	43.22000	2005-08-02 10:10:00.000
9	FAT-BELLY.COM	41.10000	2005-08-01 10:10:00.000

Figure 11-1. *First INNER JOIN*

3. We can take this a stage further and filter the rows to only list the share price row that matches the CurrentPrice in the ShareDetails.Shares table. This could be done by filtering the data on a WHERE statement, and from a performance perspective it would be better, as neither of these columns are within an index, and there could be a large number of rows for ShareDetails.SharePrices for each share as time goes on; but for this example, it demonstrates how to add a second column for the join.

```
SELECT s.ShareDesc,sp.Price,sp.PriceDate
  FROM ShareDetails.Shares s
  JOIN ShareDetails.SharePrices sp ON sp.ShareId = s.ShareId
   AND sp.Price = s.CurrentPrice
```

4. Execute the preceding code, which will return two rows as shown in Figure 11-2. As you can see, an INNER JOIN is very straightforward.

	ShareDesc	Price	PriceDate
1	ACME'S HOMEBAKE COOKIES INC	2.34125	2005-08-01 16:10:00.000
2	FAT-BELLY.COM	45.20000	2005-08-03 10:10:00.000

Figure 11-2. *INNER JOIN with multiple join columns*

5. The next join we look at is an OUTER JOIN, more specifically a LEFT OUTER JOIN. In this instance, we want to return all the rows in the left table, whether there is any data in the right table or not. The left table in this case is the ShareDetails.Shares table, as it is the left named table of the two we are concerned with. Enter the following code:

```
SELECT s.ShareDesc,sp.Price,sp.PriceDate
  FROM ShareDetails.Shares s
  LEFT OUTER JOIN ShareDetails.SharePrices sp ON sp.ShareId = s.ShareId
```

6. Once you execute this code, you should see the missing shares from the previous example listed, as you see in Figure 11-3. Notice that where no data exists in the ShareDetails.SharePrices table, the values are displayed as NULL. OUTER JOINS are a good tool when checking other queries. For example, the results in Figure 11-3 demonstrate that, quite rightly, the bottom three shares should have been missing in the first example, as they did not meet our criteria. This may not be so obvious when there are large volumes of data though.

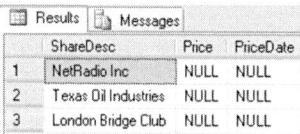

	ShareDesc	Price	PriceDate
1	ACME'S HOMEBAKE COOKIES INC	2.34125	2005-08-01 16:10:00.000
2	ACME'S HOMEBAKE COOKIES INC	2.17000	2005-08-01 14:54:00.000
3	ACME'S HOMEBAKE COOKIES INC	2.21000	2005-08-01 11:22:00.000
4	ACME'S HOMEBAKE COOKIES INC	2.41750	2005-08-01 10:16:00.000
5	ACME'S HOMEBAKE COOKIES INC	2.21250	2005-08-01 10:12:00.000
6	ACME'S HOMEBAKE COOKIES INC	2.15500	2005-08-01 10:10:00.000
7	FAT-BELLY.COM	45.20000	2005-08-03 10:10:00.000
8	FAT-BELLY.COM	43.22000	2005-08-02 10:10:00.000
9	FAT-BELLY.COM	41.10000	2005-08-01 10:10:00.000
10	NetRadio Inc	NULL	NULL
11	Texas Oil Industries	NULL	NULL
12	London Bridge Club	NULL	NULL

Figure 11-3. *LEFT OUTER JOIN*

7. To get around this problem, we can add a WHERE statement that will list those shares that do not have an item in ShareDetails.SharePrices. This is one method of achieving our goal. We will look at the other later in the chapter when we examine EXISTS. We know that when there is a missing share price, Price and PriceDate will be NULL. It is also necessary to know that Price cannot have a NULL value inserted in any rows of data. If it could, we would need to use another method, such as EXISTS.

```
SELECT s.ShareDesc,sp.Price,sp.PriceDate
  FROM ShareDetails.Shares s
  LEFT OUTER JOIN ShareDetails.SharePrices sp ON sp.ShareId = s.ShareId
WHERE sp.Price IS NULL
```

This time we will only have three rows returned, as you see in Figure 11-4.

	ShareDesc	Price	PriceDate
1	NetRadio Inc	NULL	NULL
2	Texas Oil Industries	NULL	NULL
3	London Bridge Club	NULL	NULL

Figure 11-4. *LEFT OUTER JOIN for no share prices*

8. The next example is a RIGHT OUTER JOIN. Here we expect the table on the right to return rows where there are no entries on the table in the left. In our example, such a scenario does not exist, as it would be breaking referential integrity; however, we can swap the tables around, which will show the same results as our first LEFT OUTER JOIN example. Take note that you don't have to alter the column order after ON, as it is the table definition that defines the left and right tables.

```
SELECT s.ShareDesc,sp.Price,sp.PriceDate
  FROM ShareDetails.SharePrices sp
  RIGHT OUTER JOIN ShareDetails.Shares s ON sp.ShareId = s.ShareId
```

9. Executing this code will give you the results shown in Figure 11-5.

Figure 11-5. *RIGHT OUTER JOIN*

10. If you want a LEFT OUTER JOIN and a RIGHT OUTER JOIN to be available at the same time, you need to choose the FULL OUTER JOIN. This will return rows from both the left and right tables if there are no matching rows in the other table. So to clarify, if there is a row in the left table but no match in the right table, the row from the left table will be returned with NULL values in the columns from the right table, and vice versa. This time we are going to break referential integrity and insert a share price with no share. We will then delete the row.

```
INSERT INTO ShareDetails.SharePrices
(ShareId, Price, PriceDate)
VALUES (99999,12.34,'1 Aug 2005 10:10AM')

SELECT s.ShareDesc,sp.Price,sp.PriceDate
  FROM ShareDetails.SharePrices sp
  FULL OUTER JOIN ShareDetails.Shares s ON sp.ShareId = s.ShareId
```

11. Once the preceding code has been executed, you will see the results that appear in Figure 11-6. Notice that we have rows from the ShareDetails.Shares table when there is no share price and vice versa.

Figure 11-6. *FULL OUTER JOIN*

12. The final demonstration is with a CROSS JOIN. This is a Cartesian join between our ShareDetails.Shares and ShareDetails.SharePrices table. A CROSS JOIN cannot have any filtering on it, therefore it cannot include a WHERE statement. As we are joining every row with every row, there is no need to provide an ON statement, because there is no specific row-on-row join.

```
SELECT s.ShareDesc,sp.Price,sp.PriceDate
  FROM ShareDetails.SharePrices sp
  CROSS JOIN ShareDetails.Shares s
```

13. The preceding code, when executed, generates a large amount of output. Figure 11-7 shows only a snippet of the output.

	ShareDesc	Price	PriceDate
1	ACME'S HOMEBAKE COOKIES INC	2.34125	2005-08-01 16:10:00.000
2	ACME'S HOMEBAKE COOKIES INC	2.17000	2005-08-01 14:54:00.000
3	ACME'S HOMEBAKE COOKIES INC	2.21000	2005-08-01 11:22:00.000
4	ACME'S HOMEBAKE COOKIES INC	2.41750	2005-08-01 10:16:00.000
5	ACME'S HOMEBAKE COOKIES INC	2.21250	2005-08-01 10:12:00.000
6	ACME'S HOMEBAKE COOKIES INC	2.15500	2005-08-01 10:10:00.000
7	ACME'S HOMEBAKE COOKIES INC	45.20000	2005-08-03 10:10:00.000
8	ACME'S HOMEBAKE COOKIES INC	43.22000	2005-08-02 10:10:00.000
9	ACME'S HOMEBAKE COOKIES INC	41.10000	2005-08-01 10:10:00.000
10	ACME'S HOMEBAKE COOKIES INC	12.34000	2005-08-01 10:10:00.000
11	FAT-BELLY.COM	2.34125	2005-08-01 16:10:00.000
12	FAT-BELLY.COM	2.17000	2005-08-01 14:54:00.000
13	FAT-BELLY.COM	2.21000	2005-08-01 11:22:00.000
14	FAT-BELLY.COM	2.41750	2005-08-01 10:16:00.000
15	FAT-BELLY.COM	2.21250	2005-08-01 10:12:00.000

Figure 11-7. *CROSS JOIN*

Variables

There will be times when you need to hold a value or work with a value that does not come directly from a column. Or perhaps you retrieve a value from a single row of data and a single column that you want to use in a different part of a query. It is possible to do this via a variable.

A **variable** can be declared at any time within a set of T-SQL, whether it is ad hoc or a stored procedure or trigger. However, a variable has a finite lifetime.

To inform SSE you wish to use a variable, use the following syntax:

```
DECLARE @variable_name datatype, @variable_name2 datatype
```

All variables have to be preceded with an @ sign, and as you can see from the syntax, more than one variable can be declared, although multiple variables should be separated by a comma and held on one line of code. If you move to a second line of code, you need to prefix the first variable with another DECLARE statement. All variables can hold a NULL value, and there is not an option to say that the variable cannot hold a NULL value. By default then, when a variable is declared, it will have an initial value of NULL. It is also not possible at declaration to assign a value to a variable.

To assign a value to a variable, you can use a SET statement or a SELECT statement. It is standard to use SET to set a variable value when you are not working with any tables. Let's take a look at some examples to see more of how to work with variables and their lifetime.

Try It Out: Declaring and Working with Variables

1. In this example, we define two variables; in the first, we will be placing the current date and time using the system function GETDATE(), and in the second, we are setting the value of the variable @CurrPriceInCents to the value from a column within a table with a mathematical function tagged on. Once these two have been set using SET and SELECT, we will then list them out, which can only be done via a SELECT statement.

```
DECLARE @MyDate datetime, @CurrPriceInCents money
SET @MyDate = GETDATE()
SELECT @CurrPriceInCents = CurrentPrice * 100
  FROM ShareDetails.Shares
 WHERE ShareId = 2
SELECT @MyDate,@CurrPriceInCents
```

2. Execute the code, and you will see something like the results shown in Figure 11-8, the first column showing your current data and time.

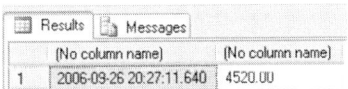

	(No column name)	(No column name)
1	2006-09-26 20:27:11.640	4520.00

Figure 11-8. *Working with our first variable*

3. If we change the query, however, into two batches, the variables in the second batch will not exist, and when we try to execute all of the code at once, we will get an error. Enter the code as it appears here; the only real change is the GO statement shown in bold.

```
DECLARE @MyDate datetime, @CurrPriceInCents money
SET @MyDate = GETDATE()
SELECT @CurrPriceInCents = CurrentPrice * 100
  FROM ShareDetails.Shares
 WHERE ShareId = 2
        GO
SELECT @MyDate,@CurrPriceInCents
```

4. The error returned when this code is executed is defined as the following results, where we are being informed that SSE doesn't know about the first variable in the last statement. This is because SSE is parsing the whole set of T-SQL before executing, rather than one batch at a time.

```
Msg 137, Level 15, State 2, Line 1
Must declare the scalar variable "@MyDate".
```

5. Remove the GO statement so you can see one more example of how variables work. You also need to remove the WHERE statement in the example so that you return all rows from the ShareDetails.Shares table. The value that will be assigned to the variable @CurrPriceInCents will be the last value returned from the query of data. The code we wish to execute is as follows. We have kept the two lines in the query, but they have now been prefixed with two dashes, --. This indicates to SSE that the lines of code have been commented out and should be ignored.

```
DECLARE @MyDate datetime, @CurrPriceInCents money
SET @MyDate = GETDATE()
SELECT @CurrPriceInCents = CurrentPrice * 100
  FROM ShareDetails.Shares
-- WHERE ShareId = 2
--GO
SELECT @MyDate,@CurrPriceInCents
```

6. If we look at the results that this produces, as shown in Figure 11-9, we can see that the value in the second column is from the last row in the `ShareDetails.Shares` table, which could also be found by performing `SELECT * FROM ShareDetails.Shares`.

Figure 11-9. *Variables and batches*

Temporary Tables

There are two types of temporary tables: local and global. These temporary tables will be created in `tempdb` and not within the database you are connected to. They also have a finite lifetime. Unlike a variable, the time such a table can "survive" is different.

A local temporary table will survive until the connection it was created within is dropped. This can happen when the stored procedure that created the temporary table completes, or when the Query Editor window is closed. A local temporary table is defined by prefixing the table name by a single hash mark, #. The scope of a local temporary table is the connection that created it only.

A global temporary table is defined by prefixing the table name by a double hash mark, ##. The scope of a global temporary table differs significantly. When a connection creates the table, it is then available to be used by any user and any connection, just like a permanent table. A global temporary table will only then be "deleted" when all connections to it have been closed.

In Chapter 8, when looking at the `SELECT` statement, you were introduced to `SELECT...INTO`, which allows a permanent table to be built from data from either another table or tables, or from a list of variables. We could make this table more transient by defining the `INTO` table to reside within the `tempdb`. However, it will still exist within `tempdb` until it is either dropped or SSE is stopped and restarted. Slightly better, but not perfect for when you just want to build an interim table between two sets of T-SQL statements.

Requiring a temporary table could happen for a number of reasons. Building a single T-SQL statement returning information from a number of tables can get complex, and perhaps could even not be ideally optimized for returning the data quickly. Splitting the query into two may make the code easier to maintain and perform better. To give an example, as our customers "age," they will have more and more transactions against their account IDs. It may be that when working out any interest to accrue, the query is taking a long time to run, as there are more and more transactions. It might be better to create a temporary table just of the transactions you are interested in, then pass this temporary table to code that then calculates the interest rather than trying to complete all the work in one pass of the data.

When it comes time to work with a temporary table, such a table can be built either by using the `CREATE TABLE` statement or by using the `SELECT...INTO` command. Let's take a look at temporary tables in action.

Try It Out: Temporary Tables

1. The first example will create a local temporary table based on the CREATE TABLE statement. We will then populate the table with some data and retrieve the data. We will then open up a different Query Editor pane and try and retrieve data from the table to show that it is local. Also of interest here is how we can use a SELECT statement in conjunction with an INSERT statement to add the values. Providing that the number of columns returned in the SELECT match either the number of columns within the table or the number of columns in the INSERT column list, using a SELECT statement is a great way of populating a table, especially temporary tables. First, create the temporary table. For the moment, just enter the code, don't execute it.

```
CREATE TABLE #SharesTmp
(ShareDesc varchar(50),
Price numeric(18,5),
PriceDate datetime)
```

2. Next we want to populate the temporary table with information from the ShareDetails.Shares and the ShareDetails.SharePrices tables. Because we are populating every column within the table, we don't need to list the columns in the INSERT INTO table part of the query. Then we use the results from a SELECT statement to populate many rows in one set of T-SQL. You can execute the code now if you want, but when we get to the third part in a moment, run the SELECT * from the same Query Editor window.

```
INSERT INTO #SharesTmp
SELECT s.ShareDesc,sp.Price,sp.PriceDate
  FROM ShareDetails.Shares s
  JOIN ShareDetails.SharePrices sp ON sp.ShareId = s.ShareId
```

3. The final part is to prove that there is data in the table.

```
SELECT * FROM #SharesTmp
```

4. When the code is executed, you should see the output that appears in Figure 11-10.

	ShareDesc	Price	PriceDate
1	ACME'S HOMEBAKE COOKIES INC	2.34125	2005-08-01 16:10:00.000
2	ACME'S HOMEBAKE COOKIES INC	2.17000	2005-08-01 14:54:00.000
3	ACME'S HOMEBAKE COOKIES INC	2.21000	2005-08-01 11:22:00.000
4	ACME'S HOMEBAKE COOKIES INC	2.41750	2005-08-01 10:16:00.000
5	ACME'S HOMEBAKE COOKIES INC	2.21250	2005-08-01 10:12:00.000
6	ACME'S HOMEBAKE COOKIES INC	2.15500	2005-08-01 10:10:00.000
7	FAT-BELLY.COM	45.20000	2005-08-03 10:10:00.000
8	FAT-BELLY.COM	43.22000	2005-08-02 10:10:00.000
9	FAT-BELLY.COM	41.10000	2005-08-01 10:10:00.000

Figure 11-10. *Temporary table*

5. Open up a fresh Query Editor pane and then try and execute the following code:

```
SELECT * FROM #SharesTmp
```

Now instead of returning a set of results like those in Figure 11-10, you will get an error message:

```
Msg 208, Level 16, State 0, Line 1
Invalid object name '#SharesTmp'.
```

6. If we change the whole query to now work with a global temporary variable, you will see a different end result. To ensure we are starting afresh, clear all the Query Editors, or execute the following DROP TABLE command in the first Query Editor.

```
DROP TABLE #SharesTmp
```

7. Enter the following code, taking note of the double hash marks, in one of the Query Editors.

```
CREATE TABLE ##SharesTmp
(ShareDesc varchar(50),
Price numeric(18,5),
PriceDate datetime)
INSERT INTO ##SharesTmp
SELECT s.ShareDesc,sp.Price,sp.PriceDate
  FROM ShareDetails.Shares s
  JOIN ShareDetails.SharePrices sp ON sp.ShareId = s.ShareId
SELECT * FROM ##SharesTmp
```

8. When you execute the code, you should see the same results as you did with the first query (refer back to Figure 11-10).

9. Move to a new Query Editor pane, ensuring that you leave the previous Query Editor pane still open. Then enter the following SELECT statement:

```
SELECT * FROM ##SharesTmp
```

10. When this is executed, you see the same results again, as shown originally in Figure 11-10.

It is not until the first Query Editor pane that defined the global table is closed or until a DROP TABLE ##SharesTmp is executed that the table will disappear.

Aggregations

An **aggregation** is where SSE performs a function on a set of data to return one aggregated value per grouping of data. This will vary from counting the number of rows returned from a SELECT statement through to figuring out maximum and minimum values. Combining some of these functions with the DISTINCT function, discussed later in the section "Distinct Values," can provide some very useful functionality. An example might be when you want to show the highest value for each distinct share to demonstrate when the share was worth the greatest amount.

Let's dive straight in by looking at different aggregation types and working through examples of each.

COUNT/COUNT_BIG

COUNT/COUNT_BIG is probably the most commonly used aggregation, and it finds out the number of rows returned from a query. You use this for checking the total number of rows in a table, or more likely the number of rows returned from a particular set of filtering criteria. Quite often this will be used to cross-check the number of rows from a query in SSE with the number of rows an application is showing to a user.

The syntax is COUNT(*) or COUNT_BIG(*). There are no columns defined, as it is rows that are being counted.

Note The difference in these two functions is that COUNT returns an integer data type, and COUNT_BIG returns a bigint data type.

Try It Out: Counting Rows

1. This example will count the number of rows in the Shares table. We know that we have only inserted five rows, and so we expect from the following code a returned value of 5.

```
SELECT COUNT(*) AS 'Number of Rows'
FROM ShareDetails.Shares
```

2. Execute the code, and you will see the results shown in Figure 11-11.

Figure 11-11. *Using COUNT()*

3. We could of course add a filter such as the following, which counts the number of shares where the price is greater than 10 dollars:

```
SELECT COUNT(*) AS 'Number of Rows'
FROM ShareDetails.Shares
WHERE CurrentPrice > 10
```

4. Execute the code, and you will now see a count of 2, as appears in Figure 11-12, as expected.

Figure 11-12. *COUNT with a filter*

SUM

If you have numeric values in a column, it is possible to aggregate them as a summation. The ideal scenario for this is to aggregate the number of transactions in a bank account to see how much the balance has changed by. This could be daily, weekly, monthly, or over any time period required. A negative amount would show that more has been taken out of the account than put in, for example.

The syntax can be shown as SUM(column1|@variable|Mathematical function). The summation does not have to be of a column, but could include a math function. One example would be to sum up the cost of purchasing shares, so you would multiply the number of shares bought multiplied by the cost paid.

Try It Out: Summing Values

1. We can do a simple SUM to add up the amount of money that has passed through the account as a withdrawal, TransactionType 1.

```
SELECT SUM(Amount) AS 'Amount Deposited'
  FROM TransactionDetails.Transactions
 WHERE CustomerId = 1
   AND TransactionType = 1
```

2. Executing this code will add up the first two rows we inserted at point 1. The results will be 100+75.67, as shown in Figure 11-13.

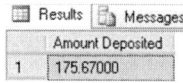

Figure 11-13. *SUMming values*

MAX/MIN

On a set of data, it is possible to get the minimum and maximum values of a column of data. This is useful if you want to see values such as the smallest share price or the greatest portfolio value, or in other scenarios outside of our example, as the maximum number of sales of each product in a period of time, or the minimum sold, so that you can see if some days are quieter than others.

Try It Out: MAX and MIN

1. In this example, you will see how to find the maximum and minimum values for a share with one statement. Enter the following code:

```
SELECT MAX(Price) MaxPrice,MIN(Price) MinPrice
FROM ShareDetails.SharePrices
WHERE ShareId = 1
```

2. Executing the code will produce the results shown in Figure 11-14.

Figure 11-14. *Finding the maximum and minimum*

AVG

As you might expect, the AVG aggregation returns the average value from the rowset of a column of data. All of the values are summed up and then divided by the number of rows that formed the underlying result set.

Try It Out: Averaging It Out

1. Our last aggregation example will produce an average value for the share prices found for share ID 1. Enter the following code:

```
SELECT AVG(Price) AvgPrice
FROM ShareDetails.SharePrices
WHERE ShareId = 1
```

2. Once you have executed the code, you should see the results shown in Figure 11-15.

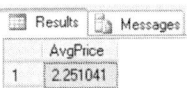

Figure 11-15. *Finding the average*

Now that we have taken a look at aggregations, we can move on to looking at grouping data. Aggregations, as you have seen, are useful, but limited. In the next section, we can expand these aggregations so that they are used with groups of data.

GROUP BY

Using aggregations, as has just been demonstrated, works well when you just wish a single row of results for a specific filtered item. If you wish to find the average price of several shares, you may be thinking you need to provide a SELECT AVG() for each share. This section will demonstrate that this is not the case. By using GROUP BY, you instruct SSE to group the data to return and provide a summary value for each grouping of data. To clarify, as you will see in the upcoming examples, we could remove the WHERE ShareId=1 statement, which would then allow you to group the results by each different ShareId.

The basic syntax for grouping is defined in the following code. It is possible to expand GROUP BY further to include rolling up or providing cubes of information; however, such a discussion falls outside the scope of this book.

```
GROUP BY [ALL] (column1[,column2,...])
```

The option ALL is a bit like an OUTER JOIN. If you have a WHERE statement as part of your SELECT statement, any grouping filtered out will still return a row in the results, but instead of aggregating the column, a value of NULL will be returned. I tend to use this as a checking mechanism. I can see the rows with values and the rows without values, and visually this will tell me that my filtering is correct.

When working with GROUP BY, the main point that you have to be aware of is that any column defined in the SELECT statement that does not form part of the aggregation **must** be contained within the GROUP BY clause and be in the same order as the SELECT statement. Failure to do this will mean that the query will give erroneous results and in many cases use a lot of resources in giving these results.

Try It Out: GROUP BY

1. This first example will demonstrate how to find maximum and minimum values for every share that has a row in the `ShareDetails.SharePrices` table where the share ID < 9999. This means that the row we added earlier when looking at joins that has no `Share` record will be excluded. The code is as follows:

```
SELECT ShareId, MIN(Price) MinPrice, Max(Price) MaxPrice
  FROM ShareDetails.SharePrices
WHERE ShareId < 9999
GROUP BY ShareId
```

2. When the code is executed, you will see the two shares listed with their corresponding minimum and maximum values, as shown in Figure 11-16.

	ShareId	MinPrice	MaxPrice
1	1	2.15500	2.41750
2	2	41.10000	45.20000

Figure 11-16. *MAX and MIN of a group*

3. If we wish to include any rows where there is a `Price` row, but the `ShareId` has a value of 9999 or greater, we would use the `ALL` option with `GROUP BY`. In the following example, we are also linking into the `ShareDetails.Shares` table to retrieve the share description.

```
SELECT sp.ShareId, s.ShareDesc,MIN(Price) MinPrice, Max(Price) MaxPrice
  FROM ShareDetails.SharePrices sp
  LEFT JOIN ShareDetails.Shares s ON s.ShareId = sp.ShareId
WHERE sp.ShareId < 9999
  GROUP BY ALL sp.ShareId, s.ShareDesc
```

4. When you execute the code, the `Price` row that is outside of the filtering returns a `NULL` value. The other rows return details as shown in Figure 11-17.

	ShareId	ShareDesc	MinPrice	MaxPrice
1	9999	NULL	NULL	NULL
2	1	ACME'S HOMEBAKE COOKIES INC	2.15500	2.41750
3	2	FAT-BELLY.COM	41.10000	45.20000

Figure 11-17. *A JOIN with a MAX and MIN group*

HAVING

When using the GROUP BY clause, it is possible to supplement your query with a HAVING clause. The HAVING clause is like a filter, but it works on aggregations of the data rather than the rows of data prior to the aggregation. Hence, it has to be included with a GROUP BY clause. It will also include the aggregation you wish to check. The code would therefore look as follows:

```
GROUP BY column1[,column2...]
HAVING [aggregation_condition]
```

The `aggregation_condition` would be where we place the aggregation and the test we wish to perform. For example, my bank charges me if I have more than 20 nonregular items pass through my account in a month. In this case, the query would group by customer ID, counting the number of nonregular transactions for each calendar month. If the count were less than or equal to 20 items, you would like this list to not include the customer in question. To clarify this, the query code would look something like the following if we were running this in August 2005:

```
SELECT CustomerId,COUNT(*)
FROM CustomerBankTransactions
WHERE TransactionDate BETWEEN '1 Aug 2005 ' AND '31 Aug 2005 '
GROUP BY CustomerId
HAVING COUNT(*) > 20
```

Try It Out: HAVING

1. In the following example, we will use the `MIN` aggregate function to remove rows where the minimum share price is greater than $10. This query is taken from our `GROUP BY ALL` example shown earlier. Although we have kept the `ALL` option within the `GROUP BY` statement, it is ignored, as it is followed by the `HAVING` clause.

```
SELECT sp.ShareId, s.ShareDesc,MIN(Price) MinPrice, Max(Price) MaxPrice
  FROM ShareDetails.SharePrices sp
  LEFT JOIN ShareDetails.Shares s ON s.ShareId = sp.ShareId
 WHERE sp.ShareId < 9999
 GROUP BY ALL sp.ShareId, s.ShareDesc
HAVING MIN(Price) > 10
```

2. The results on the executed code will only return one value, as you see in Figure 11-18, not only ignoring ACME as its share price is below $10, but also the share 99999 that has a `MinPrice` value of `NULL`.

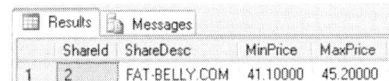

Figure 11-18. *Retrieving only certain aggregated rows*

Even if we changed `HAVING` to being less than $10, the share ID 99999 would still be ignored due to `HAVING` overriding the `GROUP BY ALL`. Not only that, `NULL` as you know is a "special" value and is neither less than or greater than any value.

Distinct Values

With some of our tables in our examples, multiple entries will exist for the same value. To clarify, in the `ShareDetails.SharePrices` table, there are multiple entries for each share as each price is stored. There may be some shares with no price, of course. But what if you wanted to see a listing of shares that did have prices, but you want to see each share listed only once? This is a simple example, and we will see more complex examples later on when we look at using aggregations within SSE. That aside, the example that follows serves it purpose well.

The syntax is to place the keyword DISTINCT after the SELECT statement and before the list of columns. The following list of columns is then tested for all the rows returned, and for each set of unique distinct values, one row will be listed.

Try It Out: Distinct Values

1. We have to join the ShareDetails.Shares and ShareDetails.SharePrices tables again so that we know we are only returning rows that have a share price. We had that code in our JOIN section earlier in the chapter. It is replicated here, and you can execute it if you wish.

```
SELECT s.ShareDesc,sp.Price,sp.PriceDate
  FROM ShareDetails.Shares s
  JOIN ShareDetails.SharePrices sp ON sp.ShareId = s.ShareId
```

2. As you know, this will return multiple rows for each share. Placing DISTINCT at the start of the column list will not make any difference, because there are different prices and different price dates.

```
SELECT DISTINCT s.ShareDesc,sp.Price,sp.PriceDate
  FROM ShareDetails.Shares s
  JOIN ShareDetails.SharePrices sp ON sp.ShareId = s.ShareId
```

3. To get a list of shares that have a value, it is necessary to remove the last two columns and only list the ShareDesc column:

```
SELECT DISTINCT s.ShareDesc
  FROM ShareDetails.Shares s
  JOIN ShareDetails.SharePrices sp ON sp.ShareId = s.ShareId
```

4. When you execute this code, you will now see the desired results, as shown in Figure 11-19.

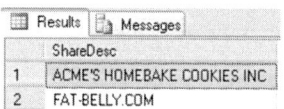

Figure 11-19. *Finding unique values*

Functions

To bring more flexibility to your T-SQL code, you can use a number of functions with the data from variables and columns. This section does not include a comprehensive list, but it does contain the most commonly used functions and the functions you will come across early in your development career. They have been split into three categories: date and time, string, and system functions. There is a short explanation for each, with some code demonstrating each function in an example with results.

Date and Time

The first set of functions involve working either with a variable that holds a date and time value or using a system function to retrieve the current data and time.

DATEADD()

If you want to add or subtract an amount of time to a column or a variable and display a new value in a rowset, or set a variable with that new value, the DATEADD() will do this. The syntax for DATEADD() is

```
DATEADD(datepart, number, date)
```

The datepart option applies to all of the date functions and details what you want to add from milliseconds to years. These are defined as reserved words and therefore will not be surrounded by quotation marks. There are a number of possible values, and they are listed in Table 11-1.

Table 11-1. *Available datepart Options*

datepart Definition	Meaning
Ms	Millisecond
ss, s	Second
mi, n	Minute
Hh	Hour
dw, w	Weekday
wk, ww	Week
dd, d	Day
dy, y	Day of year
mm, n	Month
qq, q	Quarter
yy, yyyy	Year

Taking the second option, to add the value, the number should be positive, and to subtract a number, make it negative. Moving to the final option, this can be either a value, a variable, or a column date type holding the date and time you wish to change.

Try It Out: DATEADD()

1. We will set a local variable to a date and time. After that, we will add four hours to the value and display the results, as shown in Figure 11-20.

```
DECLARE @OldTime datetime
SET @OldTime = '24 March 2006 3:00 PM'
SELECT DATEADD(hh,4,@OldTime)
```

Figure 11-20. *Adding hours to a date*

2. Taking the reverse, we will take the same variable and remove 6 hours. The results should appear as shown in Figure 11-21.

```
DECLARE @OldTime datetime
SET @OldTime = '24 March 2006 3:00 PM'
SELECT DATEADD(hh,-6,@OldTime)
```

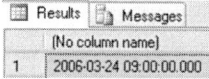

Figure 11-21. *Subtracting hours from a date*

DATEDIFF()

To find the difference between two dates, you would use the function DATEDIFF(). The syntax for this function is

```
DATEDIFF(datepart, startdate, enddate)
```

The first option contains the same options as for DATEADD(), and startdate and enddate are the two days you wish to compare. A negative number shows that the enddate is before the startdate.

Try It Out: DATEDIFF()

We will set two local variables to a date and time. After that we find the difference in milliseconds.

```
DECLARE @FirstTime datetime, @SecondTime datetime
SET @FirstTime = '24 March 2006 3:00 PM'
SET @SecondTime = '24 March 2006 3:33PM'
SELECT DATEDIFF(ms,@FirstTime,@SecondTime)
```

Figure 11-22 shows the results after executing this code.

Figure 11-22. *The difference between two dates*

DATENAME()

Returning the name of the part of the date is great for using with things like customer statements. Changing the number 6 to the word "June" makes for more pleasant reading.

The syntax is

```
DATENAME(datepart, datetoinspect)
```

You will also see this in action in DATEPART().

Try It Out: DATENAME()

In this example, we will set one date and time and then return the day of the week. We know this to be a Friday.

```
DECLARE @StatementDate datetime
SET @StatementDate = '24 March 2006 3:00 PM'
SELECT DATENAME(dw,@StatementDate)
```

Figure 11-23 shows the results after executing this code.

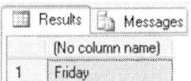

Figure 11-23. *The day name of a date*

DATEPART()

If you wish to achieve the returning part of a date from a date variable, column, or value, you can use DATEPART() within a SELECT statement.

As you may be expecting by now, the syntax has datepart as the first option, and then the datetoinspect as the second option, which will return the numeric day of the week from the date inspected.

```
DATEPART(datepart, datetoinspect)
```

Try It Out: DATEPART()

1. We need only set one local variable to a date and time. After that, we find the day of the month.

   ```
   DECLARE @WhatsTheDay datetime
   SET @WhatsTheDay = '24 March 2006 3:00 PM'
   SELECT DATEPART(dd, @WhatsTheDay)
   ```

 Figure 11-24 shows the results after executing this code.

 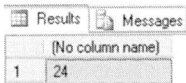

 Figure 11-24. *Finding part of a date*

2. To produce a more pleasing date and time for a statement, we can combine DATEPART() and DATENAME() to have a meaningful output. The function CAST(), which we will look at in detail shortly, is needed here, as it is a data type conversion function.

```
DECLARE @WhatsTheDay datetime
SET @WhatsTheDay = '24 March 2006 3:00 PM'
SELECT
DATENAME(dw,DATEPART(dd, @WhatsTheDay)) + ', ' +
CAST(DATEPART(dd,@WhatsTheDay) AS varchar(2)) + ' ' +
DATENAME(mm,@WhatsTheDay) + ' ' +
CAST(DATEPART(yyyy,@WhatsTheDay) AS char(4))
```

3. When this is executed, it will produce the more meaningful date shown in Figure 11-25.

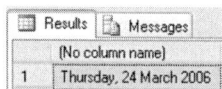

Figure 11-25. *Finding and concatenating info to provide a useful date*

GETDATE()

GETDATE() is a great function for returning the exact date and time from the system. You have seen this in action when setting up a table with a default value, and at a couple of other points in the book. There are no parameters to the syntax.

String

This next section will look at some functions that can act on those data types that are character based, such as varchar and char.

ASCII()

ASCII() will convert a single character to the equivalent ASCII code.

Try It Out: ASCII()

1. This example will return the ASCII code of the first character within a string. If the string has more than one character, only the first will be taken.

```
DECLARE @StringTest char(10)
SET @StringTest = ASCII('Robin     ')
SELECT @StringTest
```

2. Executing the code, you will see the ASCII value of the letter "R" returned, as shown in Figure 11-26.

Figure 11-26. *An ASCII value*

CHAR()

The reverse of ASCII() is the CHAR() function, which takes a numeric value and turns it into an alphanumeric character.

Try It Out: CHAR()

1. In this example, we define a local variable. Notice that the variable is a character-based data type. We then place the ASCII() value of "R" into this variable. From there, we convert back to a CHAR(). There is an implicit conversion from a character to a numeric. If the conversion results in a value greater than 255, the last value for an ASCII character, then NULL is returned. Enter the following code:

```
DECLARE @StringTest char(10)
SET @StringTest = ASCII('Robin    ')
SELECT CHAR(@StringTest)
```

2. Executing the code, you will see an "R," as shown in Figure 11-27.

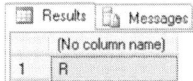

Figure 11-27. *Changing a number to a character*

3. The same result would be derived using a data type that is numeric based.

```
DECLARE @StringTest int
SET @StringTest = ASCII('12345    ')
SELECT CHAR(@StringTest)
```

LEFT()

When it is necessary to return the first *n* left characters from a string-based variable, you can achieve this through the use of LEFT(n), replacing n with the number of characters you wish to return.

Try It Out: LEFT()

1. In our next example, we take the first three characters from a local variable. Here we are taking the first three characters from Robin to return Rob:

```
DECLARE @StringTest char(10)
SET @StringTest = 'Robin    '
SELECT LEFT(@StringTest,3)
```

2. As expected, you should get the results shown in Figure 11-28 when you execute the code.

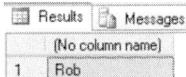

Figure 11-28. *The first LEFT characters*

LOWER()

To change alphabetic characters within a string, ensuring that all characters are in lowercase, you can use the LOWER() function.

Try It Out: LOWER()

1. Our LOWER() example combines this function along with another string function, LEFT(). Like all our functions, they can be combined together to perform several functions all at once.

```
DECLARE @StringTest char(10)
SET @StringTest = 'Robin     '
SELECT LOWER(LEFT(@StringTest,3))
```

2. As you can see, this results in displaying Rob in lowercase, as appears in Figure 11-29.

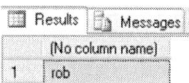

Figure 11-29. *Changing letters to lowercase*

LTRIM()

There will be times that leading spaces occur in a string that you want to remove. LTRIM() will trim these spaces on the left.

Try It Out: LTRIM()

1. To prove that leading spaces are removed by the LTRIM() function, we have to change the value within our local variable. On top of that, we have to put a string prefixing the variable to show that the variable has had the spaces removed.

```
DECLARE @StringTest char(10)
SET @StringTest = '     Robin'
SELECT 'Start-'+LTRIM(@StringTest),'Start-'+@StringTest
```

2. We produce two columns of output, as shown in Figure 11-30, the first with the variable trimmed and the second showing that the variable did have the leading spaces.

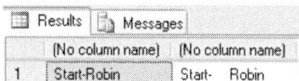

Figure 11-30. *Removing spaces from the left*

RIGHT()

The opposite of LEFT() is of course RIGHT(), and this function will return a set of characters from the right-hand side.

Try It Out: RIGHT()

1. Keep the variable used in LTRIM(), as it will allow us to return bin, which are the three right-hand-side characters now.

```
DECLARE @StringTest char(10)
SET @StringTest = '    Robin'
SELECT RIGHT(@StringTest,3)
```

2. And you can see, the results should appear as shown in Figure 11-31.

Figure 11-31. *Returning a number of characters starting from the right*

RTRIM()

When you have a CHAR() data type, no matter how many characters you enter, the variable will be filled on the right, known as **right padded**, with spaces. To remove these, use RTRIM. This will change the data from a fixed-length CHAR() to a variable-length value.

Try It Out: RTRIM()

1. This example has no spaces after Robin, and we will remove the space padding within the first column returned from the following code. The second column has the spaces trimmed.

```
DECLARE @StringTest char(10)
SET @StringTest = 'Robin'
SELECT @StringTest+'-End',RTRIM(@StringTest)+'-End'
```

2. And the results are as expected, as shown in Figure 11-32.

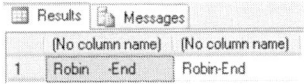

Figure 11-32. *Removing spaces from the right*

STR()

Some data types have implicit conversions. You will see later how to complete explicit conversions, but a simple conversion that will take any numeric value and convert it to a variable length string is STR(), which we look at next.

Try It Out: STR()

1. Our first example demonstrates that we cannot add a number, 82, to a string.

```
SELECT 'A'+82
```

2. When the preceding code is executed, you will see the following error:

```
Msg 245, Level 16, State 1, Line 1
Conversion failed when converting the varchar value 'A' to data type int.
```

3. Changing the example to include the STR() function will convert this numeric to a string of varying length such as varchar().

```
SELECT 'A'+STR(82)
```

4. Instead of an error, we now see the desired result, which appears in Figure 11-33. However, it isn't really desirable as there are spaces between the letter and the number. Leading zeros are translated to spaces.

Figure 11-33. *Changing a number to a string*

5. By including an LTRIM() function, we can remove those spaces:

```
SELECT 'A'+LTRIM(STR(82))
```

6. This code will now produce the correct results, as you see in Figure 11-34.

Figure 11-34. *Changing a number to a string and removing leading spaces*

SUBSTRING()

As you have seen, you can take a number of characters from the left and from the right of a string. To retrieve a number of characters that do not start with the first or last character, you need to use the function SUBSTRING(). This has three parameters: the variable or column, which character to start the retrieval from, and the number of characters to return.

Try It Out: SUBSTRING()

1. Define the variable we wish to return a substring from. Once complete, we can then take the variable, inform SSE we wish to start the substring at character position 3, and return the remaining characters.

```
DECLARE @StringTest char(10)
SET @StringTest = 'Robin     '
SELECT SUBSTRING(@StringTest,3,LEN(@StringTest))
```

2. And we have the desired result, as shown in Figure 11-35.

Figure 11-35. *Returning part of a string from within a string*

UPPER()

The final example is the reverse of the LOWER() function and will change all characters to uppercase.

Try It Out: UPPER()

1. After the declared variable has been set, we then use the UPPER() function to change the value to uppercase:

```
DECLARE @StringTest char(10)
SET @StringTest = 'Robin     '
SELECT UPPER(@StringTest)
```

2. And as you can see from Figure 11-36, Robin becomes ROBIN.

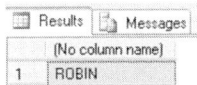

Figure 11-36. *Change the case of a string to uppercase*

System Functions

System functions are functions that provide extra functionality outside of the boundaries that can be defined as string, numeric, or date related. Three of these functions will be used extensively throughout our code, and therefore you should pay special attention to CASE, CAST, and ISNULL.

CASE WHEN...THEN...ELSE...END

We use the first function when we wish to test a condition. WHEN that condition is true, THEN we can do further processing, ELSE if it is false, then we can do something else. What happens in the WHEN

section and the THEN section can range from another CASE statement to providing a value that sets a column or a variable.

The CASE WHEN statement can be used to return a value or, if on the right-hand side of an equality statement, to set a value. Both of these scenarios are covered in the following examples.

Try It Out: CASE

1. The example will use a CASE statement to add up customers' TransactionDetails.Transactions for the month of August. If the TransactionType is 0, this is a Debit, 1 for a Credit. By using the SUM aggregation, we can add up the amounts. Combine this with a GROUP BY where the TransactionDetails. Transactions will be split between credits and debits, and we will get two rows in the results set: one for debits and one for credits.

```
SET QUOTED_IDENTIFIER OFF
SELECT CustomerId,
CASE WHEN CreditType = 0 THEN "Debits" ELSE "Credits" END
AS TranType,SUM(Amount)
  FROM TransactionDetails.Transactions t
  JOIN TransactionDetails.TransactionTypes tt ON
        tt.TransActionTypeId = t.TransactionType
 WHERE t.DateEntered BETWEEN '1 Aug 2005' AND '31 Aug 2005'
 GROUP BY CustomerId,CreditType
```

2. When the code is run, you should see the results shown in Figure 11-37.

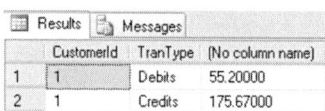

Figure 11-37. *Decisions within a string*

CAST()/CONVERT()

These are two functions used to convert from one data type to another. The main difference between them is that CAST() is ANSI SQL–92 compliant, but CONVERT() has more functionality.

The syntax for CAST() is

```
CAST(variable_or_column AS datatype)
```

This is opposed to the syntax for CONVERT(), which is

```
CONVERT(datatype,variable_or_column)
```

Not all data types can be converted between each other, such as converting a datetime to a text data type, and some conversions need neither a CAST() nor a CONVERT(). There is a grid in Books Online that provides the necessary information.

If you wish to CAST() from numeric to decimal or vice versa, you need to use CAST(); otherwise you will lose precision.

Try It Out: CAST()/CONVERT()

1. The first example will use CAST to move a number to a char(10).

```
DECLARE @Cast int
SET @Cast = 1234
SELECT CAST(@Cast as char(10)) + '-End'
```

2. Executing this code results in a left-filled character variable, as shown in Figure 11-38.

Figure 11-38. *Changing the data type of a value*

3. The second example completes the same conversion, but this time we use the CONVERT() function.

```
DECLARE @Convert int
SET @Convert = 5678
SELECT CONVERT(char(10),@Convert) + '-End'
```

4. As you can see from Figure 11-39, the only change is the value output.

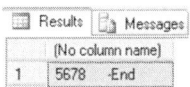

Figure 11-39. *Changing the data type of a value, using the non-ANSI standard*

ISDATE()

Although ISDATE() is a function that works with dates, this system function will take a value in a column or a variable and confirm whether it contains a valid date. The value returned is 0, or false, for an invalid date, or 1 for true if the date is okay.

Try It Out: ISDATE()

1. The first example demonstrates where a date is invalid. There are only 30 days in September.

```
DECLARE @IsDate char(15)
SET @IsDate = '31 Sep 2005'
SELECT ISDATE(@IsDate)
```

2. Execute the code, and you should get the results shown in Figure 11-40.

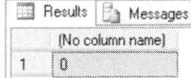

Figure 11-40. *Testing whether a value is a date*

3. Our second example is a valid date:

```
DECLARE @IsDate char(15)
SET @IsDate = '30 Sep 2005'
SELECT ISDATE(@IsDate)
```

4. This time when you run the code, you see a value of 1, as shown in Figure 11-41, denoting a valid entry.

Figure 11-41. *Showing that a value is a date*

ISNULL()

Many times so far you have seen NULL values within a column of returned data. As a value, this is very useful, as you have seen. However, you may wish to test whether a column contains a NULL or not. If there is a value, you retain it, but if there is a NULL setting, you want to convert it to a value. This could be used to cover a NULL value in an aggregation, for example. The syntax is

```
ISNULL(value_to_test,new_value)
```

where the first option is the column or variable to test whether there is a NULL value, and the second option will define what to change the value to if there is a NULL value. This change will only occur in the results and will not change the underlying data that the value came from.

Try It Out: ISNULL()

1. In this example, we define a char() variable of 10 characters in length and then set the value explicitly to NULL. The example will also work without the second line of code, which is simply there for clarity. The third line will test the variable, and as it is NULL, it will change it to a date. Note though that a date is more than 10 characters, so the value will be truncated.

```
DECLARE @IsNull char(10)
SET @IsNull = NULL
SELECT ISNULL(@IsNull,GETDATE())
```

2. And as expected, when you execute the code, you get the first 10 characters of the relevant date, as shown in Figure 11-42.

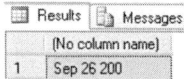

Figure 11-42. *If the value is a NULL, changing the NULL to a value*

ISNUMERIC()

This final system function tests the value within a column or variable and ascertains whether it is numeric or not. The value returned is 0, or false, for an invalid number, or 1 for true if the test is okay and can convert to a numeric.

■**Note** Currency symbols such as £ and $ will also return 1 for a valid numeric value.

Try It Out: ISNUMERIC()

1. Our first example to demonstrate ISNUMERIC() defines a character variable and contains alphabetic values. This test will fail, as shown in Figure 11-43.

```
DECLARE @IsNum char(10)
SET @IsNum = 'Robin     '
SELECT ISNUMERIC(@IsNum)
```

Figure 11-43. *Checking whether a value is a number and finding out it is not*

2. This second example places numbers and spaces in to a char field. The ISNUMERIC() test will ignore the spaces provided that there are no further alphanumeric characters.

```
DECLARE @IsNum char(10)
SET @IsNum = '1234     '
SELECT ISNUMERIC(@IsNum)
```

Figure 11-44 shows the results of running this code.

Figure 11-44. *Finding out a value is numeric*

RAISERROR

Before we look at handling errors, you need to be aware of what an error is, how it is generated and the information it generates, and finally how to generate your own errors when something is wrong. The T-SQL command RAISERROR allows us as developers to have the ability to produce our own SSE error messages when running queries or stored procedures. We are not tied to just using error messages that come with SSE; we can set up our own messages and our own level of severity for those messages. It is also possible to determine whether the message is recorded in the Windows error log or not.

However, whether we wish to use our own error message or a system error message, we can still generate an error message from SSE as if SSE itself raised it. Enterprise environments typically experience the same errors on repeated occasions, since they employ SSE in very specific ways depending on their business model. With this in mind, attention to employing RAISERROR can have big benefits by providing more meaningful feedback as well as suggested solutions for users.

By using RAISERROR, the whole SSE system will act as if SSE raised the error, as you have seen within this book.

RAISERROR can be used in one of two ways; looking at the syntax will make this clear.

```
RAISERROR ({msg_id|msg_str} {,severity,state}
          [,argument [ ,...n ] ])
          [WITH option [ ,...n ]]
```

You can either use a specific msg_id or provide an actual output string, msg_str, either as a literal or a local variable defined as string based, containing the error message that will be recorded. The msg_id references system and user-defined messages that already exist within the SSE error messages table.

When specifying a text message in the first parameter of the RAISERROR function instead of a message ID, you may find that this is easier to write than to create a new message:

```
RAISERROR('You made an error', 10, 1)
```

The next two parameters in the RAISERROR syntax are numeric and relate to how severe the error is and information about how the error was invoked. Severity levels range from 1 at the innocuous end to 25 at the fatal end. Severity levels of 2 to 14 are generally informational. Severity level 15 is for warnings, and levels 16 or higher represent errors. Severity levels from 20 to 25 are considered fatal and require the WITH LOG option, which will mean that the error is logged in the Windows error log and the connection terminated; quite simply, the stored procedure stops executing. The connection referred to here is the connection within Query Editor, or the connection made by an application using a data access method like ADO.NET. Only for a most extreme error would we set the severity to this level; in most cases, we would use a number between 1 and 18.

The last parameter within the function specifies state. Use a 1 here for most implementations, although the legitimate range is from 1 to 127. You may use this to indicate which error was thrown by providing a different state for each RAISERROR function in your stored procedure. SSE will not act on any legitimate state value, but the parameter is required.

An msg_str can define parameters within the text, and then by placing the value, either statically or via a variable, after the last parameter that you define, will replace the message parameter with that value. This is demonstrated in an upcoming example. If you do wish to add a parameter to a message string, you have to define a conversion specification. The format is

```
% [[flag] [width] [. precision] [{h | l}]] type
```

The options are as follows:

flag: A code that determines justification and spacing of the value entered:

- - (minus): Left justify the value.
- + (plus): The value will show a + or a – sign.
- 0: Prefix the output with zeros.
- #: Preface any nonzero with a 0, 0x, or 0X, depending on the formatting.
- (blank): Prefix with blanks.

width: The minimum width of the output

precision: The maximum number of characters used from the argument

h: Character types

- d or i: Signed integer
- o: Unsigned octal
- s: String
- u: Unsigned integer
- x or X: Unsigned hex

To clarify this, to place a parameter within a message string, where the parameter needs to be inserted, you would define this by a % sign followed by one of the following options: d or i for a signed integer, p for a pointer, s for a string, u for an unsigned integer, x or X for an unsigned hexadecimal, and o for an unsigned octal. Note that float, double, and single are not supported as parameter types for messages. You will see this in action in the upcoming examples.

Finally, there are three options that could be placed at the end of the RAISERROR message. These are the WITH options:

LOG will place the error message within the Windows error log.

NOWAIT sends the error directly to the client.

SETERROR will reset the error number to 50000 within the message string only.

When using any of these last WITH options, do take the greatest of care, as their misuse can create more problems than they solve. For example, you may unnecessarily use LOG a great deal, filling up the Windows error log, which leads to further problems.

There is a system stored procedure, sp_addmessage, that can create a new global error message that can be used by RAISERROR by defining the @msgnum. The syntax for adding a message is

```
sp_addmessage [@msgnum =]msg_id,
[@severity = ] severity , [ @msgtext = ] 'msg'
    [ , [ @lang = ] 'language' ]
    [ , [ @with_log = ] 'with_log' ]
    [ , [ @replace = ] 'replace' ]
```

The parameters for this system-stored procedure are as follows:

@msgnum: The number of the message, typically will be greater than 50000.

@severity: Same as the preceding, in a range of 1 to 25.

@lang: Use this if you need to define the language of the error message. Normally will be left empty.

@with_log: Set to 'TRUE' if you wish to write a message to the Windows error log.

@replace: Set to 'replace' if you are replacing an existing message and updating any of the preceding values with new settings.

▓**Note** Any message added will be specific for that database rather than the server.

It is time to move to an example that will set up an error message that will be used to say a customer is overdrawn.

<div style="background:black; color:white; text-align:center;">**Try It Out: RAISERROR**</div>

1. First of all, we want to add a new user-defined error message. To do this, we will use `sp_addmessage`. We can now add any new SSE message that we wish. Any user-defined error message must be greater than 50000 and so the first error message would normally be 50001.

   ```
   sp_addmessage @msgnum=50001,@severity=1,
   @msgtext='Customer is overdrawn'
   ```

2. We can then perform a `RAISERROR` to see the message displayed. Notice that we have to define the severity again. This is mandatory, but would be better if it was optional, and then you could always default to the severity defined.

   ```
   RAISERROR (50001,1,1)
   ```

3. When this is executed, we will see the following output:

   ```
   Customer is overdrawn
   Msg 50001, Level 1, State 1
   ```

4. This is not the friendliest of messages, as it would be better to perhaps give out the customer number as well. We can do this via a parameter. In the code that follows, we replace the message just added and now include a parameter where we are formatting with flag 0, which means we are prefixing the output with zeros; then we include the number 10, which is the precision, so that means the number will be 10 digits; and finally we indicate the message will be unsigned using the option u.

   ```
   sp_addmessage @msgnum =50001,@severity=1,
   @msgtext='Customer is overdrawn. CustomerId= %010u',@replace='replace'
   ```

5. We can then change the `RAISERROR` so that we add on another parameter. We are hard coding the customer number as customer number 243, but we could use a local variable.

   ```
   RAISERROR (50001,1,1,243)
   ```

6. Executing the code now produces output that is much better and more informative for debugging if required.

   ```
   Customer is overdrawn. CustomerId= 0000000243
   Msg 50001, Level 1, State 1
   ```

Now that you know how you can raise your own errors if scenarios crop up that need them, we can take a look at how SSE can deal with errors. We do come back to `RAISERROR` when looking at these two options next.

Error Handling

When working with T-SQL, it is important to have some sort of error handling to cater to those times when something goes wrong. Errors can be of different varieties, such as you are expecting at least one row of data returned from a query and you receive no rows. However, what we are discussing here is when SSE informs us there is something more drastically wrong. We have seen some errors throughout the book, and even in this chapter. There are two methods of error catching we can employ in such cases. The first uses a system variable, @@ERROR.

@@ERROR

This is the most basic of error handling. It has served SQL Server developers well over the years, but it can be cumbersome. When an error occurs, such as you have seen as we have gone through the book creating and manipulating objects, a global variable, @@ERROR, would have been populated with the SQL Server error message number. Similarly, if you try to do something with a set of data that is invalid, such as dividing a number by zero or exceeding the number of digits allowed in a numeric data type, SQL Server will populate this variable for you to inspect.

The downside is that the @@ERROR variable setting only lasts for the next statement following the line of code that has been executed; therefore, when you think there might be problems, you need to either pass the data to a local variable or inspect it straight away. The first example demonstrates this.

Try It Out: Using @@ERROR

1. The first example will try and divide 100 by zero, which is an error. We will then list out the error number, and then again list out the error number. Enter the following code and execute it:

```
SELECT 100/0
SELECT @@ERROR
SELECT @@ERROR
```

2. It is necessary in this instance to check both the Results and Messages tab. The first tab is the Messages tab, which will show you the error that was encountered. As expected, we see the Divide by zero error.

```
Msg 8134, Level 16, State 1, Line 1
Divide by zero error encountered.

(1 row(s) affected)

(1 row(s) affected)
```

3. Moving to the Results tab, you should see three result sets, as shown in Figure 11-45. The first, showing no information, would be where SSE would have put the division results, had it succeeded. The second result set is the number from the first SELECT @@ERROR. Notice the number corresponds to the msg number found in the Messages tab. The third result set shows a value of 0. This is because the first SELECT @@ERROR worked successfully and therefore set the system variable to 0. This demonstrates the lifetime of the value within @@ERROR.

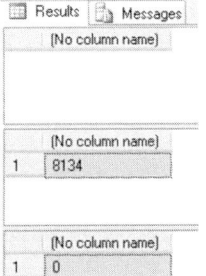

Figure 11-45. *Showing @@ERROR in multiple statements*

4. When we use the RAISERROR function, it will also set the @@ERROR variable, as we can see in the following code. However, the value will be set to 0 using our preceding example. This is because the severity level was below 11.

```
RAISERROR (50001,1,1,243)
SELECT @@ERROR
```

5. When the code is executed, you can see that @@ERROR is set to 0, as shown in Figure 11-46.

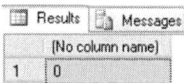

Figure 11-46. *When severity is too low to set @@ERROR*

6. By changing the severity to 11 or above, the @@ERROR setting will now be set to the message number within the RAISERROR.

```
RAISERROR (50001,11,1,243)
SELECT @@ERROR
```

7. The preceding code produces the same message as seen within our RAISERROR example, but as you can see in Figure 11-47, the error number setting now reflects that value placed in the msgnum parameter.

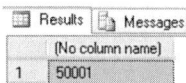

Figure 11-47. *With a higher severity, the message number is set.*

Although a useful tool, it would be better to use the next error-handling routine to be demonstrated, TRY...CATCH.

TRY...CATCH

It can be said that no matter what, any piece of code has the ability to fail and generate some sort of error. For the vast majority of this code, you will want to trap any error that occurs, check what the error is, and deal with it as best you can. As you saw previously, this could be done one statement at a time using @@ERROR to test for any error code. A new and improved functionality exists whereby a set of statements can try and execute, and if any statement has an error, it will be caught. This is known as a TRY...CATCH block.

Surrounding code with the ability to try and execute a slice of code and to catch any errors and try to deal with them has been around for quite a number of years in languages such as C++. Gladly, we now see this within SSE.

The syntax is pretty straightforward. There are two "blocks" of code. The first block, BEGIN TRY, is where there is one or more T-SQL statements that you wish to try and run. If any of the statements has an error, no further processing within that block will execute, and processing will switch to the second block, BEGIN CATCH.

```
BEGIN TRY
    { sql_statement | statement_block }
END TRY
BEGIN CATCH
    { sql_statement | statement_block }
END CATCH
```

When you generate your own error via a RAISERROR, then a bit of care has to be taken with the severity setting, as this will determine how your code works within a TRY...CATCH scenario. If you raise an error with a severity level of 0 to 10, although an error is generated and will be received by the calling program, whether that is Query Editor or a program such as C#, processing will continue without moving to the CATCH block. This can be seen as a "warning" level. Changing the severity level to 11 or above will transfer the control to the CATCH block of code. Once within the CATCH block you can raise a new error, or raise the same error by using values stored within SSE system functions.

The system functions that can be used to find useful debugging information are detailed here:

ERROR_LINE(): The line number that caused the error or performed the RAISERROR command. This will be physical rather than relative, i.e., you don't have to remove blank lines within the T-SQL to get the correct line number, unlike some software that does require this.

ERROR_MESSAGE(): The text message.

ERROR_NUMBER(): The number associated with the message.

ERROR_PROCEDURE(): If you are retrieving this within a stored procedure or trigger, the name of it will be contained here. If you are running ad hoc T-SQL code, the value will be NULL.

ERROR_SEVERITY(): The numeric severity value for the error.

ERROR_STATE(): The numeric state value for the error.

TRY...CATCH blocks can be nested, and when an error occurs, the error will be passed to the relevant CATCH section. This would be done when you wanted an overall CATCH block for "general" statements, and then you could perform specific testing and have specific error handling where you really think an error might be generated.

Not all errors are caught within a TRY...CATCH block, unfortunately. These are compile errors or when deferred name resolution takes place and the name created doesn't exist. To clarify these two points, with T-SQL code that is either ad hoc or within a stored procedure, SSE compiles the code, looking for syntax errors. However, not all code can be fully compiled and is not compiled until the statement is about to be executed. If there is an error, this will terminate the batch immediately. The second is that if you have code that references a temporary table, for example, then the table doesn't exist at run time, and so column names cannot be checked. This is known as **deferred name resolution**, and if you try to use a column that doesn't exist, this will also generate an error, terminating the batch.

There is a great deal more to TRY...CATCH blocks, but it concerns areas that are quite advanced. So now that you know the basics, let's look at some examples demonstrating what we have just discussed.

Try It Out: TRY...CATCH

1. Our first example is a straight error where we have defined an integer local variable. Within our error-handling block, after outputting a statement to demonstrate that we are within that block, we try to set a string to the variable. This is a standard error and will immediately move the execution to the CATCH block, and the last SELECT will not be executed.

```
DECLARE @Probs int
BEGIN TRY
   SELECT 'This will work'
   SELECT @Probs='Not Right'
   SELECT 10+5,
    'This will also work, however the error means it will not run'
END TRY
BEGIN CATCH
   SELECT 'An error has occurred at line ' +
      LTRIM(STR(ERROR_LINE())) +
      ' with error ' + LTRIM(STR(ERROR_NUMBER())) + ' ' + ERROR_MESSAGE()
END CATCH
```

2. When we run the code, we will see the first statement and then the SELECT statement that executes when the error is caught. We use the system functions to display relevant information, which appears in Figure 11-48.

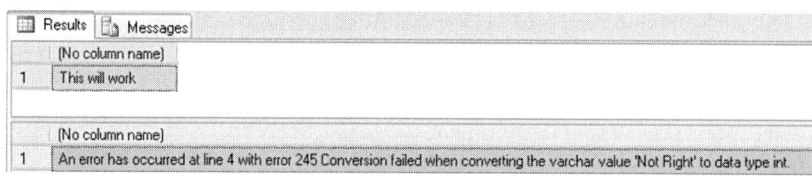

Figure 11-48. *An error is caught.*

3. Our second example demonstrates nesting TRY...CATCH blocks and how execution can continue within the outer block when an error arises within the second inner block. We keep the same error and will see the error message The second catch block. But once this is executed, processing will continue to And then this will now work.

```
DECLARE @Probs int
BEGIN TRY
   SELECT 'This will work'
   BEGIN TRY
    SELECT @Probs='Not Right'
    SELECT 10+5,
    'This will also work, however the error means it will not run'
   END TRY
   BEGIN CATCH
      SELECT 'The second catch block'
   END CATCH
   SELECT 'And then this will now work'
END TRY
BEGIN CATCH
   SELECT 'An error has occurred at line ' +
      LTRIM(STR(ERROR_LINE())) +
      ' with error ' + LTRIM(STR(ERROR_NUMBER())) + ' ' + ERROR_MESSAGE()
END CATCH
```

4. As expected, we will see three lines of output, as shown in Figure 11-49. The code in the outer CATCH block will not run, as the error was dealt with in the inner block.

Figure 11-49. *An error is caught in a nested batch.*

5. This time we will see how our T-SQL code can be successfully precompiled and execution started. Then when we try to display results from a temporary table that doesn't exist, the CATCH block does not fire, as execution terminates immediately.

```
DECLARE @Probs int
BEGIN TRY
    SELECT 'This will work'
    BEGIN TRY
        SELECT * FROM #Temp
    END TRY
    BEGIN CATCH
        SELECT 'The second catch block'
    END CATCH
    SELECT 'And then this will now work'
END TRY
BEGIN CATCH
    SELECT 'An error has occurred at line ' +
        LTRIM(STR(ERROR_LINE())) +
        ' with error ' + LTRIM(STR(ERROR_NUMBER())) + ' ' + ERROR_MESSAGE()
END CATCH
```

6. When the code is run in the Messages tab, we will see the following output, detailing one row has been returned, which comes from the first SELECT statement. We then see the SSE error. Looking at Figure 11-50, you also see just the first SELECT statement output.

```
(1 row(s) affected)
Msg 208, Level 16, State 0, Line 5
Invalid object name '#Temp'.
```

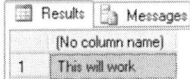

Figure 11-50. *What happens when SSE terminates execution*

7. The final example demonstrates how to reraise the same error that caused the CATCH block to fire. Recall with RAISERROR it is only possible to list a number or a local variable. Unfortunately, it is not possible to call the relevant function directly or via a SELECT statement. It is necessary to load the values into local variables.

```
DECLARE @Probs int
SELECT 'This will work'
BEGIN TRY
    SELECT @Probs='Not Right'
    SELECT 10+5,
    'This will also work, however the error means it will not run'
END TRY
BEGIN CATCH
    DECLARE @ErrMsg NVARCHAR(4000)
    DECLARE @ErrSeverity INT
    DECLARE @ErrState INT
    SELECT 'Blimey! An error'

    SELECT
        @ErrMsg = ERROR_MESSAGE(),
        @ErrSeverity = ERROR_SEVERITY(),
        @ErrState = ERROR_STATE();

    RAISERROR (@ErrMsg,@ErrSeverity,@ErrState)
END CATCH
```

Summary

Manipulating data as you have seen prior to this chapter is all well and good to a point. So far you have learned basic insertion, deletion, and updating of the data. This chapter will have advanced that knowledge to include joining of two or more tables when retrieving data. You have also looked at how to store data away from a table within a variable or in a temporary table.

Working with sets of data is the optimum way to employ SSE. With this, SSE provides functionality to aggregate a set of data through aggregation functions that, when combined with the ability to group your data and then filter based on the aggregation, allows for some powerful processing preferable to row-at-a-time processing to achieve the same results. You saw this in action over three sections within this chapter. When you then combine this with displaying distinct values, SSE shows its flexibility alongside its power in displaying differing sets of results.

The penultimate section looked at the different SSE functions that can be applied to data to allow you to manipulate data in single or multiple columns via SQL functions.

Moving to the final large section, you saw in action different methods of handling errors in your application and the potential gains and pitfalls with each method. Error handling should always be included with your code so that when something does go wrong, it can be handled.

The next chapter takes your knowledge of T-SQL to a more advanced stage: you get a chance to look at different functionality in working with data.

CHAPTER 12

■ ■ ■

Advanced T-SQL

By now you really are becoming proficient in SSE and writing code to work with the data and the objects within the database. Already you have seen some T-SQL code and encountered some scenarios that have advanced your skills as a T-SQL developer. We can now look at more advanced areas of T-SQL programming to round off your knowledge and really get you going with queries that do more than the basics.

This chapter will look at the occasions when you need a query within a query, known as a subquery. This is ideal for producing a list of values to search for, or for producing a value from another table to set a column or a variable with. It is also possible to create a transient table of data to use within a query, known as a common table expression. We look at both subqueries and common table expressions within the chapter.

From there, we explore how to take a set of data and pivot the results, just as you can do within Excel. We also take a look at different types of ranking functions where we can take our set of data and attach rankings to rows or groups of rows of data.

And finally, you will see how to deal with `varchar(max)` and `varbinary(max)` data types when loading large text, images, movies, and so on.

We will move away from our `ApressFinancial` example on occasion to the `AdventureWorks` sample that I mentioned in Chapter 3 as part of the samples database. You can install the sample databases from `http://msdn.microsoft.com/vstudio/express/sql/download/`. This will allow us to have more data to work through the examples. It is not necessary to fully appreciate this database for the examples, as it is the code that is the important point, but the text will give an overview of the tables involved as appropriate.

Subqueries

A **subquery** is a query embedded within another query or statement and can be used anywhere an expression is allowed. It can be used to check or set a value of a variable or column, or used to test whether a row of data exists in a `WHERE` clause.

To expand on this, there may be times when you wish to set a column value based on data from another query. One example we have is the `ShareDetails.Shares` table. If we had a column defined for `MaximumSharePrice` that held the highest value the share price had gone for that year, rather than doing a test every time the share price moved, we could use the `MAX` function to get the highest share price, put that value into a variable, and then set the column via that variable. The code would be similar to that defined here:

```
ALTER TABLE ShareDetails.Shares
ADD MaximumSharePrice money
DECLARE @MaxPrice money
```

```
SELECT @MaxPrice = MAX(Price)
    FROM ShareDetails.SharePrices
  WHERE ShareId = 1
SELECT @MaxPrice
UPDATE ShareDetails.Shares
SET MaximumSharePrice = @MaxPrice
WHERE ShareId = 1
```

In the preceding code, if we wished to work with more than one share, we would need to implement a loop and process each share one at a time. However, we could also perform a subquery, which implements the same functionality as shown in the code that follows. The subquery still finds the maximum price and sets the column. Notice that this time we can update all shares with one statement. The subquery joins with the main query via a WHERE clause so that as each share is dealt with, the subquery can take that ShareId and still get the maximum value.

■**Note** We call this type of subquery a **correlated subquery**.

```
SELECT ShareId,MaximumSharePrice
FROM ShareDetails.Shares
UPDATE ShareDetails.Shares
SET MaximumSharePrice = (SELECT MAX(SharePrice)
                                        FROM ShareDetails.SharePrices sp
                        WHERE sp.ShareId = s.ShareId)
FROM ShareDetails.Shares s
SELECT ShareId,MaximumSharePrice
FROM ShareDetails.Shares
```

We also came across a subquery way back in Chapter 7 when we were testing whether a backup had successfully completed or not. The code is replicated here, with the subquery section highlighted in bold. In this instance, instead of setting a value in a column, we are looking for a value to be used as part of a filtering criteria. Recall from Chapter 7 that we know the last backup will have the greatest backup_set_id. We use the subquery to find this value (as there is no system function or variable that can return this at the time of creating the backup). Once we have this value, we can use it to reinterrogate the same table, filtering out everything but the last row for the backup just created.

■**Note** Don't forget that for the FROM DISK option, you will have a different file name than the one in the following code.

```
DECLARE @BackupSet AS INT
SELECT @BackupSet = position
  FROM msdb..backupset
 WHERE database_name='ApressFinancial'
   AND backup_set_id=
                    (SELECT MAX(backup_set_id)
                    FROM msdb..backupset s
                    WHERE database_name='ApressFinancial')
IF @BackupSet IS NULL
BEGIN
    RAISERROR('Verify failed. Backup information for database
        ''ApressFinancial'' not found.', 16, 1)
END
```

```
RESTORE VERIFYONLY
FROM DISK = 'C:\Program Files\Microsoft SQL Server\MSSQL.1\MSSQL\Backup\
ApressFinancial\ApressFinancial_backup_200508061136.bak'
WITH FILE = @BackupSet,
NOUNLOAD,
NOREWIND
```

In both of these cases, we are returning a single value within the subquery, but this need not always be the case. You can also return more than one value. One value must be returned when you are trying to set a value. This is because you are using an equals (=) sign. It is possible in a WHERE clause to look for a number of values using the IN clause.

IN

If you wish to look for a number of values in your WHERE clause, such as a list of values from the ShareDetails.Shares table where the ShareId is 1, 3, or 5, then you can use an IN clause. The code to complete this example would be

```
SELECT *
  FROM ShareDetails.Shares
 WHERE ShareId IN (1,3,5)
```

Using a subquery, it would be possible to replace these numbers with the results from the subquery. The preceding query could also be written using the code that follows. The example shown here is deliberately obtuse to show how it is possible to combine a subquery and an aggregation to produce the list of ShareIds that form the IN:

```
SELECT *
  FROM ShareDetails.Shares
 WHERE ShareId IN (SELECT ShareId
                     FROM ShareDetails.Shares
                    WHERE CurrentPrice > (SELECT MIN(CurrentPrice)
                                            FROM ShareDetails.Shares)
                      AND CurrentPrice < (SELECT MAX(CurrentPrice)
                                            FROM ShareDetails.Shares))
```

■**Note** One last piece of advice: if a query starts becoming very complex, you may find that it starts performing badly. This book doesn't look at performance of queries, although it has discussed indexes and how they can help your query perform better. Always take a step back and think, "Would this work better as two queries where the first query creates a subset of data?" Writing the most complex of queries that processes all the data in one pass may not always be the best answer.

Both of these examples replace what would require a number of OR clauses within the WHERE filter such as you see in this code:

```
SELECT *
  FROM ShareDetails.Shares
 WHERE ShareId = 1
    OR ShareId = 3
    OR ShareId = 5
```

These are just three different ways a subquery can work. The fourth way involves using a subquery to check whether a row of data exists or not, which we look at next.

EXISTS

EXISTS is a clause that is very similar to IN, in that it tests a column value against a subset of data from a subquery. The difference is that EXISTS uses a join to join values from a column to a column within the subquery as opposed to IN, which compares against a comma-delimited set of values and requires no join.

Over time, our ShareDetails.Shares and ShareDetails.SharePrice tables will grow to quite a large size. If we wanted to shrink them, we could use cascading deletes so that when we delete from the ShareDetails.Shares table, we would also delete all the SharePrice rows. But how would we know which shares to delete? One way would be to see what shares are still held within our TransactionDetails.Transactions table. We would do this via EXISTS, but instead of looking for ShareIds that exist, we would use NOT EXISTS.

At present, we have no shares listed within the TransactionDetails.Transactions table, so we would see all of the ShareDetails.Shares listed. We can make life easier with EXISTS by giving tables an alias, but we also have to use the WHERE clause to make the join between the tables. However, we aren't really joining the tables as such; a better way of looking at it is to say we are filtering rows from the subquery table.

The final point to note is that you can return whatever you wish after the SELECT statement in the subquery, but it should only be one column or a value, and it is easiest to use an asterisk in conjunction with an EXISTS clause. If you do set a column or value, the value returned cannot be used anywhere within the main query and is discarded, so there is nothing to gain by returning a value from a column.

Note When using EXISTS, it is most common in SSE to use * rather than a constant like 1, as it simply returns a true or false setting.

The following code shows EXISTS in action prefixed with a NOT.

```
SELECT *
  FROM ShareDetails.Shares s
 WHERE NOT EXISTS (SELECT 1
                     FROM TransactionDetails.Transactions t
                    WHERE t.RelatedShareId = s.ShareId)
```

Note Both EXISTS and IN can be prefixed with NOT.

The APPLY Operator

It is possible to return a table as the data type from a function. The table data type can hold multiple columns and multiple rows of data as you would expect, and this is one of the main ways it differs from other data types, such as varchar, int, etc. Returning a table of data from a function allows the code invoking the function the flexibility to work with returned data as if the table permanently existed or was built as a temporary table.

To supply extensibility to this type of function, SSE provides you with an operator called APPLY, which works with a table-valued function and joins data from the calling table(s) to the data

returned from the function. The function will sit on the right-hand side of the query expression, and through the use of APPLY can return data as if you had a RIGHT OUTER JOIN or a LEFT OUTER JOIN on a "permanent" table. Before you see an example, you need to be aware that there are two types of APPLY: a CROSS APPLY and an OUTER APPLY.

- CROSS APPLY: Returns only the rows that are contained within the outer table where the row produces a result set from the table-valued function.

- OUTER APPLY: Returns the rows from the outer table and the table-valued function whether a join exists or not. This is similar to an OUTER JOIN, which you saw in Chapter 11. If no row exists in the table-valued function, you will see a NULL value in the columns from that function.

CROSS APPLY

In our example, we will build a table-valued function that accepts a CustomerId as an input parameter and returns a table of TransactionDetails.Transactions rows.

```
CREATE FUNCTION TransactionDetails.ReturnTransactions
(@CustId bigint) RETURNS @Trans TABLE
(TransactionId bigint,
CustomerId bigint,
TransactionDescription nvarchar(30),
DateEntered datetime,
Amount money)
AS
BEGIN
 INSERT INTO @Trans
 SELECT TransactionId, CustomerId, TransactionDescription,
         DateEntered, Amount
  FROM TransactionDetails.Transactions t
  JOIN TransactionDetails.TransactionTypes tt ON
         tt.TransactionTypeId = t.TransactionType
 WHERE CustomerId = @CustId
   RETURN
END
```

Now that we have the table-valued function built, we can call it and use CROSS APPLY to return only the Customer rows where there is a customer number within the table from the table-valued function. The following code demonstrates this with the first section adding in a transaction for another customer first:

```
INSERT INTO TransactionDetails.Transactions
(CustomerId,TransactionType,DateEntered,Amount,RelatedProductId)
VALUES (2,1,'18 Dec 2005',1000,1)
SELECT c.CustomerFirstName, CustomerLastName,
Trans.TransactionId,TransactionDescription,
DateEntered,Amount
FROM CustomerDetails.Customers AS c
  CROSS APPLY
  TransactionDetails.ReturnTransactions(c.CustomerId)
  AS Trans
```

The results from the preceding code are shown in Figure 12-1, where you can see that only rows from the CustomerDetails.Customers table are displayed where there is a corresponding row in the TransactionDetails.Transactions table.

Figure 12-1. *CROSS APPLY from a table-valued function*

OUTER APPLY

As mentioned previously, OUTER APPLY is very much like a RIGHT OUTER JOIN on a table, but you need to use OUTER APPLY when working with a table-valued function.

For our example, we can still use the function we built for the CROSS APPLY. With the code that follows, we are expecting those customers that have no rows returned from the table-valued function to be listed with NULL values.

```
SELECT c.CustomerFirstName, CustomerLastName,
Trans.TransactionId,TransactionDescription,
DateEntered,Amount
FROM CustomerDetails.Customers AS c
  OUTER APPLY
  TransactionDetails.ReturnTransactions(c.CustomerId)
  AS Trans
```

When this is executed, you will see the output shown in Figure 12-2.

Figure 12-2. *OUTER APPLY from a table-valued function*

Common Table Expressions

In Chapter 11, we had a look at temporary tables by defining a table in code prefixing the name with a hash mark (#). Temporary tables allow you to split a complex query or a query that, if built as one unit, would run slowly due to the complexity of the joins SSE would have to do. Therefore, creating a set of subdata in the first query would aid the performance of the query in the second. Another scenario where you may use temporary tables is when you wish to create some sort of grouping of information and then use that grouping for further analysis. As an example, you might create a temporary table that contains a sum of each day's transactions within a bank account. The second part of the query takes the temporary table and uses it to calculate the daily interest accrued.

A **common table expression** (CTE) is a bit like a temporary table. It's transient, lasting only as long as the query requires it. Temporary tables are available for use during the lifetime of the session of the query running the code or until they are explicitly dropped. The creation and use of temporary tables is a two- or three-part phase: table creation, population, and use. A CTE is built in the same code line as the SELECT, INSERT, UPDATE, or DELETE statements that use it.

■**Note** You must download the AdventureWorks sample database from Microsoft and attach the database before you can use these examples.

The best way to understand a CTE is to demonstrate an example with some code. Within the AdventureWorks database, there are a number of products held in the Production.Product table. For this example, let's say you want to know the maximum list price of stock you're holding over all the product categories. Using a temporary table, this would be a two-part process, as follows:

```
USE AdventureWorks
GO
SELECT p.ProductSubcategoryID, s.Name,SUM(ListPrice) AS ListPrice
  INTO #Temp1
  FROM Production.Product p
  JOIN Production.ProductSubcategory s ON s.ProductSubcategoryID =
    p.ProductSubcategoryID
 WHERE p.ProductSubcategoryID IS NOT NULL
 GROUP BY p.ProductSubcategoryID, s.Name

SELECT ProductSubcategoryID,Name,MAX(ListPrice)
  FROM #Temp1
  GROUP BY ProductSubcategoryID, Name
HAVING MAX(ListPrice) = (SELECT MAX(ListPrice) FROM #Temp1)

DROP TABLE #Temp1
```

However, with CTEs, this becomes a bit simpler and more efficient. In the preceding code snippet, we've created a temporary table. This table has no index on it, and therefore SSE will complete a table scan operation on it when executing the second part. In contrast, the upcoming code snippet uses the raw AdventureWorks tables. There is no creation of a temporary table, which would have used up processing time, and also existing indexes could be used in building up the query as well rather than a table scan.

The CTE is built up using the WITH keyword, which defines the name of the CTE you'll be returning, in this case ProdList, and the columns contained within it. The columns returned within the CTE will take the data types placed into it from the SELECT statement within the brackets. Of course, the number of columns within the CTE has to be the same as the table defined within the brackets. This table is built up, returned, and passed immediately into the following SELECT statement outside of the WITH block where the rows of data can then be processed as required. Therefore, the rows returned between the brackets could be seen as a temporary table that is used by the statement outside of the brackets.

```
WITH ProdList (ProductSubcategoryID,Name,ListPrice) AS
(
SELECT p.ProductSubcategoryID, s.Name,SUM(ListPrice) AS ListPrice
  FROM Production.Product p
  JOIN Production.ProductSubcategory s ON s.ProductSubcategoryID =
    p.ProductSubcategoryID
  WHERE p.ProductSubcategoryID IS NOT NULL
  GROUP BY p.ProductSubcategoryID, s.Name
)
```

```
SELECT ProductSubcategoryID,Name,MAX(ListPrice)
  FROM ProdList
  GROUP BY ProductSubcategoryID, Name
  HAVING MAX(ListPrice) = (SELECT MAX(ListPrice) FROM ProdList)
```

When the code is executed, the results should resemble the output shown in Figure 12-3.

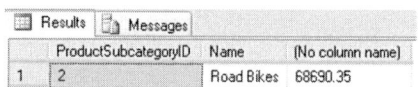

Figure 12-3. *CTE output*

Recursive CTE

A **recursive CTE** is where an initial CTE is built and then the results from that are called recursively in a UNION statement returning subsets of data until all the data is returned. This gives you the ability to create data in a hierarchical fashion as you will see in the next example.

The basis of building a recursive CTE is to build your initial query just as you saw earlier, but then append to that a UNION ALL statement with a join on the cte_name. This works so that in the "normal" CTE, data is created and built with the cte_name, which can then be referenced within the CTE from the UNION ALL. The syntax that you can see here demonstrates how this looks in its simplest form:

```
WITH cte_name ( column_name [,...n] )
AS
(
CTE_query_definition
UNION ALL
CTE_query_definition with a join on cte_name
)
```

As with all UNION statements, the number of columns must be the same in all queries that make up the recursive CTE. The data types must also match up.

■**Caution** Care **must** taken when creating a recursive CTE. It is possible to create a recursive CTE that goes into an infinite loop. While testing the recursive CTE, you can use the MAXRECURSION option, as you will see in our next example.

The following example demonstrates a recursive query that will list every employee, their job title, and the name of their manager. We have our anchor CTE, which returns the CEO of Adventure-Works. The CEO doesn't have a "boss," but we still need to return data as if he had, of the same data type as well. To resolve this dilemma, the example returns spaces that are converted to the correct data type and length. Once we have that anchor, we can then recursively call the second query, which will continue to return data, moving down the hierarchy as more data is added in until no more levels exist. So on the anchor, the EmployeeReportingStructure CTE will have the level 0, or CEO data, within it. The recursive query will then add to the CTE the level 1 employees, which then allows the recursive query to work with that data to populate level 2, and so on. From this, you should see how it is possible to create an infinite loop. To stop this from happening, as mentioned, we can put on the SELECT query that invokes the CTE and returns the data, which is an option to define how many invocations of the recursive query are made. In our example, we set this to four via OPTION (MAXRECURSION 4).

```
USE AdventureWorks
GO
WITH EmployeeReportingStructure
(ManagerID, EmployeeID, EmployeeLevel, Level,
ManagerContactId,ManagerTitle,ManagerFirst,ManagerLast,
EmployeeTitle,EmployeeFirst,EmployeeLast)
AS
(
-- Anchor member definition
    SELECT e.ManagerID, e.EmployeeID, e.Title as EmployeeLevel,
        0 AS Level,
        e.ContactID as ManagerContactId,
        CAST(' ' as nvarchar(8)) as ManagerTitle,
        CAST(' ' as nvarchar(50)) as ManagerFirst,
        CAST(' ' as nvarchar(50)) as ManagerLast,
        c.Title as EmployeeTitle,c.FirstName as EmployeeFirst,
        c.LastName as EmployeeLast
    FROM HumanResources.Employee AS e
    INNER JOIN Person.Contact c ON c.ContactID = e.ContactID
    WHERE ManagerID IS NULL
    UNION ALL
-- Recursive member definition
    SELECT e.ManagerID, e.EmployeeID, e.Title as EmployeeLevel, Level + 1,
        e.ContactID as ManagerContactId,
        m.Title as ManagerTitle,m.FirstName as ManagerFirst,
        m.LastName as ManagerLast,
        c.Title as EmployeeTitle,c.FirstName as EmployeeFirst,
        c.LastName as EmployeeLast
    FROM HumanResources.Employee AS e
    INNER JOIN Person.Contact c ON c.ContactID = e.ContactID
    INNER JOIN EmployeeReportingStructure  AS d
        ON d.EmployeeID = e.ManagerID
    INNER JOIN Person.Contact m ON m.ContactID = d.ManagerContactId
)
-- Statement that executes the CTE
SELECT ManagerID, EmployeeID,
ISNULL(ManagerTitle+' ','')+ManagerFirst+' '+ManagerLast as Manager,
EmployeeLevel,
ISNULL(EmployeeTitle+' ','')+EmployeeFirst+' '+EmployeeLast as Employee,
Level
FROM EmployeeReportingStructure
ORDER BY Level,EmployeeLast,EmployeeFirst
OPTION (MAXRECURSION 4)
```

CTEs are used not only as standalone expressions, but also within other functions, such as pivoting data, which you can see in action next.

Pivoting Data

If you have ever used Excel, you have also probably had to perform some sort of pivoting of results of the data so that rows of information are pivoted into columns of information. It is now possible to perform this kind of operation within SSE via a PIVOT operator. Pivoted data can also be changed back using the UNPIVOT operator, where columns of data can be changed into rows of data. In this next section, you will see both these operators in action. We will be using the AdventureWorks example in this section. The table involved in this example is the SalesOrderDetail table belonging to the

Sales schema. This holds details of products ordered, the quantity requested, the price they are at, and then the discount the order received.

PIVOT

Before we see PIVOT in action, we need to look at the information that we will pivot. The following code lists three products, and for each product we will sum up the amount sold, taking the discount into account:

```
USE AdventureWorks
GO
SELECT ProductID,UnitPriceDiscount,SUM(LineTotal)
  FROM Sales.SalesOrderDetail
  WHERE ProductID IN (776,711,747)
  GROUP BY ProductID,UnitPriceDiscount
  ORDER BY ProductID,UnitPriceDiscount
```

This produces one line of output for each product/discount combination, as you can see in the following results:

```
711     0.00    143788.908000
711     0.02    11421.237324
711     0.05    4384.931245
711     0.10    2679.760530
711     0.15    3131.779950
747     0.00    501788.197700
776     0.00    1198796.448000
776     0.02    23020.131792
776     0.35    32906.152500
```

By using PIVOT, we can alter this data so that we can create columns for each of the products, and each row is defined for the discount, therefore giving a cross-reference of product to discount.

```
SELECT pt.Discount,ISNULL([711],0.00) As Product711,
    ISNULL([747],0.00) As Product747,ISNULL([776],0.00) As Product776
FROM
(SELECT sod.LineTotal, sod.ProductID, sod.UnitPriceDiscount as Discount
   FROM Sales.SalesOrderDetail sod) so
PIVOT
(
SUM(so.LineTotal)
FOR so.ProductID IN ([776], [711], [747])
) AS pt
ORDER BY pt.Discount
```

Before we execute this code, let's take a look at what is happening. First of all, you need to create a subquery that contains the columns of data that the PIVOT operator can use for its aggregation (it will also be used later for displaying in the output). No filtering has been completed at this point—any columns not used in the aggregation will be ignored. The code generates a table-valued expression with an alias of so. From this table, you then instruct SSE to PIVOT the columns while completing an aggregation on a specific column, in our case a SUM of the LineTotal column from the table value expression. It is also at this point you define the columns to create via the FOR clause, in our case a column for each product of the three products. This is the equivalent to using GROUP BY for the aggregation, and it's also the equivalent of filtering data from the so table value expression. However, within the so table value expression, there's also a third column, UnitPriceDiscount. Without this, the output from the PIVOT operation would produce one row with three columns—one for

each product. So this table value expression with the PIVOT operation produces a temporary result set, which we name pt. We can then use this temporary result set to produce our output.

When you run the code, you should see output similar to what is shown in Figure 12-4.

	Discount	Product711	Product747	Product776
1	0.00	143788.908000	501788.197700	1198796.448000
2	0.02	11421.237324	0.000000	23020.131792
3	0.05	4384.931245	0.000000	0.000000
4	0.10	2679.760530	0.000000	0.000000
5	0.15	3131.779950	0.000000	0.000000
6	0.20	0.000000	0.000000	0.000000
7	0.30	0.000000	0.000000	0.000000
8	0.35	0.000000	0.000000	32906.152500
9	0.40	0.000000	0.000000	0.000000

Figure 12-4. *Pivoted data results*

UNPIVOT

The reverse of PIVOT is, of course, UNPIVOT, which will unpivot data by placing column data into rows. You can prove this by unpivoting the data just pivoted using the preceding query. The code that follows will rebuild the PIVOT operation and place the data into a temporary table. From that temporary table, you can unpivot the data.

■**Note** UNPIVOT is not the exact reverse of PIVOT. PIVOT performs an aggregation and hence merges possible multiple rows into a single row in the output. UNPIVOT does not reproduce the original table-valued expression result because rows have been merged. Besides, NULL values in the input of UNPIVOT disappear in the output, whereas there may have been original NULL values in the input before the PIVOT operation.

```
USE AdventureWorks
go
SELECT pt.Discount,ISNULL([711],0.00) As Product711,
    ISNULL([747],0.00) As Product747,ISNULL([776],0.00) As Product776
INTO #Temp1
FROM
(SELECT sod.LineTotal, sod.ProductID, sod.UnitPriceDiscount as Discount
   FROM Sales.SalesOrderDetail sod) so
PIVOT
(
SUM(so.LineTotal)
FOR so.ProductID IN ([776], [711], [747])
) AS pt
ORDER BY pt.Discount
```

UNPIVOT has similarities to PIVOT in that you build a CTE—in this case calling it up1—which you then use as the basis of unpivoting. Once the CTE is defined, you then use UNPIVOT with the column definitions of the columns to create, DiscountAppl and ProductID, where using IN defines the rows that will be produced back from the UNPIVOT.

```
SELECT ProductID,Discount, DiscountAppl
  FROM (SELECT Discount, Product711, Product747, Product776
        FROM #Temp1) up1
```

```
UNPIVOT ( DiscountAppl FOR ProductID
    IN (Product711, Product747, Product776)) As upv2
WHERE DiscountAppl <> 0
ORDER BY ProductID
```

When the preceding code is executed, the data is unpivoted, as shown in Figure 12-5.

	ProductID	Discount	DiscountAppl
1	Product711	0.00	143788.908000
2	Product711	0.02	11421.237324
3	Product711	0.05	4384.931245
4	Product711	0.10	2679.760530
5	Product711	0.15	3131.779950
6	Product747	0.00	501788.197700
7	Product776	0.00	1198796.448000
8	Product776	0.02	23020.131792
9	Product776	0.35	32906.152500

Figure 12-5. *Unpivoted data results*

Now that we have pivoted data, we can take a look at how we can rank output.

Ranking Functions

With SSE, it's now possible to rank rows of data in your T-SQL code. **Ranking functions** give you the ability to rank each row of data to provide a method of organizing the output in an ascending sequence. You can give each row a unique number or each group of similar rows the same number. You may be wondering what is wrong with other methods of ranking data, which might include IDENTITY columns. These types of columns do provide unique numbers, but gaps can form. You are also tied in to each row having its own number when you may wish to group rows together.

So then, why not GROUP BY? Well, yes, you can use GROUP BY, but what if the grouping was over more than one column? In this case, processing the data further would require knowledge about those columns. Ranking functions make it possible to provide a value that allows data to be ranked in the order required, and then that value can be used for splitting the data into further groupings.

There are four ranking functions, which will be discussed in detail in upcoming sections:

- ROW_NUMBER: Allows you to provide sequential integer values to the result rows of a query.

- RANK: Provides an ascending, nonunique ranking number to a set of rows, giving the same number to a row of the same value as another. Numbers are skipped for the number of rows that have the same value.

- DENSE_RANK: Similar to RANK, but each row number returned will be one greater than the previous setting, no matter how many rows are the same.

- NTILE: Takes the rows from the query and places them into an equal (or as close to equal as possible) number of specified numbered groups, where NTILE returns the group number the row belongs to.

Note These ranking functions can only be used with the SELECT and ORDER BY statements. Sadly, they can't be used directly in a WHERE or GROUP BY clause, but you can use them in a CTE or derived table.

```
WITH OrderedOrders AS
(SELECT SalesOrderID, OrderDate,
ROW_NUMBER() OVER (order by OrderDate)as RowNumber
FROM Sales.SalesOrderHeader )
SELECT *
FROM OrderedOrders
WHERE RowNumber between 50 and 60;
```

The syntax for ranking functions is shown as follows:

```
<function_name>() OVER([PARTITION BY <partition_by_list>]
ORDER BY <order_by_list>)
```

Taking each option as it comes, you can see how this can be placed within a SELECT statement, for example:

- function_name: Can be one of ROW_NUMBER, RANK, DENSE_RANK, or NTILE
- OVER: Defines the details of how the ranking should order or split the data
- PARTITION BY: Details which data the column should use as the basis of the split
- ORDER BY: Details the ordering of the data

ROW_NUMBER

Our first ranking function, ROW_NUMBER, allows your code to guarantee an ascending sequence of numbers to give each row a unique number. Until now, it has not been possible to guarantee sequencing of numbers, although an IDENTITY-based column could potentially give a sequence providing all INSERTs succeeded and no DELETEs took place.

This function is ideal for giving your output a reference point, for example, "Please take a look at row 10 and you'll see . . .". Another use for this function is to break the data into exact chunks for scrolling purposes in GUI systems. For example, if five rows of data are returned, row 1 could be displayed, and then "Next" would allow the application to move to row 2 easily rather than use some other method.

The following is an example that shows how the ROW_NUMBER() function can provide an ascending number for each row returned when inspecting the Employee view. This is a view in AdventureWorks that shows details of employees who work within the AdventureWorks company. The ROW_NUMBER() function is nondeterministic, and since the ORDER BY within the OVER function doesn't produce a unique sequence of data (because there might be several people with the same last name, for example), then you would need to find some other way to achieve uniqueness, if getting the same order with each execution were mandatory.

```
USE AdventureWorks
GO
SELECT ROW_NUMBER() OVER(ORDER BY LastName) AS RowNum,
    FirstName + ' ' + LastName
  FROM HumanResources.vEmployee
  WHERE JobTitle = 'Production Technician - WC60'
  ORDER BY LastName
```

When we execute the code, we see the names in last name order, as shown in Figure 12-6.

It's also possible to reset the sequence to give a unique ascending number within a section, or partition of data, using the PARTITION BY option. This would be ideal if, for example, in the same marathon you had different races, such as male, female, disabled male, disabled female, over 60, and so on. Using the category the runner is in as the basis of the partition, no matter in what order the runners cross the line, we would still have the right numbering for each category.

Figure 12-6. *Rows with row numbering*

The following example will reset the sequential number at each change of the first letter in the last name of the employees.

```
USE AdventureWorks
GO
SELECT ROW_NUMBER()
  OVER(PARTITION BY SUBSTRING(LastName,1,1)
  ORDER BY LastName) AS RowNum, FirstName + ' ' + LastName
  FROM HumanResources.vEmployee
WHERE JobTitle = 'Production Technician - WC60'
ORDER BY LastName
```

When we execute the code, as the first letter of the last name alters, we see the RowNum column, which contains the value for ROW_NUMBER(), alter, as shown in Figure 12-7.

Figure 12-7. *Rows with row numbering resetting on change of last name, first letter*

RANK

If a row of data is returned that contains the same values as another row as defined in the ORDER BY clause of your statement, the keyword RANK will give these rows the same numeric value. An internal count is kept so that on a change of value you will see a jump in the value. For example, say you watch a sport like golf, and in a golf tournament you're following Tiger Woods who wins the tournament on a score of four under par, but Colin Montgomery, Arnold Palmer, and Lee Westwood are

joint second on three under par. Finally, Michelle Wie finishes her round on two under par. Tiger would have the value 1; Colin, Arnold, and Lee would have the value 2; and Michelle would have the value 5. This is exactly what RANK does. This function would also be useful in applications that wanted to return different rankings of data but only show the data from one rank at any one time on each page.

The example to demonstrate RANK uses the vEmployeeDepartment view, in which there are more rows with the same value than in vEmployee. Here the ROW_NUMBER() function is used as it was in the previous example, enabling a cross-check where the RANK function skips to the right number. When you run the query, you'll find that there are five rows with a Department called Document Control. RANK will assign these rows the number 1. When the Department changes to Engineering, RANK will change to 6.

```
USE AdventureWorks
GO
SELECT ROW_NUMBER() OVER(ORDER BY Department) AS RowNum,
       RANK() OVER(ORDER BY Department) AS Ranking,
       FirstName + ' ' + LastName AS Employee, Department
  FROM HumanResources.vEmployeeDepartment
 ORDER BY RowNum
```

The results are shown in Figure 12-8, where you can see in the Ranking column the values remain static when the values are the same, and then skip to the correct number on a change of value.

	RowNum	Ranki...	Employee	Department
1	1	1	Tengiz Kharatishvili	Document Control
2	2	1	Zainal Arifin	Document Control
3	3	1	Sean Chai	Document Control
4	4	1	Karen Berge	Document Control
5	5	1	Chris Norred	Document Control
6	6	6	Michael Sullivan	Engineering
7	7	6	Sharon Salavaria	Engineering
8	8	6	Roberto Tamburello	Engineering
9	9	6	Gail Erickson	Engineering
10	10	6	Jossef Goldberg	Engineering
11	11	6	Terri Duffy	Engineering
12	12	12	Laura Norman	Executive
13	13	12	Ken Sánchez	Executive

Figure 12-8. *Ranking and row numbering*

DENSE_RANK

DENSE_RANK gives each group the next number in the sequence and does not jump forward if there is more than one item in a group. In our golf example from the previous section, for example, Michelle Wie would have a value of 3 instead of a value of 5 because her score is the third highest. The following code demonstrates DENSE_RANK:

```
USE AdventureWorks
GO
SELECT ROW_NUMBER() OVER(ORDER BY Department) AS RowNum,
       DENSE_RANK() OVER(ORDER BY Department) AS Ranking,
       CONVERT(varchar(25),FirstName + ' ' + LastName), Department
  FROM HumanResources.vEmployeeDepartment
 ORDER BY RowNum
```

The results can be seen in Figure 12-9. Notice that this time, on change of `Department`, the `Ranking` becomes 2 when `Document Control` becomes `Engineering`, instead of moving to 6.

	RowNum	Ranki...	(No column name)	Department
1	1	1	Tengiz Kharatishvili	Document Control
2	2	1	Zainal Arifin	Document Control
3	3	1	Sean Chai	Document Control
4	4	1	Karen Berge	Document Control
5	5	1	Chris Norred	Document Control
6	6	2	Michael Sullivan	Engineering
7	7	2	Sharon Salavaria	Engineering
8	8	2	Roberto Tamburello	Engineering
9	9	2	Gail Erickson	Engineering
10	10	2	Jossef Goldberg	Engineering
11	11	2	Terri Duffy	Engineering
12	12	3	Laura Norman	Executive
13	13	3	Ken Sánchez	Executive

Figure 12-9. *Dense ranking and row numbering*

NTILE

Batching output into manageable groups has always been tricky. For example, if you have a batch of work that needs cross-checking among a number of people, then GROUP BY has to be used, though this won't give an even split. NTILE is used to give the split a more even, although approximated, grouping. The value in parentheses after NTILE defines the number of groups to produce, so NTILE(25) would produce 25 groups of as close a split as possible of even numbers.

```
USE AdventureWorks
GO
SELECT NTILE(10) OVER(ORDER BY Department) AS NTile,
       FirstName + ' ' + LastName, Department
  FROM HumanResources.vEmployeeDepartment
```

This produces 10 groups of 29 rows each. In Figure 12-10, you will see where the first batch ends and the second batch commences.

	NTile	(No column name)	Department
24	1	David Liu	Finance
25	1	Janet Sheperdigian	Finance
26	1	Deborah Poe	Finance
27	1	Wendy Kahn	Finance
28	1	David Barber	Finance
29	1	Candy Spoon	Finance
30	2	Bryan Walton	Finance
31	2	Hao Chen	Human Resources
32	2	Mindy Martin	Human Resources
33	2	Vidur Luthra	Human Resources
34	2	Willis Johnson	Human Resources
35	2	Paula Barreto de ...	Human Resources
36	2	Grant Culbertson	Human Resources

Figure 12-10. *Batching output into groups of data*

Using the MAX Data Type

Within SSE, you can store large volumes of data or images. As we discussed when creating our database, this would either be as varchar(max) or nvarchar(max) for text-based data, or varbinary(max) for image, music, video, and other similar types of data. The easiest method for working with image and movie files is to use .NET-based code to import and retrieve the files. We don't look at this within this book (though if you care to learn more, there is an example of this in *Pro SQL Server 2005 Assemblies*, also by yours truly [Apress, 2005; ISBN: 1-59059-566-1]). It is possible in SSE to import images, etc., as I will demonstrate, but via .NET it is possible to manipulate the images, changing them to a consistent data type, and so on.

By default, SSE will only store 8,000 bytes of data within a single row on any column defined as max. If the data to insert is over 8,000 bytes, the contents are stored away from the table in a special large object (LOB) area on the server, and instead of the data within the column, a pointer value will be inserted. If the data will be consistently over 8,000 bytes and therefore will always be stored in the LOB area, you can run the following code, which will tell SSE to always store varchar(max), nvarchar(max), and varbinary(max) values in the LOB area, no matter what size they are. The first parameter defines the schema and table to set the value on; the second parameter is the option to set; and the final parameter is the value to set.

■**Note** sp_tableoption is a system-stored procedure that sets options for tables. The 'large value types out of row' text is a standard option. Others can be found here: http://msdn2.microsoft.com/en-us/ library/ms173530.aspx.

```
sp_tableoption 'TransactionDetails.Transactions',
'large value types out of row',1
```

Until now, we have not looked at the Notes column within the TransactionDetails.Transactions table, but we will make extensive use of it in a moment. First, let's switch back to holding the information in rows by executing the following code:

```
sp_tableoption 'TransactionDetails.Transactions',
'large value types out of row',0
```

Inserting text data to a max column is no different from inserting data into one defined with a specific value. However, when you wish to update values, there is new functionality we need to look at. We will also take a look at string functionality for working with the data.

To update text data, it is possible to use existing string-based functions such as SUBSTRING, but it is also possible to use a function called .WRITE(). This function has three parameters: the first is the text to update, the second is the offset to start the updating at (a value of 0 says that you are starting at the beginning, a value of 1 means the first character stays, and so on), and the third is the number of characters to remove within the existing text. The syntax is as follows:

```
SET column_name .WRITE(text,@offset,@length)
```

■**Note** Notice that there is a space between column_name and .WRITE.

To clarify, the third parameter defines the number of characters to remove from the existing text that is contained within the column you are updating and is not the length of the data you are inserting. Therefore, the following would remove 15 characters of text from position 10 and replace the text defined in the first parameter at position 10:

```
SET column_name .WRITE(text,10,15)
```

Try It Out: Working with Text LOB Data

1. First of all, let's insert a text value into the `TransactionDetails.Transactions` table. Enter and execute the following code:

```
UPDATE TransactionDetails.Transactions
SET Notes = 'From online auction sale of Redskins memorabilia, helmet'
WHERE TransactionId = 1
```

2. We can see that the insertion has worked by executing the following code and looking at the results in Figure 12-11:

```
SELECT * FROM TransactionDetails.Transactions
```

	TransactionId	CustomerId	TransactionType	DateEntered	Amount	ReferenceDetails	Notes
1	1	1	1	2005-08-01 00:00:00.000	100.00000	NULL	From online
2	2	1	1	2005-08-03 00:00:00.000	75.67000	NULL	NULL
3	3	1	2	2005-08-05 00:00:00.000	35.20000	NULL	NULL
4	4	1	2	2005-08-06 00:00:00.000	20.00000	NULL	NULL
5	5	1	2	2006-09-18 20:25:49.000	100.00000	NULL	NULL
6	7	1	3	2006-09-19 20:29:08.950	200.00000	NULL	NULL
7	8	2	1	2005-12-18 00:00:00.000	1000.00000	NULL	NULL

Figure 12-11. *Notes added to a varchar(max) column*

3. If we realize that we have made a mistake and in fact it wasn't Redskins memorabilia but Dolphins memorabilia we're after, and we want to change several other items of information, then we could use the following to update the whole of the Notes column:

```
UPDATE TransactionDetails.Transactions
SET Notes .WRITE(
'From online auction sale of Dolphins memorabilia, helmet',
0,8000)
WHERE TransactionId = 1
```

4. If we then list the details again, we can see the updates as shown in Figure 12-12.

```
SELECT * FROM TransactionDetails.Transactions WHERE TransactionId = 1
```

	TransactionId	CustomerId	TransactionType	DateEntered	Amount	ReferenceDetails	Notes
1	1	1	1	2005-08-01 00:00:00.000	100.00000	NULL	From online auction sale of Dolphins memorabili...

Figure 12-12. *Notes modified*

5. The last demonstration will show how we update part of a piece of text. In this case, we still didn't get it right, as it is Raiders memorabilia we're really after. Here we update column 28 and remove 8 characters, the number of characters in the word "Dolphins."

```
UPDATE TransactionDetails.Transactions
SET Notes .WRITE('Raiders',28,8)
WHERE TransactionId = 1
```

6. If we then list the details again, we can see the updates in Figure 12-13.

```
SELECT * FROM TransactionDetails.Transactions WHERE TransactionId = 1
```

	TransactionId	CustomerId	TransactionType	DateEntered	Amount	ReferenceDetails	Notes
1	1	1	1	2005-08-01 00:00:00.000	100.00000	NULL	From online auction sale of Raiders memorabilia,...

Figure 12-13. *Notes modified at a specific location*

Now that we have looked at text, we can turn our attention to images.

Image LOB

Inserting and updating images, movies, Excel spreadsheets, or any data that can be accessed via an OLE DB data provider into rows of data uses a T-SQL function called OPENROWSET. OLE DB is a method used to access data information from a program. This can be Excel files, text files, image files, and so on. Some OLE DB providers offer basic functionality, whereas others allow you to work with the data within your program as if you were using the raw data's own program. For example, there is an OLE DB provider that allows you to work with Excel data in, for example, a C# program as if you were in Excel. We won't look at the OPENROWSET function extensively, as it is quite an advanced area, but bulk loading binary data located locally on our computer doesn't need any provider or connection information, and so it is very straightforward.

▓**Note** Remember that although we are using an image here, it could be music files, movies, or any data at all that could be loaded.

The OPENROWSET syntax that we will be looking at is shown next, and it has to be used as if we were returning data from a table; therefore, we would need to use the INSERT...SELECT FROM OPENROWSET functionality.

```
FROM OPENROWSET(BULK 'name_of_file', type_of_operation) column_alias
```

The first option is the name_of_file. This will define the location, the name, and the file type of the file we wish to bulk load. The second option allows us three different types of bulk loading:

- SINGLE_BLOB: This reads the file and creates one row and column of data, which is defined as varbinary(max).

- SINGLE_CLOB: Same as SINGLE_BLOB, except the data is varchar(max) using the collation of the database the query is run on.

- SINGLE_NCLOB: Same as SINGLE_CLOB, but the data is Unicoded and therefore is nvarchar(max).

Once the data has been bulk loaded, we need to give the column a column alias so we can refer to it within our SELECT statement. So the full statement would look like this:

```
INSERT INTO table_name (column_in_table)
SELECT column_alias.*
FROM OPENROWSET(BULK 'name_of_file ',type_of_operation) column_alias
```

To update a row, you would need to use OPENROWSET as a subquery to set the column you wish to modify.

Try It Out: Working with Image LOB Data

1. In this example, we will insert a new TransactionDetails.Transactions row and insert an image into the Notes column. This image could be from a document scanner of the share purchase made, a copy of the written check from the bank account, or even a copy of the credit card receipt. As security is the byword in SSE, we won't be showing this, and I will use a picture found on my computer.

2. Enter the following code noting that we have a column alias of PhotoToLoad that is defined from the BULK operation and then used in the SELECT:

```
INSERT INTO TransactionDetails.Transactions
(CustomerId, TransactionType, DateEntered,Amount,Notes, RelatedProductId)
SELECT 2, 1, GETDATE(),1000,PhotoToLoad.*, 1
FROM OPENROWSET
    (BULK 'c:\temp\photo.tif', SINGLE_BLOB) PhotoToLoad
```

3. When you execute this code, you should see the following results:

```
(1 row(s) affected)
```

4. To update the row, first of all find the TransactionId of the row entered previously; in my case this was 9. Then we create a subquery in which we only return one column and one row. We can use this to set the value within a column.

```
UPDATE TransactionDetails.Transactions
SET Notes = (
SELECT PhotoToLoad.*
FROM OPENROWSET
    (BULK 'c:\Temp\photo2.jpg', SINGLE_BLOB) PhotoToLoad)
WHERE transactionId = 9
```

5. When you execute this code, you should see the following results:

```
(1 row(s) affected)
```

Note Images cannot be retrieved for viewing within SSE. It is necessary to use some sort of graphical program or tool to display the image.

Summary

Subqueries are one of the most commonly used areas of T-SQL, but common table expressions are also useful, so knowing these areas well will enable you to move forward at a rapid pace. Combine this knowledge with the functionality covered in Chapter 11, especially when working with JOINs, and you should start to see how powerful T-SQL can be at working with time-based data.

Now that your T-SQL knowledge is quite advanced, we can move on to look at triggers.

CHAPTER 13

■■■

Triggers

Now that you're quite proficient in using SQL Server Express, you ought to know about one last aspect of it. Triggers are that one last step, and this chapter is the missing link in the foundation of your knowledge and skill set.

At times a modification to data somewhere in your database requires action on data elsewhere, either in the same database, another database, or elsewhere within SSE; a trigger is the object that will do this for you. When a modification to your data occurs, SSE will fire a trigger, which is a specialized stored procedure that will run, performing the actions that you desire. Triggers are similar to constraints but more powerful, and they require more system overhead, which can affect performance. Triggers are most commonly used to perform business rule validation, cascade data modifications (changes in one table requiring changes in other tables), keep track of changes for each row (audit trailing), or any other processing required when data in a specific table is modified.

The goals of this chapter are as follows:

- Describe what a trigger is
- Detail potential problems surrounding triggers
- Show the CREATE TRIGGER T-SQL syntax
- Discuss when to use a constraint and when to use a trigger
- Show the system tables and functions specific to triggers
- Demonstrate the creation of a trigger through a template and T-SQL
- Discuss image data types and the problems that surround updating image columns and firing a trigger

First of all, let's see just what constitutes a trigger.

What Is a Trigger?

A **trigger** is a specialized stored procedure that will be executed either on a data modification, known as a Data Modification Language (DML) trigger, or on a data model action, such as CREATE TABLE, known as a Data Definition Language (DDL) trigger. DML triggers are pieces of code attached to a specific table that are set to automatically run in response to an INSERT, a DELETE, or an UPDATE statement. A DDL trigger is attached to an action that occurs within either a database or a server. The first part of the chapter will look at DML triggers, followed by an investigation of DDL triggers.

■**Note** Unlike stored procedures, you cannot manually execute a trigger or use parameters or return values with triggers.

The DML Trigger

Triggers have many uses. Perhaps the most common for a DML trigger is to enforce a business rule; for example, when a customer places an order, to check that he or she has sufficient funds or that you have enough stock. If any of these checks fail, you can complete further actions or return error messages and roll back the update.

DML triggers can be used as a form of extra validation, for example, to perform quite complex checks on data that a constraint could not achieve. Keep in mind that using constraints instead of triggers gives you better performance, but triggers are the better choice when dealing with complex data validation. Another use for a DML trigger is to make changes in another table based on what is about to happen in the triggered table. For example, when you add an order, you could fire a DML trigger that would reduce the quantity of items in stock. Finally, DML triggers can be used to create an automated audit trail that records a change history for each row.

We can create separate triggers for any table action except SELECT, or triggers that will fire on any combination of table actions. Obviously, as no table modifications occur on a SELECT statement, it is impossible to create such a trigger. There are three main types of triggers:

- INSERT trigger
- DELETE trigger
- UPDATE trigger

You can also have a combination of the three types of triggers.

Triggers can update tables within other databases if desired, and it is also possible for triggers to span servers as well, so don't think the scope of triggers is limited to the current database.

It is possible for a trigger to fire a data modification, which in turn will execute another trigger, which is known as a **nested trigger**. For example, imagine you have Table A, which has a trigger on it to fire a modification within Table B, which in turn has a trigger on it that fires a modification within Table C. If a modification is made to Table A, Table A's trigger will fire, modifying the data in Table B, which will fire the trigger in Table B, thus modifying data in Table C. This nesting of triggers can go up to 32 triggers deep before you reach the limit set within SSE; however, if you start getting close to that sort of level, you either have a very complex system, or perhaps you have been overly zealous with your creation of triggers!

It is possible to switch off any nesting of triggers so that when one trigger fires, no other trigger can fire; however, this is not usually the norm. Be aware that your performance will suffer greatly when you start using nested triggers; use them only when necessary.

■**Note** There is one statement that will stop a DELETE trigger from firing. If you issue a TRUNCATE TABLE T-SQL command, it is as if the table has been wiped without any logging. This also means that a DELETE trigger will not fire, as it is not a deletion per se that is happening.

As with stored procedures, do take care when building triggers: you don't want to create a potentially endless loop in which a trigger causes an update, which fires a trigger already fired earlier within the loop, thereby repeating the process.

CREATE TRIGGER Syntax for DML Triggers

The creation of a trigger through T-SQL code can be quite complex if you use the full trigger syntax. However, the reduced version that I cover here is much more manageable and easier to demonstrate.

When building a trigger, it can be created for a single action or for multiple actions. To expand on this, a trigger can be for only insertion of a row, or it could cover inserting and updating the row.

■**Note** Although this chapter will demonstrate DML triggers on tables, a trigger can also be placed on a view as well, so that when data is modified through a view, it too can fire a trigger if required.

Here is the syntax for creating a basic trigger:

```
CREATE TRIGGER trigger_name
ON {table | view}
[WITH ENCRYPTION]
{
{{FOR {AFTER | INSTEAD OF} {[INSERT] [,] [UPDATE] [,] [DELETE]}
AS
[{IF [UPDATE (column)
[{AND | OR} UPDATE (column)]] ]
COLUMNS_UPDATE()]
sql_statements}}
```

Let's explore the options in this syntax more closely:

CREATE TRIGGER trigger_name: First of all, as ever, you need to inform SSE what you are attempting to do, and in this instance, you wish to create a trigger. The name for the trigger must also follow the SSE standards for naming objects within a database. In this chapter, you name the triggers starting with tg to indicate the object is a trigger, followed by the type of trigger (ins for insert, del for delete, and upd for update), and then the name of the root table the trigger will be associated with.

ON {table | view}: It is then necessary to give the name of the single table or view that the trigger relates to, which is named after the ON keyword. Each trigger is attached to one table only.

[WITH ENCRYPTION]: As with views and stored procedures, you can encrypt the trigger using the WITH ENCRYPTION options so that the code cannot be viewed by prying eyes.

{FOR | AFTER | INSTEAD OF}:

- The FOR | AFTER trigger will run the code within the trigger after the underlying data is modified. Therefore, if you have any constraints on the table for cascading changes, the table and these cascades will complete before the trigger fires. You either specify FOR or AFTER.

- INSTEAD OF: The most complex of the three options to understand, as a trigger defined with this option will run the T-SQL within the trigger rather than allow the data modification to run. This includes any cascading. To clarify, if you have an INSTEAD OF trigger that will execute on a data INSERT, the insertion will not take place.

{[INSERT] [,] [UPDATE] [,] [DELETE]}: This section of the syntax determines on what action(s) the trigger will execute. This can be an INSERT, an UPDATE, or a DELETE T-SQL command. As mentioned earlier, the trigger can fire on one, two, or three of these commands, depending on what you wish the trigger to do. Therefore, at this point, you need to mention which combination of commands, separated by a comma, you wish to work with.

AS: The keyword AS defines that the trigger code has commenced, just as the AS keyword defines the start of a stored procedure. After all, a trigger is just a specialized stored procedure.

`[{IF UPDATE (column) [{AND | OR} UPDATE (column)]]`: This option can be used within a trigger that is not available within a stored procedure, and that is the test to check whether a specific column has been modified or not. This happens through the use of the `UPDATE()` keyword. By placing the name of the column to test in between the parentheses, a logical `TRUE` or `FALSE` will be returned depending on whether the column has been updated or not. The deletion of a row will not set the `UPDATE` test to `TRUE` or `FALSE`, as you are removing an item and not updating it. An `INSERT` or an `UPDATE` will set the `UPDATE` test to the necessary value.

`COLUMNS_UPDATE()`: This has functionality similar to `UPDATE()`, but instead of testing a specific named column, it tests multiple columns in one test.

`sql_statements`: At this point you code the trigger just like any other stored procedure.

The main thought that you must keep in mind when building a trigger is that a trigger fires after each statement that modifies data is executed, but before the modification is actually made to rows in the table. Therefore, if you have a statement that updates many rows, the trigger will fire after all the updates have been requested to be done, not when each of the rows has been dealt with.

■**Note** Keep in mind, the `FOR` trigger executes before the underlying data is modified; therefore, a trigger can issue a `ROLLBACK` for that particular action if so desired.

Now that you know how to create a trigger, we'll look at which situations they best apply to, as opposed to constraints.

Why Not Use a Constraint?

There is nothing stopping you from using a constraint to enforce a business rule, and in fact, constraints should be used to enforce data integrity. Constraints also give you better performance than triggers. However, they are limited in what they can achieve and what information is available to them to complete their job.

Triggers are more commonly used for validation of business rules, or for more complex data validation, which may or may not then go on to complete further updates of data elsewhere within SSE.

A constraint is only able to validate data that is within the table the constraint is being built for or a specified value entered at design time. This is in contrast to a trigger, which can span databases, or even servers, and check against any data set at design time or built from data collected from other actions against any table. This can happen if the necessary access rights are given to all objects involved.

However, constraints are the objects to use to ensure that data forming a key is correct, or when referential integrity needs to be enforced through a foreign key constraint.

At times a fine line will exist between building a constraint and a trigger, when the trigger is meant to perform a very simple validation task. In this case, if the decision deals with any form of data integrity, then use a constraint, which will give you better performance than using a trigger. If the object to be built is for business rules and may require complex validation, needs to handle multiple databases or servers, or requires advanced error handling, then build a trigger. For example, a trigger must be used if you need a change on one table to result in an action (update, delete, etc.) on a table that is located in another database. You might have this situation if you keep an audit trail (change history) database separate from your production database. It is doubtful that you would want to use a trigger if you are doing something simple like verifying that a date field only contains values within a certain range.

Deleted and Inserted Logical Tables

When a table is modified, whether this is by an insertion, an update, or a deletion, an exact copy of the row of data is held in two system logical tables called DELETED and INSERTED. When a row is inserted into a table, a full copy of the inserted row is placed into the INSERTED table. Every item of information placed into each column for the insertion is then available for checking. If a deletion is performed, a copy of the row is placed in the DELETED table. Finally, when an update occurs, a copy of the row before the modification is placed in the DELETED table, and then a copy of the row of data after the modification is placed in the INSERTED table.

The INSERTED and DELETED tables will hold one row from each table for each modification. Therefore, if you perform an UPDATE that updates 100 rows, the DELETED logical table is populated with the 100 rows prior to the UPDATE. The modification then takes place, and the INSERTED table is populated with 100 rows. Finally, the trigger will fire. Once the trigger has completed, the data for that table is removed from the relevant logical tables.

These tables do not actually physically exist within SSE, so it is not possible to complete any further processing on these tables, such as creating an index, and the data can only be interrogated via a SELECT statement and cannot be modified. Therefore, you can access these tables within a trigger to find out which rows have been inserted, updated, or deleted.

To check what columns have been modified, it would be possible to compare each and every column value between the two tables to see what information had been altered. Luckily, as was discussed when we examined the syntax, there is a function, UPDATE(), that can test whether a column has been modified.

Now that you are fully up to date as to what a DML trigger is and how it works, it is time to create and test our first trigger within the database.

Creating a DML FOR Trigger

The first trigger we will look at is a DML trigger. The following example will demonstrate how to create a trigger on a data insertion, but also what happens to that INSERT when there is a problem in the trigger itself. As we are near the end of the book, our T-SQL within the trigger will be more advanced than some of the code so far.

Try It Out: Creating a Trigger in Query Editor

The purpose of our example is to change a customer's account balance when a financial transaction occurs as defined by an INSERT in the TransactionDetails.Transactions table. We want to change the balance AFTER the row has been inserted into the TransactionDetails.Transactions table. This is so we do not change the customer's account balance if later in the INSERT of the row a problem occurs and the INSERT does not complete.

1. Ensure that Query Editor is running and that you are logged in with an ID that can insert objects into the database. First of all, it is necessary to give the trigger a meaningful name. Then you define the table that the trigger will be attached to, which in this case is the TransactionDetails.Transactions table. The final part of the start of the trigger will then define the type of trigger and on what actions the trigger will execute. This will be a FOR AFTER trigger on an INSERT on the TransactionDetails.Transactions table. The first part of the code looks as follows:

```
CREATE TRIGGER trgInsTransactions
ON TransactionDetails.Transactions
AFTER INSERT
AS
```

2. It is now time to enter the remainder of the code for the trigger. We need to retrieve the Amount and TransactionType from the INSERTED table to be able to use these in the update of the CustomerDetails.Customers table. We can JOIN from the INSERTED table to the TransactionDetails.TransactionTypes table to find out whether we are dealing with a credit or a debit transaction. If it is a debit, through the use of a subquery and a CASE statement we can alter the Amount by multiplying it by -1 so that we are reducing a customer's balance. Notice the subquery includes a WHERE statement so that if we are entering a transaction type that does not affect the cash balance, such as recording a share movement, the ClearedBalance will not be altered. The final action is to update the customer's balance, which we will do via an UPDATE statement. There is a great deal to take in, so take time over the code. Also, the two examples of running this trigger should clear up any queries you will have.

Note This trigger does have a deliberate bug, which is included so that you can see a little later in this section what happens when a trigger has a bug.

```
UPDATE CustomerDetails.Customers
   SET ClearedBalance = ClearedBalance +
      (SELECT CASE WHEN CreditType = 0
                   THEN i.Amount * -1
                   ELSE i.Amount
                   END
        FROM INSERTED i
        JOIN TransactionDetails.TransactionTypes tt
          ON tt.TransactionTypeId = i.TransactionType
        WHERE AffectCashBalance = 1)
   FROM CustomerDetails.Customers c
   JOIN INSERTED i ON i.CustomerId = c.CustomerId
```

3. Execute the code to create the trigger in the database. We can test the trigger now by inserting a cash withdrawal or deposit relating to the two transaction types we currently have. We will list the customer balance before executing the INSERT into the TransactionDetails.Transactions table, and then we will add the row and look at the balance again to show that it has changed. Enter the following code, which inserts a withdrawal of $200 from Vic McGlynn's account:

```
SELECT ClearedBalance
  FROM CustomerDetails.Customers
 WHERE customerId=1

INSERT INTO TransactionDetails.Transactions (CustomerId,TransactionType,
Amount,RelatedProductId, DateEntered)
VALUES (1,2,200,1,GETDATE())

SELECT ClearedBalance
  FROM CustomerDetails.Customers
 WHERE customerId=1
```

4. Execute the code. As you see in Figure 13-1, the results should show that the balance has dropped by $200 as expected. You could also double-check that the transaction exists in the TransactionDetails. Transactions table.

Figure 13-1. *Balance reduction after trigger action*

5. So our next test is to simulate a noncash transaction that has been recorded. For example, if you bought some shares, there would be the cash transaction removing the funds from your bank account, `ProductId=1`, and then a second row entered on `ProductId=2`, which is the equities product showing the addition of shares. This is a simple accounting procedure of one debit and one credit. Enter the following code:

```
SELECT ClearedBalance
  FROM CustomerDetails.Customers
 WHERE customerId=1

INSERT INTO TransactionDetails.Transactions (CustomerId,TransactionType,
Amount,RelatedProductId, DateEntered)
VALUES (1,3,200,1,GETDATE())

SELECT ClearedBalance
  FROM CustomerDetails.Customers
 WHERE customerId=1
```

6. Now execute the code. Instead of seeing two rows where the balance hasn't altered, we see the following error message and on the Results tab only one set of output, as shown in Figure 13-2. Our trigger has a bug in that the subquery will return a `NULL` value where the transaction type does not affect a cash balance, and we have not accounted for that scenario. There are two reasons for showing you an error: the first is to demonstrate how to alter a trigger, the second, more importantly, is to determine whether the `INSERT` statement succeeded or failed.

```
(1 row(s) affected)
Msg 515, Level 16, State 2, Procedure trgInsTransactions, Line 6
Cannot insert the value NULL into column 'ClearedBalance', table
'ApressFinancial.CustomerDetails.Customers'; column does not allow nulls.
UPDATE fails.
The statement has been terminated.
```

Figure 13-2. *Balance not updated*

7. To reiterate, the `INSERT` statement is correct and would normally work. However, as the trigger has a bug, the transaction did not insert the data and was rolled back. You can see this by inspecting the `TransactionDetails.Transactions` table with the following code and the results shown in Figure 13-3.

```
SELECT *
  FROM TransactionDetails.Transactions
 WHERE CustomerId=1
```

	TransactionId	CustomerId	TransactionType	DateEntered	Amount	ReferenceDetails	Notes
1	1	1	1	2005-08-01 00:00:00.000	100.00000	NULL	NULL
2	2	1	1	2005-08-03 00:00:00.000	75.67000	NULL	NULL
3	3	1	2	2005-08-05 00:00:00.000	35.20000	NULL	NULL
4	4	1	2	2005-08-06 00:00:00.000	20.00000	NULL	NULL
5	5	1	2	2006-09-19 20:25:49.000	200.00000	NULL	NULL

Figure 13-3. *Transaction table listing*

8. We can change a trigger using the `ALTER TRIGGER` command. The changes to the code occur in the sub-query: we surround the single column we will have returned with an `ISNULL()` test. If the result is NULL, we transpose this with the value of 0, as the cash balance is not to alter. The code we need to change is in **bold**.

```
ALTER TRIGGER TransactionDetails.trgInsTransactions
ON TransactionDetails.Transactions
AFTER INSERT
AS
UPDATE CustomerDetails.Customers
    SET ClearedBalance = ClearedBalance +
        ISNULL((SELECT CASE WHEN CreditType = 0
                    THEN i.Amount * -1
                    ELSE i.Amount
                    END
          FROM INSERTED i
          JOIN TransactionDetails.TransactionTypes tt
            ON tt.TransactionTypeId = i.TransactionType
          WHERE AffectCashBalance = 1),0)
    FROM CustomerDetails.Customers c
    JOIN INSERTED i ON i.CustomerId = c.CustomerId
```

9. Once the changes have been completed, we can then execute the code to alter the trigger. We can now rerun our test, which will add a row to the `TransactionDetails.Transactions` table without altering the balance. If you wish, you could also list the `TransactionDetails.Transactions` table to prove that the `INSERT` succeeded this time, as Figure 13-4 demonstrates.

```
SELECT ClearedBalance
  FROM CustomerDetails.Customers
 WHERE customerId=1

INSERT INTO TransactionDetails.Transactions (CustomerId,TransactionType,
Amount,RelatedProductId, DateEntered)
VALUES (1,3,200,1,GETDATE())

SELECT ClearedBalance
  FROM CustomerDetails.Customers
 WHERE customerId=1
```

Figure 13-4. *Transactions table with no balance change*

Checking Specific Columns

It is possible to check whether a specific column or set of columns has been updated via the UPDATE()
or COLUMNS_UPDATED() functions available within a trigger. This can reduce the amount of processing
within the trigger and therefore speed up your batch and transactions. Checking columns and only
performing specific T-SQL code if a column is altered will reduce trigger overheads. As you will see,
only when an amount or type of transaction has altered do you really need to perform an UPDATE on
the CustomerDetails.Customers table.

The first statement we will look at is UPDATE().

Using UPDATE()

The UPDATE() function is a very simple, yet powerful tool to a developer who is building a trigger. It is
possible to check against a specific column, or a list of columns, to see whether a value has been
inserted or updated within that column. It is not possible to check whether a value has been deleted
for a column, because, quite simply, you cannot delete columns, you can only delete whole rows of
data. If you wish to check more than one column at the same time, place the columns one after
another with either an AND or an OR depending on what you wish to happen. Each individual UPDATE()
will return TRUE if a value has been updated. If there are a number of columns, each column will have
to be defined separately. For example:

```
IF UPDATE(column1) [AND | OR UPDATE(column2)]
```

You can use this function to deal with updates to the TransactionDetails.Transactions table. For
example, there will be times that a transaction row has been incorrectly inserted. The trigger we created
previously would have to be modified to deal with an UPDATE to alter the CustomerDetails.Customers
ClearedBalance. The UPDATE would remove the value within the DELETED table and then apply the value
within the INSERTED table. However, what if the alteration has nothing to do with any transaction that
would alter the cash balance? For example, we were changing the date entered. By simply checking
each column as necessary, it is possible to see whether an update is required to the CustomerDetails.
Customers table. The two columns that would interest us are Amount and TransactionType.

Try It Out: UPDATE() Function

1. Within Query Editor, let's alter our trigger to deal with an UPDATE first of all before moving to the UPDATE()
 function. The first part of the alteration is to tag an UPDATE to the AFTER statement.

```
ALTER TRIGGER TransactionDetails.trgInsTransactions
ON TransactionDetails.Transactions
AFTER INSERT,UPDATE
AS
```

2. Then we need to deal with the undoing of the amount in the DELETED table row from the CustomerDetails.
 Customers table. The actions on the ClearedBalance need to be the opposite of the addition.

```
UPDATE CustomerDetails.Customers
   SET ClearedBalance = ClearedBalance -
      ISNULL((SELECT CASE WHEN CreditType = 0
                     THEN d.Amount * -1
                     ELSE d.Amount
                     END
         FROM DELETED d
         JOIN TransactionDetails.TransactionTypes tt
           ON tt.TransactionTypeId = d.TransactionType
         WHERE AffectCashBalance = 1),0)
      FROM CustomerDetails.Customers c
      JOIN DELETED d ON d.CustomerId = c.CustomerId
```

3. The remainder of the trigger is the same. Once you have added in the following code, execute it so that the
 trigger is altered.

```
UPDATE CustomerDetails.Customers
   SET ClearedBalance = ClearedBalance +
      ISNULL((SELECT CASE WHEN CreditType = 0
                     THEN i.Amount * -1
                     ELSE i.Amount
                     END
         FROM INSERTED i
         JOIN TransactionDetails.TransactionTypes tt
           ON tt.TransactionTypeId = i.TransactionType
         WHERE AffectCashBalance = 1),0)
      FROM CustomerDetails.Customers c
      JOIN INSERTED i ON i.CustomerId = c.CustomerId
```

4. We can test that the trigger works by reducing the amount of a withdrawal, in this case TransactionId
 number 12, which currently sits at a value of $200. The following code will list the transactions for
 CustomerId=1 and the current cleared balance. We then alter the amount of the withdrawal from $200
 down to $100. The final actions will list the tables to prove the update to the TransactionDetails.
 Transactions and CustomerDetails.Customers tables has succeeded.

```
SELECT *
  FROM TransactionDetails.Transactions
 WHERE CustomerId = 1
SELECT ClearedBalance
  FROM CustomerDetails.Customers
 WHERE CustomerId = 1
UPDATE TransactionDetails.Transactions
   SET Amount = 100
 WHERE TransactionId = 5
SELECT *
  FROM TransactionDetails.Transactions
 WHERE CustomerId = 1
SELECT ClearedBalance
  FROM CustomerDetails.Customers
 WHERE CustomerId = 1
```

5. Once you execute the code, the transactions amount and cleared balances are altered as shown in
 Figure 13-5. So now we know the trigger has worked and will do these actions no matter what happens
 to the TransactionDetails.Transactions table.

Figure 13-5. *Transactions and balances*

6. We are now going to alter the trigger to test the `Amount` and `TransactionType` columns. If there is an update, we will complete the actions described previously; if not, we will skip this processing. We will prove which path the trigger takes by using the system function `RAISERROR`, which you saw in the discussion of error handling in Chapter 11. Each section of the `IF` statement will have an appropriate `RAISERROR`.

7. We will now alter the trigger to update the `CustomerDetails.Customers` table only if `Amount` or `TransactionType` is altered. If we execute this code, we will have a `RAISERROR` saying this is what we have done. Similarly, if we don't update the table, we will have an appropriate but different `RAISERROR`. The trigger is defined in the following code with the alterations shown in **bold**. Once you have made the same changes, execute the code to alter the trigger.

```
ALTER TRIGGER TransactionDetails.trgInsTransactions
ON TransactionDetails.Transactions
AFTER INSERT,UPDATE
AS
IF UPDATE(Amount) OR Update(TransactionType)
BEGIN
UPDATE CustomerDetails.Customers
   SET ClearedBalance = ClearedBalance -
      ISNULL((SELECT CASE WHEN CreditType = 0
                  THEN d.Amount * -1
                  ELSE d.Amount
                  END
         FROM DELETED d
         JOIN TransactionDetails.TransactionTypes tt
           ON tt.TransactionTypeId = d.TransactionType
         WHERE AffectCashBalance = 1),0)
   FROM CustomerDetails.Customers c
   JOIN DELETED d ON d.CustomerId = c.CustomerId
```

```
UPDATE CustomerDetails.Customers
   SET ClearedBalance = ClearedBalance +
       ISNULL((SELECT CASE WHEN CreditType = 0
                     THEN i.Amount * -1
                     ELSE i.Amount
                     END
         FROM INSERTED i
         JOIN TransactionDetails.TransactionTypes tt
           ON tt.TransactionTypeId = i.TransactionType
         WHERE AffectCashBalance = 1),0)
   FROM CustomerDetails.Customers c
   JOIN INSERTED i ON i.CustomerId = c.CustomerId
RAISERROR ('We have completed an update',10,1)
END
ELSE
RAISERROR ('Updates have been skipped',10,1)
```

8. We can now test out the example, which will not update the Amount or TransactionType but will alter the DateEntered.

```
SELECT *
  FROM TransactionDetails.Transactions
 WHERE TransactionId=5
SELECT ClearedBalance
  FROM CustomerDetails.Customers
 WHERE CustomerId = 1
UPDATE TransactionDetails.Transactions
   SET DateEntered = DATEADD(dd,-1,DateEntered)
 WHERE TransactionId = 5
SELECT *
  FROM TransactionDetails.Transactions
 WHERE TransactionId=5
SELECT ClearedBalance
  FROM CustomerDetails.Customers
 WHERE CustomerId = 1
```

9. Once we have run this code, we will see the Results tab showing the DateEntered being altered but the ClearedBalance not, as Figure 13-6 illustrates. However, at this point we don't know if this is because we have removed and then re-added the amount, giving a null effect.

	TransactionId	CustomerId	TransactionType	DateEntered	Amount	ReferenceDetails	Notes	Related
1	5	1	2	2006-09-19 20:25:49.000	100.00000	NULL	NULL	NULL

	ClearedBalance
1	4211.22

	TransactionId	CustomerId	TransactionType	DateEntered	Amount	ReferenceDetails	Notes	Related
1	5	1	2	2006-09-18 20:25:49.000	100.00000	NULL	NULL	NULL

	ClearedBalance
1	4211.22

Figure 13-6. *Details where updates have been skipped*

10. Moving to the Messages tab, we can see the RAISERROR that occurred when we skipped updating the CustomerDetails.Customers table. There are also fewer "row(s) affected" messages.

```
(1 row(s) affected)
(1 row(s) affected)
Updates have been skipped
(1 row(s) affected)
(1 row(s) affected)
(1 row(s) affected)
```

This brings us to the end of looking at the UPDATE() function. Let's move on to COLUMNS_UPDATED().

Using COLUMNS_UPDATED()

Instead of working with a named single column, the COLUMNS_UPDATED() function can work with multiple columns. It does this through the use of bit flags rather than naming columns. There are 8 bits in a byte, and a bit can be either off (a value of 0) or on (a value of 1).

COLUMNS_UPDATED() checks the bits of a single byte, which is provided by SSE, to see whether a column has been updated. It can do this by correlating a bit with a column in the underlying table. So to clarify, the TransactionDetails.Transactions table has nine columns. The first column, TransactionId, would relate to the first bit within the byte. The Amount column is the fifth column and therefore would relate to the fifth bit within the byte. If the first bit is on (a value of 1), the TransactionId column has been updated. Similarly, if the fourth bit is on, the Amount column has been updated.

■**Note** Confusingly, when talking about bits, the first bit is known as bit 0, the second bit is known as bit 1, and the byte is made up of bits 0 through 7. Therefore, the TransactionId column is bit 0, and the Amount column is bit 4. We will use this convention from this point on.

The bit flag settings are based on the column order of the table definition. To test for a bit value, you use the ampersand (&) operator to test a specific bit or multiple set of bits. Before we discuss how this works, inspect Table 13-1. A bit value increases by the power of 2 as you progress down the bit settings, as you can see.

Table 13-1. *Bit Settings to Values Cross-Reference*

Bit	Value
0	1
1	2
2	4
3	8
4	16
5	32
6	64
7	128

Note Another point about bits is that they work from right to left. For example, 00000010 shows bit 1 is set and therefore has a value of 2.

Now if bits 2 and 4 are switched on within a byte, in other words, they have a setting of true (00010100), the value is 4 + 16, which equates to 20. Therefore, to test whether the third and fifth columns of our table have *both* been altered, we would use the following syntax:

```
IF COLUMNS_UPDATE() & 20 = 20
```

This is a great deal to take in and understand, so I have included the following code to help you to understand this further. Here we have a byte data type variable. We then set the variable to a value; in this case, we believe that bits 0 and 1 will be set. By using the & operator we can check this. To reiterate, slightly confusingly, it's not the bit position we have to test, but the corresponding bit value, so bit 0 has a value of 1.

```
DECLARE @BitTest varbinary
SET @BitTest = 3
SELECT @BitTest & 1,@BitTest & 2,@BitTest & 4,@BitTest & 8,@BitTest & 16
```

As a byte contains 8 bits, COLUMNS_UPDATED() can only test the first eight columns on this basis. Obviously, tables will contain more than eight columns, as you have seen with the TransactionDetails.Transactions table we have just been using.

Once a table has more than eight columns, things change. Instead of being able to test COLUMNS_UPDATED() & 20 > 0 to check whether columns 3 or 5 have been updated, it is necessary to SUBSTRING() the value first. Therefore, to test columns 3 or 5, the code needs to read as follows:

```
IF (SUBSTRING(COLUMNS_UPDATED(),1,1) & 20) > 0
```

However, even this is not the correct solution, although we are almost there. It is necessary to substring the COLUMNS_UPDATED() into 8-bit chunks for each set of eight columns. However, we need to involve the power() function to get the correct value to test for. The syntax for the power() section of the test is as follows:

```
power(2,(column_to_test - 1))
```

Therefore, if you wish to test whether column 9 has been updated, the statement would be as follows, where we take the second set of eight columns using the SUBSTRING character 2, and then test the first column of the second set of eight; in other words column 8 + 1 = 9.

```
IF (SUBSTRING(COLUMNS_UPDATED(),2,1)=power(2,(1-1)))
```

The following tests columns 1, 4, and 10 to see whether any of them has changed:

```
IF (SUBSTRING(COLUMNS_UPDATED(),1,1)=power(2,(1-1))
OR SUBSTRING(COLUMNS_UPDATED(),1,1)=power(2,(4-1))
OR SUBSTRING(COLUMNS_UPDATED(),2,1)=power(2,(2-1)))
```

We can use this function to deal with updates to the TransactionDetails.Transactions table. For example, there will be times that a transaction row has been incorrectly inserted. The trigger we created previously would have to be modified to deal with an UPDATE that alters the customer's ClearedBalance. The UPDATE would remove the value within the DELETED table and then apply the value within the INSERTED table. However, what if the alteration has nothing to do with any transaction that would alter the cash balance? For example, say we were changing the date entered. By simply checking each column as necessary, it is possible to see whether an update is required to the CustomerDetails.Customers table. The two columns that would interest us are Amount and TransactionType.

Try It Out: COLUMNS_UPDATED()

The example in this section will take the same example as UPDATE() and convert it to use COLUMNS_UPDATED(). It is a two-line change. The following test will see whether either the TransactionType OR the Amount has altered by checking the two column settings using the power() function.

```
ALTER TRIGGER TransactionDetails.trgInsTransactions
ON TransactionDetails.Transactions
AFTER UPDATE,INSERT
AS
    IF (SUBSTRING(COLUMNS_UPDATED(),1,1) = power(2,(3-1))
    OR SUBSTRING(COLUMNS_UPDATED(),1,1) = power(2,(5-1)))
BEGIN
UPDATE CustomerDetails.Customers
   SET ClearedBalance = ClearedBalance -
      ISNULL((SELECT CASE WHEN CreditType = 0
                    THEN d.Amount * -1
                    ELSE d.Amount
                    END
         FROM DELETED d
         JOIN TransactionDetails.TransactionTypes tt
           ON tt.TransactionTypeId = d.TransactionType
         WHERE AffectCashBalance = 1),0)
   FROM CustomerDetails.Customers c
   JOIN DELETED d ON d.CustomerId = c.CustomerId

UPDATE CustomerDetails.Customers
   SET ClearedBalance = ClearedBalance +
      ISNULL((SELECT CASE WHEN CreditType = 0
                    THEN i.Amount * -1
                    ELSE i.Amount
                    END
         FROM INSERTED i
         JOIN TransactionDetails.TransactionTypes tt
           ON tt.TransactionTypeId = i.TransactionType
         WHERE AffectCashBalance = 1),0)
   FROM CustomerDetails.Customers c
   JOIN INSERTED i ON i.CustomerId = c.CustomerId
RAISERROR ('We have completed an update ',10,1)
END
ELSE
RAISERROR ('Updates have been skipped',10,1)
```

Now that we have covered DML triggers, we can take a look at DDL triggers.

DDL Triggers

Checking whether an action has happened on an object within SSE either on a database or within the server is not code that you will write every day. As more and more audit requirements are enforced on companies to ensure that their data is safe and has not been amended, auditors are now also turning their attention to areas that may cause objects to be altered. A DDL trigger is like a data trigger, as it can execute on the creation, deletion, or modification of rows within system tables rather than on user tables. So how does this help you?

I am sure we can all recall specific stories involving major institutions having a program running that removed funds or stock. My favorite is one in which a developer wrote a program that calculated interest on clients' accounts. Obviously, there needed to be roundings, so the bank always rounded down to the nearest cent. However, all the "down roundings" added up each month to a fairly substantial amount of money. Of course, auditors saw that the data updates were correct, as the amount on the Transactions table matched the amount in the client's account. The interest calculation stored procedure also passed QA at the time. However, once it was live, the developer altered the stored procedure so that all the down roundings were added up in a local variable, and at the end of the process, the amount was added to a "hidden" account. It was a simple stored procedure that never went wrong, and of course it was encrypted so nobody could see what the developer had done easily at a glance (they would have had to unobfusticate the code). If the stored procedure needed an update, it was the "old" correct code that went live, and the developer simply waited until the time was right and reapplied his code. Auditors could not figure out why at a global level thousands of dollars could not be accounted for over time. Of course, eventually they did, but if they had a DDL trigger so that whenever a stored procedure was released they received an e-mail or some other notification, they could have immediately seen two releases of the stored procedure and been asking "Why?" within minutes. Our example will demonstrate this in action.

First of all, let's look at database-scoped events.

DDL_DATABASE_LEVEL_EVENTS

This section presents a listing of all the events that can force a DDL trigger to execute. Similar to DML triggers that can execute on one or more actions, a DDL trigger can also be linked to one or more actions. However, a DDL trigger is not linked to a specific table or type of action. Therefore, one trigger could execute on any number of unrelated transactions. For example, the same trigger could fire on a stored procedure being created, a user login being dropped, and a table being altered. I doubt if you will create many if any triggers like this, but it is possible.

There are two ways that you can create a trap for events that fire. It is possible to either trap these events individually (or as a comma-separated list) or as a catchall. You will see how to do this once we have looked at what events are available.

Database-scoped Events

Table 13-2 lists all the DDL database actions that can be trapped. This is quite a comprehensive list and covers every database event there is. Many of the actions you will recognize from previous chapters, although the commands have spaces between words rather than underscores.

Table 13-2. *Database-scoped Events*

CREATE_TABLE	ALTER_TABLE	DROP_TABLE
CREATE_VIEW	ALTER_VIEW	DROP_VIEW
CREATE_SYNONYM	DROP_SYNONYM	CREATE_FUNCTION
ALTER_FUNCTION	DROP_FUNCTION	CREATE_PROCEDURE
ALTER_PROCEDURE	DROP_PROCEDURE	CREATE_TRIGGER
ALTER_TRIGGER	DROP_TRIGGER	CREATE_EVENT_NOTIFICATION
DROP_EVENT_NOTIFICATION	CREATE_INDEX	ALTER_INDEX
DROP_INDEX	CREATE_STATISTICS	UPDATE_STATISTICS
DROP_STATISTICS	CREATE_ASSEMBLY	ALTER_ASSEMBLY
DROP_ASSEMBLY	CREATE_TYPE	DROP_TYPE

CREATE_USER	ALTER_USER	DROP_USER
CREATE_ROLE	ALTER_ROLE	DROP_ROLE
CREATE_APPLICATION_ROLE	ALTER_APPLICATION_ROLE	DROP_APPLICATION_ROLE
CREATE_SCHEMA	ALTER_SCHEMA	DROP_SCHEMA
CREATE_MESSAGE_TYPE	ALTER_MESSAGE_TYPE	DROP_MESSAGE_TYPE
CREATE_CONTRACT	ALTER_CONTRACT	DROP_CONTRACT
CREATE_QUEUE	ALTER_QUEUE	DROP_QUEUE
CREATE_SERVICE	ALTER_SERVICE	DROP_SERVICE
CREATE_ROUTE	ALTER_ROUTE	DROP_ROUTE
CREATE_REMOTE_SERVICE_BINDING	ALTER_REMOTE_SERVICE_BINDING	DROP_REMOTE_SERVICE_BINDING
GRANT_DATABASE	DENY_DATABASE	REVOKE_DATABASE
CREATE_SECEXPR	DROP_SECEXPR	CREATE_XML_SCHEMA
ALTER_XML_SCHEMA	DROP_XML_SCHEMA	CREATE_PARTITION_FUNCTION
ALTER_PARTITION_FUNCTION	DROP_PARTITION_FUNCTION	CREATE_PARTITION_SCHEME
ALTER_PARTITION_SCHEME	DROP_PARTITION_SCHEME	

DDL Statements with Server Scope

Database-level events are not the only events that can be trapped within a trigger; server events can also be caught.

Table 13-3 lists the DDL statements that have the scope of the whole server. Many of these you may not come across for a while, if at all, so we will concentrate on database-scoped events.

Table 13-3. *Server-Scoped Events*

CREATE_LOGIN	ALTER_LOGIN	DROP_LOGIN
CREATE_HTTP_ENDPOINT	DROP_HTTP_ENDPOINT	GRANT_SERVER_ACCESS
DENY_SERVER_ACCESS	REVOKE_SERVER_ACCESS	CREATE_CERT
ALTER_CERT	DROP_CERT	

A DDL trigger can also accept every event that occurs within the database and, within the T-SQL code, decide what to do with each event, from ignoring upwards; however, catching every event results in an overhead on every action.

Note It is not possible to have a trigger that fires on both server and database events; it's either one or the other.

The syntax for a DDL trigger is very similar to that for a DML trigger:

```
CREATE TRIGGER trigger_name
ON {ALL SERVER | DATABASE}
[WITH ENCRYPTION]
{
{{FOR | AFTER } {event_type,...}
AS
sql_statements}}
```

The main options that are different are as follows:

- `ALL SERVER | DATABASE`: The trigger will fire either for the server or the database you are attached to when creating the trigger.

- `Event_type`: This is a comma-separated list from either the database or server list of DDL actions that can be trapped.

Note You can also catch events that can be grouped together. For example, all table and view events can be defined with a group, or this group can be refined down to just table events or view events. The only grouping we will look at is how to catch every database-level event.

Dropping a DDL Trigger

Removing a DDL trigger from the system is not like removing other objects where you simply say `DROP object_type object_name`. With a DDL trigger, you have to suffix this with the scope of the trigger.

```
DROP TRIGGER trigger_name ON {DATABASE | ALL SERVER}
```

EVENTDATA()

As an event fires, although there are no `INSERTED` and `DELETED` tables to inspect what has changed, you can use a function called `EVENTDATA()`. This function returns an XML data type containing information about the event that fired the trigger. The basic syntax of the XML data is as follows, although the contents of the function will be altered depending on what event fired.

```
<SQLInstance>
    <PostTime>date-time</PostTime>
    <SPID>spid</SPID>
    <ComputerName>name</ComputerName>
</SQLInstance>
```

I won't detail what each event will return in XML format; otherwise, we will be here for many pages. However, in one of the examples that follow we will create a trigger that will fire on every database event, trap the event data, and display the details.

Database-level events have the following base syntax, different from the previously shown base syntax:

```
<SQLInstance>
    <PostTime>date-time</PostTime>
    <SPID>spid</SPID>
    <ComputerName>name</ComputerName>
    <DatabaseName>name</DatabaseName>
    <UserName>name</UserName>
    <LoginName>name</LoginName>
</SQLInstance>
```

The XML elements can be described as follows:

- PostTime: The date and time of the event firing
- SPID: The SSE process ID that was assigned to the code that caused the trigger to fire
- ComputerName: The name of the computer that caused the event to fire
- DatabaseName: The name of the database that caused the event to fire
- UserName: The name of the user who caused the event to fire
- LoginName: The login name of the user who caused the event to fire

It's time to see a DDL trigger in action.

Try It Out: DDL Trigger

1. This first example will create a trigger that will execute when a stored procedure is created, altered, or dropped. When it finds this action, it will check the time of day, and if the time is during the working day, the action will be disallowed and rolled back. On top of this, we will raise an error listing the stored procedure. This will allow you to see how to retrieve information from the EVENTDATA() function. The final action is to roll back the changes if an action is happening during the working day.

```
CREATE TRIGGER trgSprocs
ON DATABASE
FOR CREATE_PROCEDURE, ALTER_PROCEDURE, DROP_PROCEDURE
AS
IF DATEPART(hh,GETDATE()) > 9 AND DATEPART(hh,GETDATE()) < 17
BEGIN
    DECLARE @Message nvarchar(max)
    SELECT @Message =
      'Completing work during core hours. Trying to release - '
      + EVENTDATA().value
        ('(/EVENT_INSTANCE/TSQLCommand/CommandText)[1]','nvarchar(max)')
    RAISERROR (@Message, 16, 1)
    ROLLBACK
END
```

2. We can now test the trigger. Depending on what time of day you run the code, the following will either succeed or fail.

```
CREATE PROCEDURE Test1
AS
SELECT 'Hello all'
```

3. Try running the preceding code between 9 a.m. and 5 p.m. so that it is possible to see the creation fail. Running the code in the afternoon provided me with the following error:

```
Msg 50000, Level 16, State 1, Procedure trgSprocs, Line 11
Completing work during core hours.
Trying to release - CREATE PROCEDURE Test1
AS
SELECT 'Hello all'
Msg 3609, Level 16, State 2, Procedure Test1, Line 3
The transaction ended in the trigger. The batch has been aborted.
```

4. It is necessary to drop the preceding trigger so we can move on, unless of course you are now outside of the prohibited hours and you wish the trigger to remain:

```
DROP TRIGGER trgSprocs ON DATABASE
```

5. We can create our second DDL trigger. This time we will not look for any specific event but wish this trigger to execute on any action that occurs at the database. This will allow us to see the XML data generated on any event we want to.

```
CREATE TRIGGER trgDBDump
ON DATABASE
FOR DDL_DATABASE_LEVEL_EVENTS
AS
    SELECT EVENTDATA()
```

6. This trigger can be tested by successfully creating the stored procedure we couldn't in our first example.

```
CREATE PROCEDURE Test1
AS
SELECT 'Hello all'
```

7. Check the results window. You should see results that you have not seen before. What is returned is XML data, and the results window displays the data as shown in Figure 13-7.

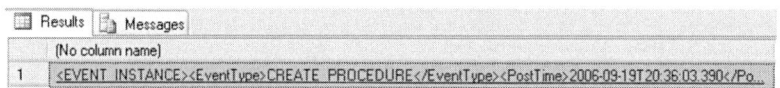

Figure 13-7. *Event data XML*

8. If you click the row, a new Query Editor pane opens after a few moments, and the XML data is transposed into an XML document layout. Each of the nodes can be inspected just like the `CommandText` node was earlier.

```
<EVENT_INSTANCE>
  <EventType>CREATE_PROCEDURE</EventType>
  <PostTime>2006-09-19T20:36:03.390</PostTime>
  <SPID>55</SPID>
  <ServerName>XP-HOME</ServerName>
  <LoginName>XP-HOME\rdewson</LoginName>
  <UserName>dbo</UserName>
  <DatabaseName>ApressFinancial</DatabaseName>
  <SchemaName>dbo</SchemaName>
  <ObjectName>Test1</ObjectName>
  <ObjectType>PROCEDURE</ObjectType>
  <TSQLCommand>
    <SetOptions ANSI_NULLS="ON" ANSI_NULL_DEFAULT="ON" ANSI_PADDING="ON"
QUOTED_IDENTIFIER="ON" ENCRYPTED="FALSE" />
    <CommandText>CREATE PROCEDURE Test1
AS
SELECT 'Hello all'
</CommandText>
  </TSQLCommand>
</EVENT_INSTANCE>
```

Summary

DML triggers should be seen as specialized and specific stored procedures set up to help your system maintain data integrity, cascade updates throughout a system, or enforce business rules. Discounting the two system tables, INSERTED and DELETED, and that you can check what columns have been modified, the whole essence of a trigger is that it is a stored procedure that runs automatically when a specified event occurs on a specific table.

DDL triggers are built mainly for security or reporting of system changes to compliance departments and the like. With the EventData() XML information available to a trigger, a great deal of useful information can be inspected and used further.

Coding a trigger is just like coding a stored procedure with the full control of flow, error handling, and processing that is available to you within a stored procedure.

The aim of this chapter was to demonstrate how a trigger is fired, and how to use the information that is available to you within the system to update subsequent tables or to stop processing and roll back the changes.

The DML triggers in this chapter demonstrated how to use the virtual tables, as well as how to determine whether a column has been modified. The DDL triggers demonstrated how you can trap events and determine what has been changed within a database or a server.

We are now at the end of the book if you are running on a Windows XP Home operating system. However, in the next chapter, we take a look at Reporting Services, which allows the production of reports from your SSE data. This has to be completed on a machine with Internet Information Services (IIS) installed, which is not available on Windows XP Home.

CHAPTER 14

■ ■ ■

SQL Server 2005 Reporting Services

SQL Server Reporting Services is available for SSE, but only if you have installed SSE on Windows XP Pro or another version of Windows that has Internet Information Services (IIS) installed. IIS is not available on Windows XP Home.

Reporting Services was introduced as an add-on with SQL Server 2000 but now comes as part of the full installation, which you can choose to install like other components. This flexible reporting tool makes it possible to export reports in PDF or Excel or even to e-mail them to someone. Reporting Services was designed and developed purely for SQL Server, therefore removing the generic format that other tools that work with different types of database have to provide. This chapter is an overview of Reporting Services, as it does require a whole book to use this tool well. In this chapter, you'll learn the following:

- What Reporting Services is
- Elements of the architecture both for neophytes and for more advanced programmers
- Creating a basic report using Report Wizard
- Creating a report from scratch
- Building and using a report model

First, let's take a look at Reporting Services.

■**Tip** To learn more about Reporting Services, please see Landrum and Voytek, *Pro SQL Server 2005 Reporting Services* (Apress, 2005).

What Is Reporting Services?

The SQL Server 2005 Business Intelligence solution comes with the promise of delivering the right data to the right people at any time. The promise is fulfilled using three main products: Integration Services, Analysis Services, and Reporting Services.

Reporting Services is used to produce reports to make information available to anyone who needs it, at any time, and in the right format (paper, PDF, Excel spreadsheet, etc.). If you prefer the official definition, Reporting Services is a server-based platform for creating, managing, and delivering both traditional and interactive reports.

Prior to Reporting Services, the only option was to use a third-party report builder such as Business Objects' Crystal Reports. This is a generic reporting tool that works with many different data sources and is quite powerful when it comes to the content and type of reports it can produce. Crystal Reports can also be automated via VB .NET, C#, VBA, and so on, so there is still a market for this type of product. However, Reporting Services is built purely for SQL Server and thus is optimized for it.

For SSE, it is necessary to download the SQL Server Express toolkit and to install Business Intelligence Development Studio to build reports easily. This is one of the downloads found on the SQL Server Express download page you saw in Chapter 1.

The next section will add more details in terms of architecture.

Reporting Services Architecture

Reporting Services components can be grouped in three layers: an application layer, a server layer, and a data layer:

- The application layer includes report tools, configuration utilities, custom applications, browsers, etc.

- The server layer is represented by your report server and is the primary component of Reporting Services. You will deploy your reports to the report server that will handle reports delivery, scheduling, data processing, etc.

- Finally, two SQL Server databases, ReportServer and ReportServerTempDB, make the data layer.

There are three stages in a typical reporting life cycle: authoring, management, and delivery.

In the authoring stage, you create the reports and, actually, you create an XML file. Every report is defined using Report Definition Language (RDL), an extensible open schema that contains the layout and data information for a report. After creating a report, you will publish it to a report server to make it available to end users.

The next stage is management, where you manage the objects stored on a report server. You can set access security, schedule operations, manage data sources, etc.

In the last stage, the reports will get to those who need them using various formats (PDF, XLS, etc.), various devices (e-mail, file shares, etc.), and two delivery methods (**push**, known also as **on-demand delivery**, and **pull**, or **event-based delivery**). There is another option for report delivery, a local delivery, if you use the new Report Viewer controls. The new Report Viewer controls allow you to embed reports in both Windows Forms applications and web applications. The controls can deliver reports through your application without a connection to a report server.

Reporting Services Architecture: A Closer Look

The architecture of Reporting Services can be divided in three layers: a server layer, an application layer, and a data layer, as shown in Figure 14-1.

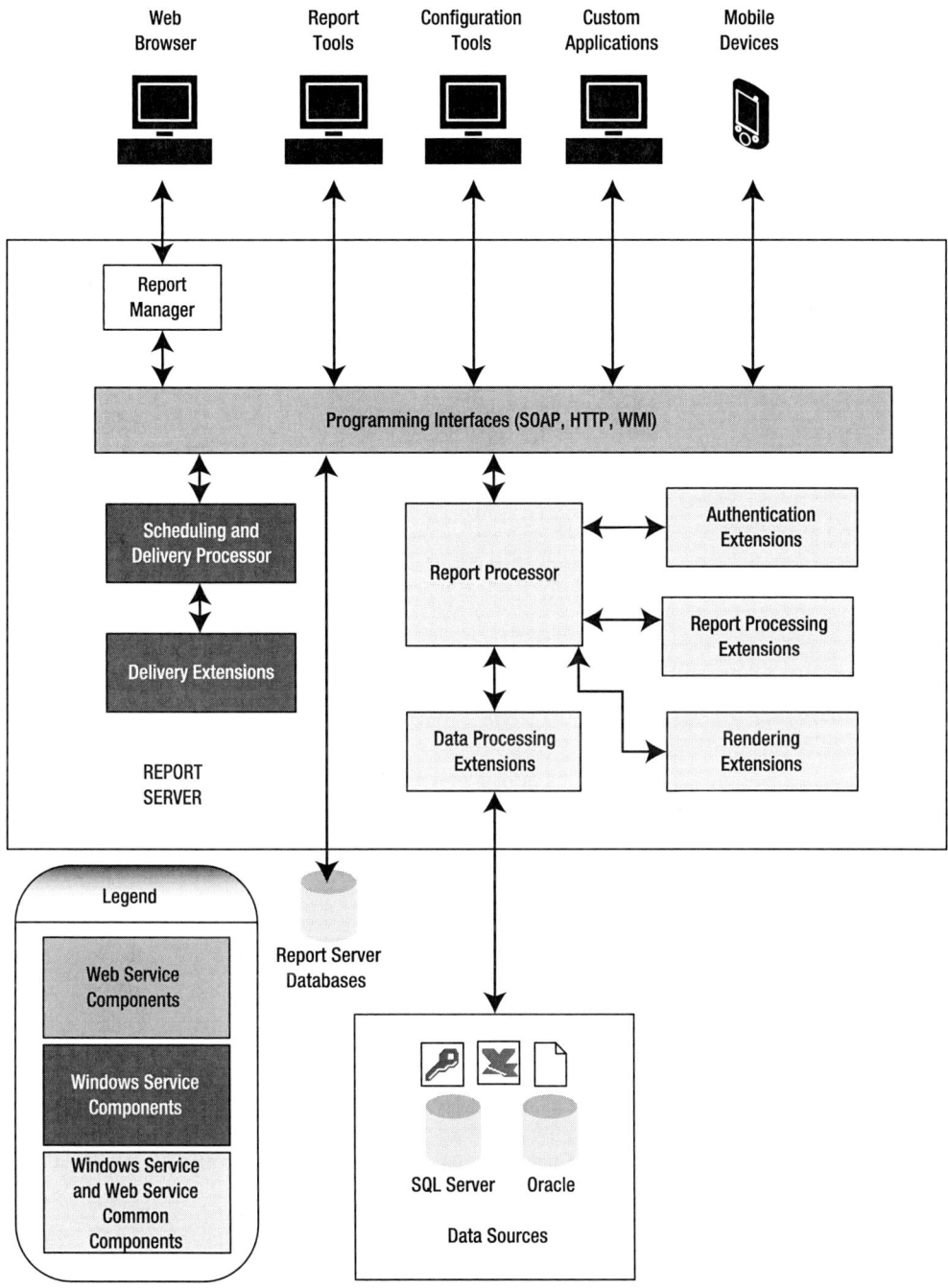

Figure 14-1. *Reporting Services architecture*

The Application Layer

At the application layer level, you will find report design tools, configuration tools, and custom applications that use Reporting Services. Let's see how they're used.

Web Browser

The simplest way to access a report server is using a browser. On the report server, all the items (such as reports, report models, and shared data sources) are organized using a folder hierarchy. The folder hierarchy looks like a file system, but actually all folders and items are stored in SQL Server databases. You can navigate through this hierarchy and view reports using a browser. The path to a report will be included in the URL used by the browser to display the report. For example, one of the reports we will create in the demo section "Building a Report Model" can be viewed with a browser using this URL: `http://localhost/Reports/Pages/Report.aspx?ItemPath=%2fAd+Hoc+Report`.

Another use for a web browser is web-based management using Report Manager. Report Manager is an ASP.NET web application used as a management tool for Reporting Services.

Report Tools

A new option in SQL Server 2005 Reporting Services is the ability to create report models and ad hoc reports. A report model hides the complexity of a report. Based on a model, a business user can create with limited technical knowledge ad hoc reports. You will see them at work in the examples later in the chapter.

To create reports, report models, or ad hoc reports, you have several options from Notepad to third-party tools.

Business Intelligence Development Studio

Based on a Visual Studio 2005 shell, Business Intelligence Development Studio is the tool that allows you to create Business Intelligence projects. For Reporting Services, you have three project templates available: Report Model Project, Report Server Project, and Report Server Project Wizard. Using the project templates will actually start the Report Designer and the Model Designer tools, which are design tools for creating the reports and report models.

Configuration Tools

For Reporting Services configuration and management, you have several tools at your disposal. You have seen some of these tools already within the book, but let's take a moment to see how they fit within Reporting Services.

SQL Server Configuration Manager

If you want to configure SQL Server 2005 components services or network libraries, you will use this tool.

Surface Area Configuration

This tool will help you to reduce the options of a potential attacker by stopping or disabling unused components. For Reporting Services, you can enable or disable the following features:

Web services request and HTTP access: If you don't use a client application that requires Reporting Services web service, you can disable this feature.

Scheduled events and report delivery: If you only require "on-demand" report delivery, you can disable this feature.

Windows Integrated Security: In order to access data sources, you need some credentials. If you disable the Windows Integrated Security feature, you will reduce the probability that a user's identity may be used without the user's consent. To access data, you will have several options for obtaining credentials such as prompted credentials, stored credentials, or no credentials.

Report Manager

I already mentioned this web-based application that allows you to manage a report server. If you don't want to install any client utilities on your machine, the best management tool for you is Report Manager.

Reporting Services Configuration Tool

This tool allows you to configure and modify the settings for a Reporting Services installation. You can configure the virtual directories used by a report server or by the Report Manager application; you can also configure service accounts, report server databases, or e-mail delivery. If you want to scale out reporting services, you need this tool.

SQL Server Management Studio Express

You should already be familiar with this tool, which allows the central management for all SQL Server component servers. It can replace Report Manager.

Command-Line Tools

In addition to graphical tools, Reporting Services comes with several command-line utilities such as rs (for scripted operations), rsconfig (for configuring the report server connection to the report server database), rskeymgmt (for encryption key management), and sac (the command-line version of Surface Area Configuration).

Custom Applications

The extensible architecture of all Reporting Services components gives you the possibility to develop custom applications for the complete reporting life cycle. You can build new report designers, management tools, or just use reports in your applications. I want to remind you that you can use Report Viewer controls, a set of two freely redistributable controls that can work in local and remote mode. In local mode, you can display reports without connecting to a report server.

The Server Layer

At the server layer, we have the report server—a perfect marriage between a Windows service and a web service.

Web Service

The web service provides a set of programmatic interfaces that allow communication with the report server. It is required for on-demand delivery and for the use of client tools like Report Manager, Report Builder, and SSMSE.

Windows Service

The Windows service is responsible for scheduling and delivery services, server maintenance, encryption operations, and initialization services. It is required for the push delivery method of reports. For scheduling operations, the SQL Server Agent Service is required. If you need just push delivery reporting, you can run only the Report Server Windows service. However, you will have available just a part of the Reporting Services functionality.

Programming Interfaces

Applications can use Simple Object Access Protocol (SOAP), URL endpoints, and Windows Management Instrumentation (WMI) to talk to report servers. If you are not familiar with web services, they simply allow you to use programmatic functionality (functions, procedures, T-SQL stored procedures, etc.) over the Web, by including the call to the functionality in your request to the web server and the result of the call in the web server's response. The communication protocol used by web services is SOAP. SOAP messages have a simple XML form. I captured the SOAP message used by Business Intelligence Development Studio to deploy a report. You can see that it is not so complicated. A report named ContactList is created in a parent folder named `Advance Report`. The existing report, if any, will be overwritten.

```
<?xml version="1.0" encoding="utf-8" ?>
<soap:Envelope xmlns:soap="http://schemas.xmlsoap.org/soap/envelope/"
xmlns:xsi="http://www.w3.org/2001/XMLSchema-instance"
xmlns:xsd="http://www.w3.org/2001/XMLSchema">
<soap:Body>
<CreateReport xmlns="http://schemas.microsoft.com/sqlserver/2005/06/30/reporting/
reportingservices">
  <Report>ContactList</Report>
  <Parent>/Advanced Report</Parent>
  <Overwrite>true</Overwrite>
  <Definition>...</Definition>
</CreateReport>
</soap:Body>
</soap:Envelope>
```

The importance of SOAP is that anyone can talk SOAP, which is XML. To receive SOAP messages, the Reporting Services web service uses URL endpoints, to which you send your requests to work with reports. SQL Server 2005 Reporting Services comes with two new endpoints—for report execution and for report management. You can use both of them to build custom tools, though most likely you will be just as happy with the existing set of tools.

Processors

The only components of Reporting Services that cannot be extended are the Report Processor and the Scheduling and Delivery Processor. The former is responsible for combining a report layout with data and rendering a report in a requested format. The latter is responsible for delivering reports and supports scheduled operations.

Extensions

One of the main benefits of Reporting Services is extensibility. You can add custom extensions for authentication, data processing, report processing, rendering, and delivery.

Authentication Extensions

Reporting Services uses a role-based security model in which users and groups are mapped to roles. Roles are just a collection of tasks that can be performed. For example, the Report Builder Role allows a user to build and edit reports with Report Builder. You can also define your own custom roles. The default authentication uses Internet Information Services (IIS) to authenticate Windows users and groups. You can use a custom authentication extension to get users from other directory services like Novell Directory Services, for example.

Data Processing Extensions

Data processing extensions are used to get data from various sources like SQL Server, Analysis Services, Oracle, OLE DB, and ODBC data sources.

Rendering Extensions

Rendering extensions work with Report Processor to transform the data and the report layout into a specific format. Available are the following rendering extensions: HTML (to generate HTML using UTF-8 encoding), Excel (to generate reports in a native file format for Excel—Binary Interchange File Format), CSV (to generate Comma-Separated Value files), Image (to generate image formats like BMP, GIF, JPEG, TIFF, EMF, and WMF), PDF, and XML.

Report Processing Extensions

If you need custom report items, you can add new report processing extensions for them.

Delivery Extensions

The delivery extensions are used by the Scheduling and Delivery Processor to deliver reports to locations such as e-mails or file shares. You can even send short notices (not the actual report) to mobile devices like pagers and phones.

The Data Layer

At the data layer level, we have two SQL Server databases and a lot of options for data sources.

Report Server Databases

SQL Server 2005 Reporting Services uses two SQL Server databases (ReportServer, ReportServerTempDB) to store the information used by the report server. The ReportServer database stores static metadata such as report definitions, data sources, users, roles, subscriptions, and schedule definitions. The ReportServerTempDB database stores temporary objects such as work tables or session data. To store Reporting Services databases, you can use another server and even another version of SQL Server (SQL Server 2000).

Data Sources

The report data sources can come from SQL Server, Analysis Services, Excel, Access, Oracle, flat files or any OLE DB, and ODBC data sources. Using data processing extensions, you can add new sources of data.

Building Your First Report Using Report Wizard

In this section, we will call the Report Wizard—the perfect tool for a beginner or for an urgent situation. The end result will be a simple report containing a list of contacts from the CustomerDetails.Customers table.

Try It Out: Using the Report Wizard

1. From the Start menu, select Programs ➤ Microsoft SQL Server 2005 ➤ SQL Server Business Intelligence Development Studio.

2. Once Business Intelligence Development Studio has opened, on the File menu, select New ➤ Project.

3. In the Project Types list, click Business Intelligence Projects.

4. In the Templates list, click Report Server Project Wizard.

5. Type **ApressFinancial Reports** in the Name text box as shown in Figure 14-2 and leave the default location as it is. Click OK to start the Report Wizard.

Figure 14-2. *Selecting the report template and report name*

6. On the Welcome page of the wizard, you will see a short description of the wizard. You may read it and then click Next.

7. On the Select the Data Source page, make sure that Microsoft SQL Server is selected in the Type drop-down list.

8. This step defines the source of data for our report. We will use the ApressFinancial database from our local server. Click the Edit button to open the Connection Properties dialog box and specify XP-PRO for Server Name and ApressFinancial in Select or Enter a Database Name as shown in Figure 14-3.

Figure 14-3. *Defining the report data source*

An alternative is to remain in the Data Source page and type **Data source= XP-PRO; Initial Catalog= ApressFinancial** in the Connection String text box as you see in Figure 14-4. When you are done, click Next.

Figure 14-4. *Defining the data source using a connection string*

Tip Remember to change the data source to the SQL Server instance you are connecting to, such as (local).

9. In the Design the Query page, shown in Figure 14-5, you can use the Query Designer to build your query. This example is very simple, so just use the Query String text box to enter the following `SELECT` statement and click Next:

```
SELECT CustomerFirstName, CustomerLastName
FROM CustomerDetails.Customers
```

Figure 14-5. *Design the Query window*

10. The next step is to select a report type. Select Tabular and click Next.

11. To make the report really simple, you will not use any kind of grouping. In the Design the Table page, shown in Figure 14-6, you can select fields by which to group or fields to display in the Report Details section. Select the `CustomerFirstName` and `CustomerLastName` fields and click the Details button to add them to the details section. Click Next.

12. The Choose the Table Style page allows you to apply a style to the report by selecting a style template. I prefer the style template named `Corporate`. Select one template and click Next.

13. The next page, Choose the Deployment Location, allows you to configure the report server and the folder to which the report will be published. The Report Wizard will not publish the report. You can do that yourself later on.

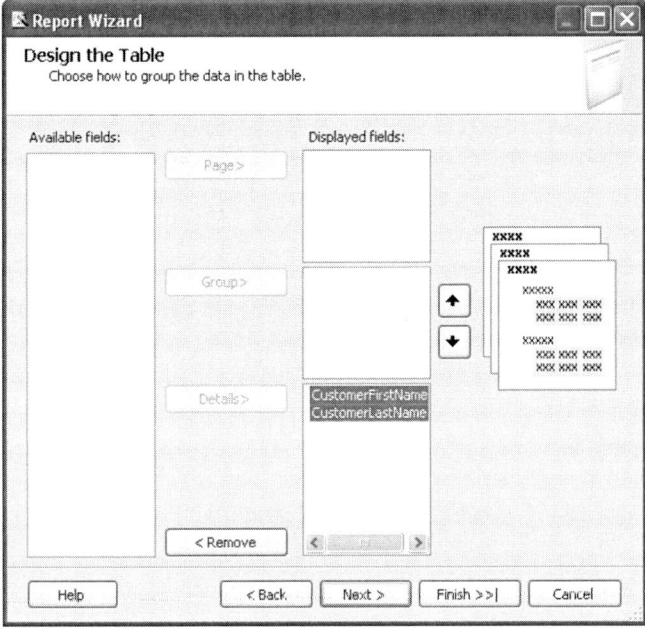

Figure 14-6. *Design the Table window*

14. The last step gives you a final chance to verify the information entered. You have the option to rename the report, but for this demo leave the default name—Report1. Select the Preview Report check box to preview the report as soon as the wizard finishes. Click Finish.

You should now see the preview of your report, which should resemble Figure 14-7.

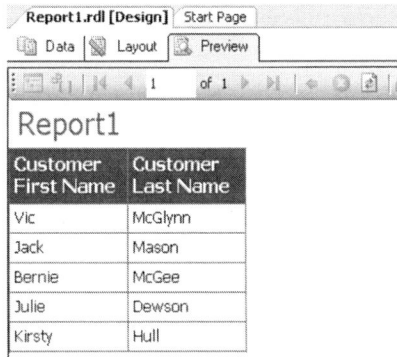

Figure 14-7. *The preview of your report*

You can see the two columns selected in step 11, the tabular format of data, because of the choice you selected in step 10, and of course, the great name Report1 that you can change if you choose to.

In the next section, I will tell you about various report items.

Building a Report from Scratch

In this section, you will learn to build a report without the help of Report Wizard. That means you will have to define a data source and build a design for your report. Once the report is built, I will show you how to add some interactivity by using parameters. Before we do the actual demo, let's take a closer look at some of the elements we can use.

Data-Related Elements

The utility of a report is given mainly by its data. In order to have data in a report, you have to define at least a data source and a dataset. The data view of the graphical designer interface allows you to create datasets and data sources.

Data Source

A data source represents the necessary information to get to the source of data. Such information may include a server name, a database name, and user credentials. The information included depends on the source of data; for example, if you use a text file, you will specify the path to the text file and not a server name or a database name. A report can use multiple data sources. There are two types of data source: shared (can be used by multiple reports) or report specific (used by a particular report). Another difference between shared and report-specific data sources is their storage location. A shared data source is stored in a separate file on the report server, while the report-specific data source is stored in the report definition.

Dataset

Once we have defined the source of data, we will create a dataset using a query. You can imagine a dataset as a set of rows with data. I would mention that a report can also have multiple datasets.

Fields

Every row of a dataset has one or more fields that typically refer to columns returned by the query that populated the dataset. You can find all fields of a dataset in the Fields window. Using the drag-and-drop method, you can include any field in your report.

Custom Fields

Besides the existing fields of a dataset, you can create new ones using several methods. You can create calculated fields based on existing fields, or you can use aggregate functions such as Sum, Avg, and Count.

Report Items

There are two types of report items: data regions and independent items.

Data Regions

Data regions display data from datasets, with each data region being associated to one dataset. You can use the following types of data regions:

Table: A table is a data region that uses static columns and rows to display data.

Matrix: A matrix (or crosstab) is a more complex data region having columns and rows that can contain other columns or rows.

List: A list allows you to display data in a free-form way. It basically repeats each row (or group) in a dataset.

Chart: A chart data region displays the data in a graphical form, allowing you to use charts like bar charts, pie charts, and other types of charts.

Independent Items

Independent report items are not linked to a dataset but can be used to display data or graphical elements. The following types are available:

Text box: A text box is a control that allows displaying of text data. They can be placed anywhere on a report.

Line: A line is a graphical element that can be placed anywhere in your report. You can apply different styles to it, you can make it horizontal, vertical, diagonal, etc.

Image: An image control allows you to display images from a file, URL, or database field.

Rectangle: A rectangle is a graphical element that can be used to group other items, and as the name says its shape is rectangular.

Subreport: A subreport is an item that references the body of another report.

Reports Structure

A report can contain different sections such as report header and footer (that appear on the first and the last pages, respectively), page headers and page footers (that appear on all pages), and table headers and footers (displayed at the beginning and the end of a table). If you have groups, you can use group headers and group footers.

That should be enough to give you an overview on what you need to create a report. Let's start the demo.

Try It Out: Creating a Report Structure

1. You will use a new simple table for this report. Open SSMSE and run the following script:

```
USE ApressFinancial
GO
CREATE TABLE CustomerDetails.Title
(CustomerTitleId int primary key,TitleDescription varchar(16))
GO
INSERT INTO CustomerDetails.Title VALUES (1,'Mr')
INSERT INTO CustomerDetails.Title VALUES (2,'Ms')
INSERT INTO CustomerDetails.Title VALUES (3,'Dr')
```

2. Open Business Intelligence Development Studio and on the File menu, select New ➤ New Project.

3. Click Business Intelligence Projects in the Project Types list. Click the Report Server Project template.

4. In the Name text box, type **Advanced Report** and then click OK to create the project.

5. Make sure the Solution Explorer window is visible. To display it, from the View menu click Solution Explorer.

6. In Solution Explorer, right-click Reports, and select Add ➤ New Item.

7. In Add New Item, click Report.

8. In Name text box, type **ContactList.rdl** as shown in Figure 14-8 and then click Add. The new report will open in Data view.

Figure 14-8. *Naming our report*

Next, we will create a data source for the report and two datasets.

9. In the Data view, expand the Dataset drop-down list, and select New Dataset as shown in Figure 14-9 to open the Data Source dialog box.

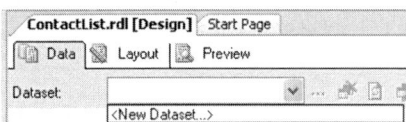

Figure 14-9. *Specifying a new dataset*

10. In the Name text box, type **ApressFinancial**. In the Type drop-down menu, select Microsoft SQL Server.

11. In the Connection String text box, type the following:

```
Data source=XP-PRO; Initial Catalog=ApressFinancial
```

12. Your screen should look like what you see in Figure 14-10. Click OK to add the ApressFinancial dataset.

Figure 14-10. *Using the Data Source dialog box to define our dataset*

13. Type the following query into the SQL pane that opened:

```
SELECT C.CustomerFirstName, C.CustomerLastName, T.TitleDescription
FROM CustomerDetails.Customers C
JOIN CustomerDetails.Title T
ON C.CustomerTitleId = T.CustomerTitleId
```

14. Click the Run (!) button on the Query Designer toolbar to see the results of the query, as shown in Figure 14-11.

CustomerFirstNa...	CustomerLastNa...	TitleDescription
Vic	McGlynn	Mr
Jack	Mason	Dr
Bernie	McGee	Dr
Julie	Dewson	Ms
Kirsty	Hull	Mr

Figure 14-11. *Our query results*

15. To create a second dataset, expand the Dataset drop-down list, and select New Dataset.

16. In the Dataset dialog box, type **TitleList** in the Name text box, leave ApressFinancial as the data source, make sure that in the Command Type drop-down list Text is selected, and in the Query string text box type the following:

```
SELECT TitleDescription FROM CustomerDetails.Title
```

17. Make sure your screen looks like Figure 14-12, and click OK to add the TitleList dataset. Optionally, you can click the Run (!) button on the Query Designer toolbar to see the results of the query.

Figure 14-12. *Creating our TitleList dataset*

Next, we will add a table data region to the report layout.

18. Click the Layout tab to display the Layout view.

19. If the Toolbox window is not displayed, from the View menu click Toolbox. It may be just a small tab on the top left of your design area, so it may not be very visible.

20. In the Toolbox window, click the Table control and then click the design surface. A table with three columns is generated.

21. In the Datasets Explorer, which should be on the left of the design pane, expand the ApressFinancial dataset to see the fields. If it is not visible, you can bring it up by selecting the last option in the View menu.

22. Drag the CustomerFirstName field to the detail row (the middle row) of the first column.

23. Drag the CustomerLastName field to the detail row (the middle row) of the second column.

24. Drag the TitleDescription field to the detail row (the middle row) of the third column. Your columns should now resemble those shown in Figure 14-13.

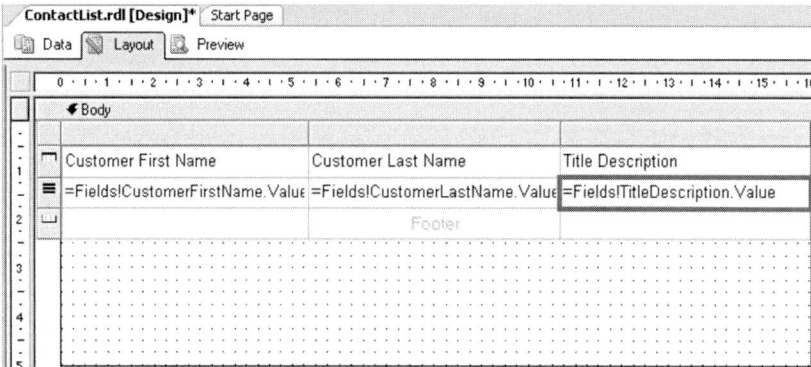

Figure 14-13. *The design surface*

25. Use the File menu and click Save All. Click the Preview tab to see the report, as shown in Figure 14-14.

Figure 14-14. *Our newly generated report*

Try It Out: Enhancing the Report

In this section, we will change the design of the report we have just created by adding a new column, and we will add some interactivity to our report using a parameter.

1. Click the Layout tab to display the Layout view.

2. Click the first column of the first row in the table that displays the text Customer First Name. Delete the word "Customer."

3. Repeat the step for the second column.

4. Delete the word "Description" from the third column using the same method.

5. Click the table to display the column and row handles. Right-click the handle of the third column (Title) and select Insert Column to the Right, as shown in Figure 14-15.

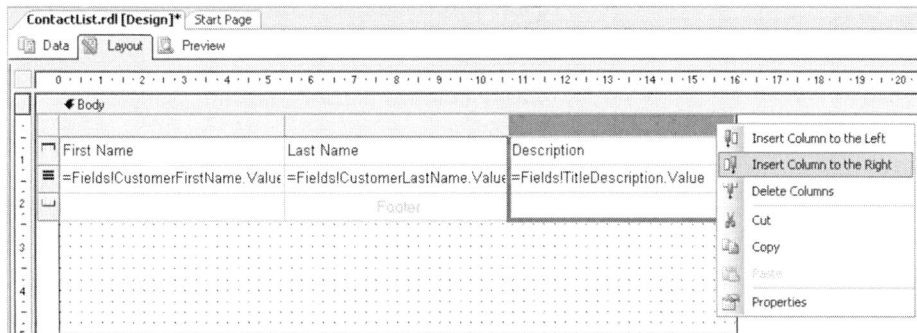

Figure 14-15. *Adding a new column*

6. Click the detail row for the new column (the fourth). Enter the following expression that will calculate the initials:

```
=Left(Fields!CustomerFirstName.Value,1) & " " &
Left(Fields!CustomerLastName.Value,1)
```

7. Click the first row for the new column and type **Initials**. Click the Preview tab to see the results, as shown in Figure 14-16.

Figure 14-16. *Our report after adding a new column*

Try It Out: Deploying the Report (Optional)

In this section, we will deploy the report created in the last two "Try It Out" sections to the local report server and use Report Manager to view the report.

1. Once you have saved the report using Save All, in the Solution Explorer, right-click Advanced Report solution, and select Properties.

2. In the TargetServerURL text box, type **http://localhost/ReportServer** as shown in Figure 14-17 and click OK.

Figure 14-17. *Specifying our target server*

3. The Reporting Services service has to be running. It should be running, but in case you need to get to the service because you have switched it to start manually, from the Control Panel, select Administrative Tools ➤ Services. Find SQL Server Reporting Services, right-click, and click Start, as shown in Figure 14-18.

Figure 14-18. *Starting Reporting Services*

4. In the Solution Explorer, right-click Advanced Report solution, and select Deploy.

5. Open Microsoft Internet Explorer and in the address bar, type the Report Manager URL: **http://localhost/Reportserver/**. This will redirect you to the report pages folder list.

6. Click the `Advanced Report` folder. Click the Contact List report (see Figure 14-19).

Figure 14-19. *The Contact List report*

And that concludes this demonstration.

Summary

This brings us to the end of *Beginning SQL Server 2005 Express for Developers*, demonstrating how to report on the data within the example we have been building up throughout the book. You have seen an overview of how Reporting Services works and is architected, and you have learned how to build a simple report and preview the results. The next move for you would be to read Rizzo et al., *Pro SQL Server 2005* (Apress, 2005), and also Dewson and Skinner, *Pro SQL Server 2005 Assemblies* (Apress, 2005).

Glossary of Terms

The aim of this glossary is to provide you with a quick and easy-to-use reference to many of the terms that you will come across in this book.

When beginning to study a new area of interest (especially in specialized areas), whether it be computing or something else, it is very easy for a book, or any discussion for that matter, to get bogged down in new terminology, making it difficult for new readers to advance in their knowledge of the subject at hand. Essentially, that's where this glossary comes in. Hopefully, it will help you come to grips with SQL Server Express, and you may also find this to be a practical reference guide for use in meetings, or indeed any other situation while you are starting out with SSE. So, without further ado, let's begin.

A

alias

A substituted name for a database, table, column, or other database object. If you have, for example, a table or a column with a long name, or if you are joining two tables together and both have a column name that is the same, or if the name of a column doesn't lend itself to describing the contents that it holds well enough for output displayed, you could use an alias on the column name. In this way, you can assign a more descriptive name to the column, via the alias, thereby making it easier to see just what the data in the column actually represents. The alias on the column name would typically be used within SELECT statements or WHERE clauses of SQL commands.

You can also alias table names to make it easier when building queries joining two tables, so that the alias is used in defining the join conditions, rather than a long table name.

ANSI

Acronym for American National Standards Institute, a body of representatives from industry and business based in America that defines standards in many areas, one of which deals with databases.

authentication

When logging in to SSE, whether using Windows authentication or SQL Server authentication, this refers to the process of verifying that the submitted user ID is valid for a given instance of SSE, and then allocating permissions to that user based upon his or her user profile.

B

backup device

A hardware device such as a tape drive used to back up an SSE database. If a tape drive is used, the tape drive must be attached directly to the computer that the SSE database resides on, and cannot be a tape drive found on a remote computer elsewhere on the network. This can also refer to a file on the same computer, or another server if required.

batch

A set of T-SQL statements forming one group, or batch, of actions. In Query Editor, you define the end of a batch by employing a GO statement, or another statement set via options within Query Editor. A batch allows you to put together a set of tasks that either has to be a batch because of the way SSE works, or you wish to "batch" together a set of tasks as one single unit. Finally, certain T-SQL commands must be placed in a batch on their own with no other T-SQL commands. These are statements like CREATE TABLE.

BEGIN TRAN[SACTION]

This statement is used to denote the start of a transaction, in which modifications will be made to data that can either be placed into the database using a COMMIT TRAN[SACTION] or rejected using a ROLLBACK TRAN[SACTION].

C

checkpoint

This is a system function that will write all the confirmed transaction changes from memory to disk.

clustered index

An index on a table (or view) that defines the physical order in which the data will be stored on the table (or view).

COMMIT TRAN[SACTION]

Used to inform SSE that any data modifications performed within a transaction, by using the BEGIN TRAN[SACTION] statement, are to be accepted as valid and are to be committed to the database (that is, permanently written to the database).

constraint

A constraint can be a check placed against a table column to ensure that the data entered is valid or of a default value; a foreign key constraint identifies the relationship between two tables, while a primary key constraint identifies the column(s) that make a unique primary key on a table.

control-of-flow language

The T-SQL structures contained within a stored procedure, trigger, or batch of queries that determines the flow of execution of T-SQL statements. Contains commands such as BEGIN...END and IF...ELSE, among others.

D

data integrity

Ensuring that all the data stored within your database is valid, and that there are no inconsistencies between rows or tables of any information. Maintaining data integrity is paramount to the success of your database solution, and can be achieved through the use of constraints, keys, and triggers.

database

A repository of objects holding information that makes up a single unit of information. A database will hold not only rows and columns of information within a table, but also objects that let you work with that data, such as stored procedures, views, etc.

database diagram

A graphical representation generated by SSE that displays all or a subset of the tables and relationships within an SSE database. A default database diagram can be specifically generated by a developer, and is displayed within SQL Server Management Studio Express. However, there can be as many diagrams as desired containing any tables within the database.

deadlock

Deadlocks occur when there is a cyclic dependency between two or more threads for some set of resources. This can be seen when there are two transactions trying to complete an update simultaneously. For example, when transaction 1 has updated one table (TableA) and is trying to update another table (TableB), the second transaction has updated TableB, and is waiting to update TableA. After a period of time, SSE will choose a deadlock victim, which is the update to be canceled and rolled back.

Note Deadlock is a condition that can occur on any system with multiple threads, not just on a relational database management system.

default database

When a user is created in SSE, a default database is automatically assigned to him or her (this is master by default). This can be changed to any other database within the server. When no specific database is defined in a connection made by programs written in languages such as C#, Java, etc., the default database assignment is used to denote which database will be connected to.

default instance

An instance of SSE installed on a server with no instance name specified. The name of the instance is thus just the computer name. Multiple instances of SSE can be installed on the same computer; however, only one can be the default instance. The other instances are referred to as named instances and have the form computername\instancename.

delimiter

Characters that denote the start and end of object names. Delimiter characters are either double quotation marks or square brackets.

differential database backup

When a backup of a database is made, only the changes made to the database since the last full database backup are reflected.

F

foreign key (FK)

A column or set of columns that match the definition of a primary key or a unique key from another table. A foreign key is used to establish relationships between two tables through the correspondence between the foreign key and the primary key.

foreign table

If a table has a foreign key, the table is known as a foreign table. The use of this term is most common when discussing relationships between tables.

full backup

When a complete backup of all the data within an SSE database is taken. All data, including indexes, are backed up.

G

GRANT statement

When you wish to give a user or, indeed, a group of users defined within a role permissions to access or work with a database object, you need to grant permission to them to perform the desired task using the GRANT statement.

I

identity column

A specialized column within a table that has its value automatically inserted by SSE each time a new record is inserted. This value cannot be altered. The value is unique for the table and is based upon a starting point and an increment for each insertion.

index

By creating an index on a table, you can potentially speed up access and data retrieval from a table or a view. An index consists of one or more columns from a table.

instance

An installation of SSE on a local or remote computer. It is possible to have multiple instances of SSE installed and running simultaneously on the same computer, with some or all instances registered within SQL Server Management Studio Express.

ISO

Stands for the International Organization for Standardization. This is one of the two international standards bodies responsible for developing international data communication standards. The other organization is the International Electrotechnical Commission (IEC). ISO and IEC are responsible for the SQL-92 standard for SQL. If a database is defined as SQL-92 compliant, this means it meets the requirements described in this standard, which is set by these two organizations.

J

JOIN condition

A T-SQL clause or a condition forms a relationship that is used to relate two tables and make them appear as one. There are different types of joins such as `INNER`, `OUTER`, and `CROSS`.

K

key

A single column or combination of columns that defines a single row as unique, such as a primary key or unique key, or defines a relationship between two tables, such as a foreign key.

L

local server

An example of a local server is the type of server used throughout this book, where the instance of SSE is running on the same computer as the application.

local variable

Within a stored procedure or query, when wishing to store values without wanting to place them within a table, you can place them within a user-defined, locally scoped variable using the `DECLARE` statement. You can then assign a value using a `SELECT` or `SET` statement. As soon as the batch of work is complete, the local variable is no longer valid, and the values within will have been destroyed.

lock

When updating data, a lock is placed on the data stopping any other connection from being able to modify the data that has been locked, until such time as the process is finished and releases the lock. A lock is released at the end of a transaction when its work has been processed. Locks should be held for as short a time as possible to avoid a deadlock situation. There are different levels of locking from row-level locking to full table-level locking.

M

many-to-many relationship

Such a relationship exists between two tables where there are many records in one table that can be related, or linked, to many records in the other table.

many-to-one relationship

A many-to-one relationship exists when two tables are linked and there are many records in one table that can be linked to one record in another table. An example of this would be when looking at customer records and retrieving the country of residence from a cross-reference number within the customer's details. Also known as a one-to-many relationship.

master database

This is the most important database within SSE, as it serves as the database that controls logins, environment settings, and system error messages. This database should not be altered manually. Using commands or system stored procedures within SSE will alter the database, but it is only through these system stored procedures, such as sp_configure, that the master database should be altered. This database also holds the location of the files of all other databases.

media set

All the media, whether it is tapes or disks, involved in the process of making a backup.

model database

Used as the template from which all other databases and database objects are built. Details from within this database are copied when creating a new database, and when a new object is created, the details of the object initially come from the model database. Alter with care.

msdb

A system-defined database used by SSE Agent when dealing with automated jobs and alerts.

N

nonclustered index

An index where the columns listed within the index do not define the physical order of the data in the table the index is defined on (as opposed to a clustered index).

NULL

A NULL value indicates that no data is stored in a column. This is a special entry within a column, as it can be placed in any data type and yet means nothing. If you find this value within a column, you cannot compare it through a comparison with any other value, other than another NULL, which is a special comparison. In other words, it is not less than or greater than any other value, or equal to any other value, even another NULL value.

nullability

Expresses whether or not a column or a parameter can accept NULL values.

O

object

An object is any component contained within a database. As such, an object can range from a database table right through to a stored procedure.

one-to-many relationship

Similar to a many-to-one relationship, wherein there is one row in the master table being related to many rows in a child table. However, the difference here is it is the master table that is driving the relationship. An example is a table for orders (one) with a table of order details (many).

one-to-one relationship

This relationship is rarely found, and exists when every record in one table has a relationship with a single record in another table, and vice versa. This could be when a single row in a table becomes split in two where data in one table is referenced frequently, and data in the other table is not so frequently referenced. This will then speed up data retrieval on the frequently referenced table.

P

precision

The number of digits found in a noninteger number both to the left and to the right of the decimal point.

primary key (PK)

A single column or a set of columns from a single table that can uniquely identify a row of data within that table or view. No two rows can have the same value within a primary key, and no column defined for the primary key can contain a NULL value.

Q

query

A single or set of T-SQL statements that deals with any aspect of data manipulation or retrieval. Queries can be run once, or run many times, and quite often are stored within stored procedures.

R

referential integrity (RI)

A state wherein all the relationships between tables are valid, commonly achieved using constraints and keys, to ensure that data integrity, hence referential integrity, remains valid. Preserving referential integrity will ensure that all data within the database is in alignment.

ROLLBACK TRAN[SACTION]

If, when working within a transaction, you decide that you no longer wish for any data modifications within the transaction to be committed to SSE, you would issue a `ROLLBACK TRAN[SACTION]` statement that will restore the original values to all of the columns and all the rows that have been modified in the most recent transaction.

row

A single, "horizontal" set of values that have come from one or more tables through a query, or a single "horizontal" set of values from all columns within a table. This can also be called a record.

S

schema

A method for organizing database entities with similar logical entities. Logins can then be assigned access rights at the schema level and also can be assigned default schemas.

SET

Used to define the value of a variable rather than using `SELECT`.

SQL (Structured Query Language) query

An action that will either manipulate or retrieve data from objects within SSE or any other database that supports SQL, for example, Oracle or Sybase.

SQL Server authentication

A method for validating login attempts to SSE using a user ID and password that are defined with SSE.

stored procedure

A set of T-SQL statements grouped together and stored as a compiled object within SSE, associated to a particular table. Can contain control of flow, error statements, etc.

subquery

A `SELECT` statement that is used to aid data selection by being nested within another SQL query.

T

table

A database object that contains rows and columns of data. Each column will have a predefined data type and may also have constraints, indexes, and keys associated with it.

table scan

Occurs when SSE scans every row within a table while performing a SQL command, rather than using an index. It does this because it is faster to scan the table rather than look at an index, then retrieve the data from the table, and then move back to get the next index item, etc., and tends to occur when there are no indexes that return a low percentage of the data. To clarify, an ideal index should return a low number of rows; otherwise, a table scan is highly likely to occur.

tempdb

Transient database that will hold any temporary tables, indexes, and any temporary storage needed by a query, a stored procedure, or any system process.

temporary table

A table placed in the tempdb, which is then lost at the end of the session that built it. It could also be a table defined with a single hash, #, or double hash, ##.

Transact-SQL (T-SQL)

A language extension to the SQL-92 defined standards for a database to allow administrating data and objects within SSE.

transaction

A logical unit of work, such that if it contains any data modifications, these modifications can be committed or rolled back depending upon a decision that can be made at any time within the transaction. Related to BEGIN TRAN[SACTION], COMMIT TRAN[SACTION], and ROLLBACK TRAN[SACTION].

transaction log

A file separate from the database, which holds all of the data modifications made within a transaction. Used by SSE for recovery purposes.

trigger

A specialized stored procedure that is executed when data in the table associated with the trigger is modified. Used to enforce referential integrity or business rules.

U

unique index

An index that defines that no two rows within a table are the same.

user-defined data type

A data type based upon an SSE data type created by the user or built using .NET code, which will not be based on an SSE base data type.

user-defined function (UDF)

A function that can be created by a user to perform frequently used, or business-related, logic. User-defined functions are different from stored procedures because they are like code snippets rather than full-blown procedures and can form a column in a result set. These user-defined functions can be written in T-SQL or .NET code.

V

variables

Holders of values used in queries, stored procedures, etc.; although they can hold information from a column of data from a table, they are not actually part of any table. Variables are defined by using the DECLARE command and are prefixed with the @ sign. Values are placed into variables using the SET or the SELECT statement.

view

A database object that can be used as a security measure when dealing with data that is sensitive, or to make data schemas more friendly and usable for user-defined data queries. Acts in a similar fashion to a table, and although called a view, data can be updated and deleted providing specific conditions are met.

W

Windows authentication

An SSE database will have either Windows authentication or SQL Server authentication, or both, as the basis of any connections to the database. Windows authentication uses the Windows user ID and logon as the basis of its connection to SSE. This is more secure than SQL Server authentication for connecting to SSE.

X

XML

Short for eXtensible Markup Language, this is a technique for building self-describing data. This can be used to pass data and information between systems or to a web browser. Data can be retrieved from SSE in an XML format. There is also a specific data type for XML data.

Index

Find it faster at http://superindex.apress.com/

You Need the Companion eBook

Your purchase of this book entitles you to buy the companion PDF-version eBook for only $10. Take the weightless companion with you anywhere.

We believe this Apress title will prove so indispensable that you'll want to carry it with you everywhere, which is why we are offering the companion eBook (in PDF format) for $10 to customers who purchase this book now. Convenient and fully searchable, the PDF version of any content-rich, page-heavy Apress book makes a valuable addition to your programming library. You can easily find and copy code—or perform examples by quickly toggling between instructions and the application. Even simultaneously tackling a donut, diet soda, and complex code becomes simplified with hands-free eBooks!

Once you purchase your book, getting the $10 companion eBook is simple:

❶ Visit **www.apress.com/promo/tendollars/**.

❷ Complete a basic registration form to receive a randomly generated question about this title.

❸ Answer the question correctly in 60 seconds, and you will receive a promotional code to redeem for the $10.00 eBook.

2560 Ninth Street • Suite 219 • Berkeley, CA 94710

eBookshop